Advances in Industrial Control

Other titles published in this Series:

Robust Control of Diesel Ship Propulsion
Nikolaos Xiros

Hydraulic Servo-systems
Mohieddine Jelali and Andreas Kroll

Strategies for Feedback Linearisation
Freddy Garces, Victor M. Becerra, Chandrasekhar Kambhampati and Kevin Warwick

Robust Autonomous Guidance
Alberto Isidori, Lorenzo Marconi and Andrea Serrani

Dynamic Modelling of Gas Turbines
Gennady G. Kulikov and Haydn A. Thompson (Eds.)

Control of Fuel Cell Power Systems
Jay T. Pukrushpan, Anna G. Stefanopoulou and Huei Peng

Fuzzy Logic, Identification and Predictive Control
Jairo Espinosa, Joos Vandewalle and Vincent Wertz

Optimal Real-time Control of Sewer Networks
Magdalene Marinaki and Markos Papageorgiou

Process Modelling for Control
Benoît Codrons

Computational Intelligence in Time Series Forecasting
Ajoy K. Palit and Dobrivoje Popovic

Modelling and Control of mini-Flying Machines
Pedro Castillo, Rogelio Lozano and Alejandro Dzul

Rudder and Fin Ship Roll Stabilization
Tristan Perez

Measurement, Control, and Communication Using IEEE 1588
John Eidson

Piezoelectric Transducers for Vibration Control and Damping
S.O. Reza Moheimani and Andrew J. Fleming
Publication due March 2006

Windup in Control
Peter Hippe
Publication due April 2006

Manufacturing Systems Control Design
Stjepan Bogdan, Frank L. Lewis, Zdenko Kovačić and José Mireles Jr.
Publication due May 2006

Practical Grey-box Process Identification
Torsten Bohlin
Publication due May 2006

Nonlinear H_2/H_∞ Constrained Feedback Control
Murad Abu-Khalaf, Jie Huang and Frank L. Lewis
Publication due May 2006

Ben M. Chen, Tong H. Lee, Kemao Peng
and Venkatakrishnan Venkataramanan

Hard Disk Drive Servo Systems

2nd Edition

With 124 Figures

Ben M. Chen, PhD
Department of Electrical and Computer Engineering
National University of Singapore
4 Engineering Drive 3
Singapore 117576

Tong H. Lee, PhD
Department of Electrical and Computer Engineering
National University of Singapore
4 Engineering Drive 3
Singapore 117576

Kemao Peng, PhD
Department of Electrical and Computer Engineering
National University of Singapore
4 Engineering Drive 3
Singapore 117576

Venkatakrishnan Venkataramanan, PhD
Mechatronics and Recording Channel Division
Data Storage Institute
DSI Building, 5 Engineering Drive 1
Singapore 117608

British Library Cataloguing in Publication Data
Hard disk drive servo systems. - 2nd ed. - (Advances in industrial control)
1.Servomechanisms 2.Data disk drives - Design 3.Hard disks (Computer science)
I.Chen, Ben M., 1963-
629.8'323

Advances in Industrial Control series ISSN 1430-9491
e-ISBN 1-84628-305-1 Printed on acid-free paper
ISBN-13: 978-1-84996-575-0 e-ISBN-13: 978-1-84628-305-5

First published 2002
Second edition 2006

Printed in Germany

9 8 7 6 5 4 3 2 1

Springer Science+Business Media
springer.com

Advances in Industrial Control

Professor Emeritus O.P. Malik
Department of Electrical and Computer Engineering
University of Calgary
2500, University Drive, NW
Calgary
Alberta
T2N 1N4
Canada

Professor K.-F. Man
Electronic Engineering Department
City University of Hong Kong
Tat Chee Avenue
Kowloon
Hong Kong

Professor G. Olsson
Department of Industrial Electrical Engineering and Automation
Lund Institute of Technology
Box 118
S-221 00 Lund
Sweden

Professor A. Ray
Pennsylvania State University
Department of Mechanical Engineering
0329 Reber Building
University Park
PA 16802
USA

Professor D.E. Seborg
Chemical Engineering
3335 Engineering II
University of California Santa Barbara
Santa Barbara
CA 93106
USA

Doctor K.K. Tan
Department of Electrical Engineering
National University of Singapore
4 Engineering Drive 3
Singapore 117576

Doctor I. Yamamoto
Technical Headquarters
Nagasaki Research & Development Center
Mitsubishi Heavy Industries Ltd
5-717-1, Fukahori-Machi
Nagasaki 851-0392
Japan

To our families

Series Editors' Foreword

The series *Advances in Industrial Control* aims to report and encourage technology transfer in control engineering. The rapid development of control technology has an impact on all areas of the control discipline. New theory, new controllers, actuators, sensors, new industrial processes, computer methods, new applications, new philosophies…, new challenges. Much of this development work resides in industrial reports, feasibility study papers and the reports of advanced collaborative projects. The series offers an opportunity for researchers to present an extended exposition of such new work in all aspects of industrial control for wider and rapid dissemination.

Hard disk drive systems are ubiquitous in today's computer systems and the technology is still evolving. There is a review of hard disk drive technology and construction in the early pages of this monograph that looks at the characteristics of the disks and there it can be read that: "bit density… continues to increase at an amazing rate", "spindle speed… the move to faster and faster spindle speeds continue", "form factors… the trend…is downward… to smaller and smaller drives", "performance… factors are improving", "redundant arrays of inexpensive disks… becoming increasingly common, and is now seen in consumer desktop machines", "reliability… is improving slowly… it is very hard to improve the reliability of a product when it is changing rapidly" and finally "interfaces… continue to create new and improved standards… to match the increase in performance of the hard disks themselves". To match this forward drive in technology, control techniques need to progress too and that is the main reason why Professor Chen and his co-authors T.H. Lee, K. Peng and V. Venkataramanan have produced this second edition of their well-received *Advances in Industrial Control* monograph *Hard Disk Drive Servo Systems*.

The monograph opens with two chapters that create the historical context and the system modelling framework for hard disk drive systems. These chapters are followed by the control and applications content of the monograph. Hard disk drive systems are beset by nonlinear effects arising from friction, high-frequency mechanical resonances and actuator saturation so any control methods used have to be able to deal with these physical problems. Furthermore, there are two operational modes to contend with: track seeking and track following each with thors emerges from the interplay between the desire to mitigate the nonlinear effects and yet find a control strategy to unify the control of the two operational modes. To reveal the strategy developed in this Foreword would be like prematurely revealing the ending of a fascinating mystery story.

The monograph also has other valuable features: Chapter 3 contains succinct presentations of five different control methods with formulas given for both continuous and discrete forms. Two chapters on nonlinear control follow that covering linear control techniques. These chapters review classical time-optimal control and introduce the relatively new composite nonlinear feedback (CNF) control method. Again, presentations are given in both the continuous-time and discrete-time domains for completeness.

The second part of the monograph comprises five applications studies presented over five chapters. Whilst the first three of these chapters test out the control methods discussed in earlier chapters, the last two chapters introduce new applications hardware into the hard disk drive servo system problem: micro drive systems and piezoelectric actuators; nonlinear system effects are prominent in these new hardware systems.

Overall, it is an excellent monograph that exemplifies the topicality of control engineering problems today. Many lecturers will find invaluable material within this monograph with which to enthuse and motivate a new generation of control engineering students. Right at the end of this monograph, Professor Chen and his coauthors have extracted a benchmark control design problem for a typical hard disk drive system. The authors present their solution and "invite interested readers to challenge our design", so happy reading and computing!

M.J. Grimble and M.A. Johnson
Industrial Control Centre
Glasgow, Scotland, U.K.

Preface

Nowadays, it is hard for us to imagine what life would be like without computers and what computers would be like without hard disks. Hard disks provide an important data-storage medium for computers and other data-processing systems. Many of us can still recall that the storage medium used on computers in the 1960s and 1970s was actually paper, which was later replaced by magnetic tapes. The key technological breakthrough that enabled the creation of the modern hard disk drives (HDDs) came in the 1950s, when a group of researchers and engineers in IBM made the very first production hard disk, IBM 305 RAMAC (random access method of accounting and control). The first generation of hard disks used in personal computers in the early 1980s had a capacity of 10 megabytes. Modern hard disks have a capacity of several hundred gigabytes.

In modern HDDs, rotating disks coated with a thin magnetic layer or recording medium are written with data that are arranged in concentric circles or tracks. Data are read or written with a read/write (R/W) head, which consists of a small horseshoe-shaped electromagnet. It is suggested that, on a disk surface, tracks should be written as closely spaced as possible so that we can maximize the usage of the disk surface. This means an increase in the track density, which subsequently means a more stringent requirement on the allowable variations of the position of the head from the true track center. The prevalent trend in hard disk design is towards smaller drives with increasingly larger capacities. This implies that the track width has to be smaller, leading to lower error tolerance in the positioning of the head. As such, it is necessary to introduce more advanced control techniques to achieve tighter regulation in the control of the HDD servomechanism.

The scope of this second edition remains the same. It is to provide a systematic treatment on the design of modern HDD servo systems. We particularly focus on the applications of some newly developed control theories, namely the robust and perfect tracking (RPT) control, and the composite nonlinear feedback (CNF) control. Emphasis is made on HDD servo systems with either a single-stage voice-coil-motor (VCM) actuator or a dual-stage actuator in which an additional microactuator is attached to a conventional VCM actuator to provide faster response and hence higher bandwidth in the track-following stage. New design considerations and techniques, which have drastically improved the overall performance of our HDD servo systems, are introduced in this new edition. We also take this opportunity to extend the CNF control technique to systems with external disturbances and to include a comprehensive modeling and compensation of friction and nonlinearities as well as a complete servo system design of a microdrive.

The intended audience of this book includes practicing engineers in hard disk and CD-ROM drive industries and researchers in areas related to servo systems and engineering. An appropriate background for this monograph would be some senior level and/or first-year graduate level courses in linear systems and multivariable control. Some knowledge of control techniques for systems with actuator nonlinearities would certainly be helpful.

We have the benefit of the collaboration of several coworkers, from whom we have learned a great deal. Many of the results presented in this monograph are the results of our collaboration. Among these coworkers are Professor Chang C. Hang of the National University of Singapore, Dr Siri Weerasooriya, Dr Tony Huang, Mr Wei Guo and Dr Guoxiao Guo of the Data Storage Institute of Singapore. We are indebted to them for their contributions.

The authors of this monograph are particularly thankful to Guoyang Cheng for his help in proofreading the whole manuscript. The first two authors would also like to thank their current and former graduate students, especially Yi Guo, Xiaoping Hu, Lan Wang, Teck-Beng Goh, Kexiu Liu, Zhongming Li, Chen Lin and Guoyang Cheng, for their help and contributions.

We are grateful to Professor Zongli Lin of the University of Virginia, for his invaluable comments and discussions on the subject related to the composite nonlinear feedback control technique of Chapter 5. This technique, originally proposed by Zongli and his coworkers and later enhanced by us, has emerged as an effective tool in designing HDD servo systems. We are also indebted to Professor Iven Mareels of the University of Melbourne and Professor Frank Lewis of the University of Texas at Arlington, who were visiting our department here at the National University of Singapore, for many beneficial discussions on related subjects.

We would like to acknowledge the National University of Singapore for providing us with the funds for three research projects on the development of HDD servo systems. We are also grateful to people in the Design Technology Institute and the Data Storage Institute of Singapore for their support to our projects.

Last, but certainly not the least, we owe a debt of gratitude to our families for their sacrifice, understanding and encouragement during the course of preparing this monograph. It is very natural that we once again dedicate this second edition to our families.

Kent Ridge, Singapore
October 2005

Ben M. Chen
Tong H. Lee
Kemao Peng
V. Venkataramanan

Contents

Notation

We adopt the following notation and abbreviations throughout this monograph.

$\mathbb{R}$	the set of real numbers
$\mathbb{C}$	the entire complex plane
$\mathbb{C}^{\odot}$	the set of complex numbers inside the unit circle
$\mathbb{C}^{\otimes}$	the set of complex numbers outside the unit circle
$\mathbb{C}^{\circ}$	the unit circle in the complex plane
$\mathbb{C}^{-}$	the open left-half complex plane
$\mathbb{C}^{+}$	the open right-half complex plane
$\mathbb{C}^{0}$	the imaginary axis in the complex plane
I	an identity matrix
I_k	an identity matrix of dimension $k \times k$
X'	the transpose of X
X^{H}	the complex conjugate transpose of X
$\mathrm{Im}\,(X)$	the range space of X
$\mathrm{Ker}\,(X)$	the null space of X
$X^{\dagger}$	the Moore–Penrose (pseudo) inverse of X
$\lambda(X)$	the set of eigenvalues of X
$\lambda_{\max}(X)$	the maximum eigenvalue of X
$\sigma_{\max}(X)$	the maximum singular value of X
$\|X\|$	the usual 2-norm of a matrix X
$\|\|G\|\|_2$	the H_2-norm of a stable system $G(s)$ or $G(z)$
$\|\|g\|\|_2$	the l_2-norm of a signal $g(t)$ or $g(k)$
L_2	the set of all functions whose l_2 norms are finite
$\|\|g\|\|_p$	the l_p-norm of a signal $g(t)$ or $g(k)$

L_p	the set of all functions whose l_p-norms are finite
$\|\|G\|\|_\infty$	the H_∞-norm of a stable system $G(s)$ or $G(z)$
$\dim(\mathcal{X})$	the dimension of a subspace $\mathcal{X}$
$\mathcal{X}^\perp$	the orthogonal complement of a subspace $\mathcal{X}$ of $\mathbb{R}^n$
ARE	algebraic Riccati equation
CNF	composite nonlinear feedback
DSA	digital signal analyzer
DSP	digital signal processor
GB	gigabytes
HDD	hard disk drive
LDV	laser Doppler vibrometer
LQG	linear quadratic Gaussian
LQR	linear quadratic regulator
LTR	loop transfer recovery
MB	megabytes
MSC	mode-switching control
N/RRO	non-/repeatable runouts
PES	position error signal
PID	proportional-integral-derivative
PTOS	proximate time-optimal servomechanism
RPT	robust and perfect tracking
R/W	read/write
TMR	track misregistration
TOC	time-optimal control
TPI	tracks per inch (kTPI = kilo TPI)
VCM	voice-coil-motor
ZOH	zero-order hold

Also, $C^{-1}\{\mathcal{X}\} := \{x \mid Cx \in \mathcal{X}\}$, where $\mathcal{X}$ is a subspace and C is a matrix. Finally, we append a $\Diamond$ at the end of a proof or a result statement.

Part I

Introduction and Background Material

1

Introduction

1.1 Introduction

Hard disk drives (HDDs) provide an important data-storage medium for computers and other data-processing systems. In most commercial HDDs, rotating disks coated with a thin magnetic layer or recording medium are written with data that are arranged in concentric circles or tracks. Data are read or written with a read/write (R/W) head, which consists of a small horseshoe-shaped electromagnet. Figure 1.1 shows a simple illustration of a typical hard disk servo system with a voice-coil-motor (VCM) actuator.

The two main functions of the R/W head-positioning servomechanism in disk drives are track seeking and track following. Track seeking moves the R/W head from the present track to a specified destination track in minimum time using a bounded control effort. Track following maintains the head as close as possible to the destination track center while information is being read from or written to the disk. Track density is the reciprocal of the track width. It is suggested that, on a disk surface, tracks should be written as closely spaced as possible so that we can maximize the usage of the disk surface. This means an increase in the track density, which subsequently means a more stringent requirement on the allowable variations of the position of the heads from the true track center.

The prevalent trend in hard disk design is towards smaller hard disks with increasingly larger capacities. This implies that the track width has to be smaller, which leads to lower error tolerance in the positioning of the head. The controller for track following has to achieve tighter regulation in the control of the servomechanism. Basically, the servo system of an HDD can be divided into three stages, *i.e.* the track-seeking, track-settling and track-following stages (see Figure 1.2 for a detailed illustration). Current HDDs use a combination of classical control techniques, such as the proximate time-optimal control technique in the track-seeking stage, and lead-lag compensators, proportional-integral-derivative (PID) compensators in the track-following stage, plus some notch filters to reduce the effects of high-frequency resonance modes (see, *e.g.*, [1–16] and references cited therein). These classical methods can no longer meet the demand for HDDs of higher performance. Thus, many con-

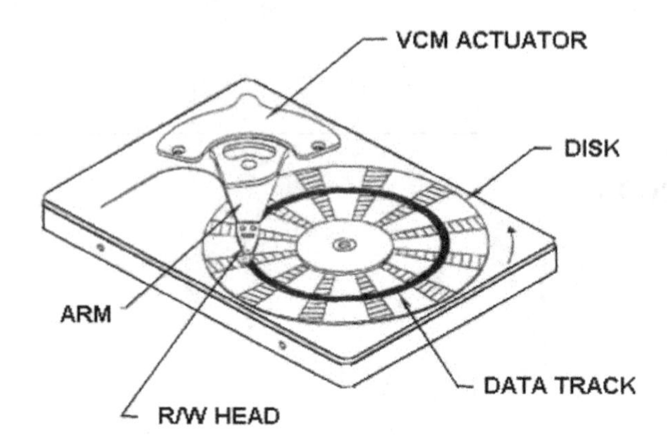

Figure 1.1. A typical HDD with a VCM actuator servo system

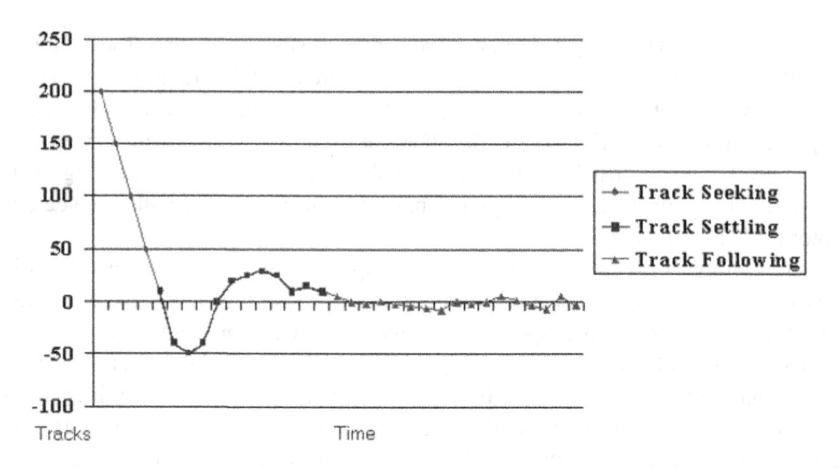

Figure 1.2. Track seeking and following of an HDD servo system

trol approaches have been tried, such as the linear quadratic Gaussian (LQG) with the loop transfer recovery (LTR) approach (see, *e.g.*, [17–19], H_∞ control approach (see, *e.g.*, [20–26], and adaptive control (see, *e.g.*, [27–30]) and so on. Although much work has been conducted to date, more studies need to be done to achieve better performance in HDDs.

The scope of this book is to provide a systematic treatment on the design of modern HDD servo systems. In particular, we focus on the applications of some newly developed results in control theory, *i.e.* robust and perfect tracking (RPT) control, which is suitable for track following, and composite nonlinear feedback (CNF) control, which is for track seeking and following. The emphasis is on HDD servo systems with either a single-stage VCM actuator or a dual-stage actuator in which an additional microactuator is attached to a conventional VCM actuator to provide faster response and hence higher bandwidth in the track-following stage. Most of the results presented in this book are from research carried out by the authors and their coworkers over the last few years. The purpose of this book is to discuss various aspects of the subject under a single cover.

1.2 Historical Development

The first generation of hard disks used in PCs had a capacity of 10 megabytes (MB) and cost over $100 per MB. Modern hard disks have capacities approaching 100 gigabytes (GB) and cost less than 1 cent per MB. This represents an improvement of 1000000% in less than 20 years and now it is cumulatively improving at 70% per year. At the same time, the speed of the hard disk and its interfaces has also increased dramatically.

Some of the very earliest computers had no storage at all. Each time a program had to be run it would have to be entered manually. It was realized then that to utilize the power of computers fully there was a need for permanent storage.

During the initial search for permanent storage, paper played a major role in human life. The computer scientists were also psychologically influenced by *paper*. This led to the use of paper as the first storage medium on computers, though magnetic storage had already gained momentum by that time. Programs and data were recorded using holes punched into paper tapes or punch cards to represent a "1", and paper blocks to represent a "0" (or *vice versa*). This type of storage was used for many years until the creation of magnetic tapes. However, these tapes also lost their place when random access to the data was needed for quick and efficient usage of data stored. Thus, an improvement needed to be found. Disk drive development took an eventful spin when IBM announced, in May 1955, a product that offered unprecedented random-access storage to 5 million characters each of 7-bit.

These early prototypes had the heads of the hard disk in contact with the disk surface. This was done to allow the low-sensitivity electronics to be able to better read the magnetic fields on the disk surface. However, owing to the fact that manufacturing techniques were not nearly as sophisticated as they are now, it was not possible to produce a disk surface that was smooth enough for the head to slide smoothly over it at high speed while in contact with the surface. As a result, the heads and the magnetic coating on the surface of the disk would wear out over time. Thus the problem of reliability was not addressed.

IBM engineers working under R. Johnson at IBM in San Jose, California, between 1952 and 1954 realized that, with the proper design, the R/W heads could be suspended above the disk surface and read the bits as they passed underneath. This critical discovery, that contact with the surface of the disk was no longer necessary, was implemented as IBM 305 RAMAC (random access method of accounting and control), introduced on September 13, 1956. This early version stored 5 million characters on 50 disks, each 24″ in diameter. The capacity was approximately 5 MB. Its bit density was about 2000 bits per square inch and the data transfer rate was an impressive 8800 bytes per second. Over the succeeding years, the technology improved incrementally; bit density, capacity and performance all increased.

Next, we summarize the interesting history of the hard disk. In what follows, we present lists of some historical "firsts" and new trends in the development of HDDs. These lists are generated from the following sources on the net: www.pcguide.com, www.storage.ibm.com, www.storagereview.com and www.mkdata.dk [31, 32].

1.2.1 Chronological List of HDD History

There have been a number of important "firsts" in the world of hard disks over their first 40 years or so. The following is a list, in chronological order, of some of the products developed during the past half-century that introduced key or important technologies in HDDs.

- FIRST HARD DISK (1956): IBM 305 RAMAC was introduced. It had a capacity of about 5 MB, stored on fifty 24″ disks. Its bit density was a mere 2000 bits per square inch and its data throughput was about 8800 bytes per second.
- FIRST AIR-BEARING HEADS (1962): IBM's model 1301 lowered the flying height of the R/W heads to 250 microinches. It had a 28-MB capacity with half as many heads as the original RAMAC, and increased both bit density and throughput by about 1000%.
- FIRST REMOVABLE DISK DRIVE (1965): IBM's model 2310 was the first disk drive with a removable disk pack. While many PC users think of removable hard disks as being a modern invention, in fact they were very popular in the 1960s and 1970s.
- FIRST FERRITE HEADS (1966): IBM's model 2314 was the first hard disk to use ferrite core heads, the first type later used on PC hard disks.
- FIRST MODERN HARD DISK DESIGN (1973): IBM's model 3340, nicknamed the *Winchester*, was introduced. With a capacity of 60 MB, it introduced several key technologies that led to it being considered by many as the ancestor of the modern disk drives.
- FIRST THIN-FILM HEADS (1979): IBM's model 3370 was the first with thin-film heads, which would for many years be the standard in the HDD industry.
- FIRST 8″ FORM FACTOR DISK DRIVE (1979): IBM's model 3310 was the first disk drive with 8″ platters, greatly reduced in size from the 14″ that had been the standard for over a decade.
- FIRST 5.25″ FORM FACTOR DISK DRIVE (1980): Seagate's ST-506 was the first drive in the 5.25″ form factor, used in the earliest PCs.
- FIRST 3.5″ FORM FACTOR DISK DRIVE (1983): Rodime introduced RO352, the first disk drive to use the 3.5″ form factor, which became one of the most important industry standards.
- FIRST EXPANSION CARD DISK DRIVE (1985): Quantum introduced the Hardcard, a 10.5-MB hard disk mounted on an industry standard architecture (ISA) expansion card for PCs that were originally built without a hard disk. This product put Quantum "on the map" so to speak.

- FIRST VOICE-COIL ACTUATOR 3.5″ DRIVE (1986): Conner Peripherals introduced CP340, the first disk drive to use a voice-coil actuator.
- FIRST "LOW-PROFILE" 3.5″ DISK DRIVE (1988): Conner Peripherals introduced CP3022, which was the first 3.5″ drive to use the reduced 1″ height, now called *low profile* and the standard for modern 3.5″ drives.
- FIRST 2.5″ FORM FACTOR DISK DRIVE (1988): PrairieTek introduced a drive using 2.5″ platters. This size later became a standard for portable computing.
- FIRST DRIVE WITH MR HEADS AND PARTIAL RESPONSE AND MAXIMUM LIKELIHOOD (PRML) DATA DECODING (1990): IBM's model 681 (Redwing), an 857 MB drive, was the first to use MR heads and PRML data decoding.
- FIRST THIN-FILM DISKS (1991): IBM's *Pacifica* mainframe drive was the first to replace oxide media with thin-film media on the platter surface.
- FIRST 1.8″ FORM FACTOR DISK DRIVE (1991): Integral Peripherals' 1820 was the first hard disk with 1.8″ platters, later used for PC-card disk drives.
- FIRST 1.3″ FORM FACTOR DISK DRIVE (1992): Hewlett Packard's C3013A is the first 1.3″ drive.
- FIRST 1″ HIGH 1 GB DISK DRIVE (1993): IBM unveiled the world's first 1″ high 1 GB disk drive, storing 354 million bits per square inch.
- FIRST 7200 RPM ULTRA ATA-INTERFACE DISK DRIVE (1997): Industry's first of this kind for desktop computers from Seagate Technology.
- FIRST 10000 RPM DISK DRIVE (1998): Seagate Technology introduced the first 10000 rpm drives, *i.e.* the 9.1-GB (ST19101) and 4.55-GB (ST34501) Cheetah family.
- FIRST ULTRA ATA/100 DISK DIVES (2000): Seagate announced the first Ultra ATA/100 interface on its Barracuda ATA II disk drive, the industry's fastest desktop PC disk drive.
- LARGEST HDD (2000): At the time of the preparation of the first edition, Seagate's Barracuda 180 was the largest single drive in the world. It had a capacity of 180 GB.
- FIRST 100 GB/IN^2 DISK DRIVE (2001): Seagate Technology demonstrated a disk drive of more than 100 billion data bits per square inch.
- FIRST 60-GB/PLATTER DISK DRIVE (2002): Seagate Technology launched the Barracuda ATA V disc drive; it's the first hard drive to achieve 120 GB using only two discs.
- FIRST 2.5″/10,000 RPM ENTERPRISE DISK DRIVE (2004): Seagate Technology introduced Savvio, the world's first family of 2.5″ enterprise-class hard disk drives.
- LARGEST 1″ DISK DRIVE (2005): At the time of the preparation of this second edition, Seagate Technology produces the largest capacity 1″ disk drive, which has a capacity of 8 GB.
- LARGEST HDD (2005): At the time of the presentation of this second edition, Barracuda 7200.9 of Seagate Technology is the largest drive in the world, which has a capacity of 500 GB.

1.2.2 Trends in Advances of HDD Systems

In spite of a slow change in the basic design of hard disks over the years, accelerated improvements in terms of their capacity, storage, reliability and other characteristics have been made. In what follows, the various trends are highlighted.

- BIT DENSITY: The bit density of hard disk platters continues to increase at an amazing rate, even exceeding some of the optimistic predictions of a few years ago. Densities in the laboratory are now approaching 1000 Gbits per square inch, and modern disks pack as much as 60 GB of data onto a single 3.5″ platter.
- CAPACITY: Hard disk capacity continues to increase at an accelerating rate. From 10 MB in 1981, the normal capacity is now well over 400 GB. Consumer drives would most likely have a capacity of 1 TB within a couple of years.
- SPINDLE SPEED: The move to faster and faster spindle speeds continues. Since increasing the spindle speed improves both random access and sequential performance, this is likely to continue. 7200 rpm spindles are now standard on mainstream IDE/ATA drives. A 15000 rpm SCSI drive was announced by Seagate in 2000.
- FORM FACTOR: The trend in form factors is downward: to smaller and smaller drives. 5.25″ drives have now all but disappeared from the mainstream market, with 3.5″ drives dominating the desktop and server segment. In the mobile world, 2.5″ drives are the standard, with smaller sizes becoming more prevalent; IBM in 1999 announced its Microdrive, a tiny 170 MB or 340 MB device, only 1″ in diameter and less than 0.25″ thick. Over the next few years, desktop and server drives are likely to make a transition to the 2.5″ form factor as well. The primary reasons for this "shrinking trend" include the enhanced rigidity of smaller platters, reduction of mass to enable faster spin speeds, and improved reliability due to enhanced ease of manufacturing.
- PERFORMANCE: Both positioning and transfer performance factors are improving. The speed with which data can be pulled from the disk is increasing more rapidly than the improvement of positioning performance, suggesting that, over the next few years, addressing seek time and latency will be the areas of greatest value to hard disk engineers.
- REDUNDANT ARRAYS OF INEXPENSIVE DISKS (RAID): In the province of only high-end servers, the use of multiple disk arrays to improve performance and reliability is becoming increasingly common, and is now seen even in consumer desktop machines.
- RELIABILITY: The reliability of hard disks is improving slowly as manufacturers refine their processes and add new reliability-enhancing features, but this characteristic is not changing nearly as rapidly as the others above. It is simply very hard to improve the reliability of a product when it is changing rapidly.
- INTERFACES: Despite the introduction to the PC world of new interfaces, such as the IEEE-1394 and universal serial bus (USB), the mainstream interfaces are the same as they were through the 1990s: IDE/ATA and SCSI. The interfaces

themselves continue to create new and improved standards with higher maximum transfer rates, to match the increase in performance of the hard disks themselves.

1.3 Overview of HDD Servo Systems

1.3.1 Mechanical Structure of an HDD

The physical structure of a typical modern hard disk drive is depicted in Figure 1.3. The authors of this book are thankful to Seagate Technology for granting permission to use this figure in our work. A brief description (see also, *e.g.*, [33]) is given below:

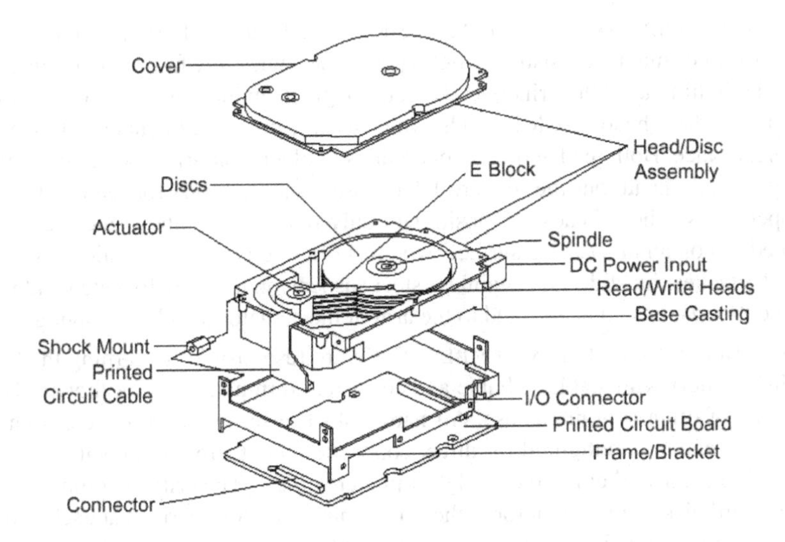

Figure 1.3. Mechanical structure of a typical HDD

1. DEVICE ENCLOSER. This is the most important component, as it determines the reliability of the disk drives. It helps to keep the contamination low. With the aid of recirculation and a breather filter, it keeps out dust and other contamination that could enter between the R/W heads and the platters over which they float, and reduces the possibility of head crashes. The two major parts, the base casting and top cover, are sealed with a gasket. The base casting provides supports for the spindle, actuator, VCM yoke and electronics card.
2. DISK. Every hard disk has one or more flat rotating disks, each with two magnetic surfaces, called platters. These are made of either an Al–Mg alloy substrate material electroless plated with Ni–P, or a mixture of glass and ceramic. The magnetic material, to allow data storage, is applied as a thin coating on both

sides of each platter together with a carbon overcoat. The surfaces of each platter are precision machined and treated to remove any imperfections, and attention is paid during the manufacturing process to ensure a very smooth surface.

3. ACTUATOR ASSEMBLY. This consists of a VCM, data flex cable or printed circuit cable, actuator arms and crash-stops at both ends of travel. The data are read/written from/to the platters using the R/W heads mounted on the top and bottom surfaces of each platter. The heads are supported by the actuator arm. The actuator in HDDs, *i.e.* the VCM actuator, is so named as it works like a loudspeaker. The electrical input to the VCM is supplied through a flex cable. The coil of the VCM actuator extends between a yoke/magnets. The write-driver/preamplifier is often part of the actuator assembly, which is mounted on a data flex cable.
4. HEAD/SUSPENSION ASSEMBLY. The R/W heads are of ferrite, metal-in-gap, thin-film or magnetoresistive (MR) types. Older types, *i.e.* ferrite, metal-in-gap, and thin film, used the principle of electromagnetic induction, whereas the modern disk drive heads use MR heads, which use the principle of change of magnetoresistance. Both read and write operations in older disk drives were performed by a single head, but the modern HDDs use separate heads for read and write operations. These heads are positioned only microinches above the recording medium on an air-bearing surface, which is often referred to as a slider. A gimbal attaches the slider to a stainless steel suspension to allow for pitch and roll, and the suspension is attached to the arm of the actuator by a ball swaging.
5. SPINDLE AND MOTOR ASSEMBLY. These are responsible for turning the hard disk platters with stable, reliable and consistent turning power for thousands of hours of often continuous use. All hard disks use servo-controlled DC spindle motors and are configured for direct connection, *i.e.* there are no belts or gears used to connect them to the hard disk platter spindle. The critical component of the hard disk's spindle motor is the set of spindle motor bearings at each end of the spindle shaft. These bearings are used to turn the platters smoothly. The disk clamper and spacers are other important parts of this assembly.
6. ELECTRONICS CARD: This provides an interface to the host personal computer (PC). The most common interfaces used are the integrated drive electronics (IDE), the advanced technology attachment (ATA), and the small computer systems interface (SCSI), which all use integrated electronic circuits. These integrated circuits have a power driver for the spindle motor, VCM, R/W electronics, servo demodulator, controller chip for timing control and control of interface, microcontroller/digital signal processor (DSP) for servo control and control interface, and ROM and RAM for microcode and data transfer.

Lastly, we note that a fairly complete report on the basic mechanical and electrical structures of hard and floppy disk drives used in the 1970s and early 1980s can be found in Zhang [34].

1.3.2 Issues on Control System Design

For smaller drives with larger capacities, the control system has to achieve tighter regulation in the control of the servomechanism. To read (or write) the data reliably from (or to) the disk, the absolute track-following error with respect to the target track center, which is commonly called track misregistration (TMR), must be less than 10% of the track pitch. For example, for a 3.5″ HDD with 25 kTPI, the track pitch is about 1 μm and its TMR must be less than 0.1 μm. Thus, for 70 kTPI, TMR must be less than 0.036 μm. This requires rigorous analysis of the sources of TMR and development of advanced techniques to overcome or eliminate these sources to meet the increasing demand for higher TPI. Figure 1.4 shows a typical disk drive servo channel indicating the various sources of disturbances and errors. Some of the larger components of TMR are due to the following error sources, listed roughly in the order of impact (see, *e.g.*, [35] and references cited therein):

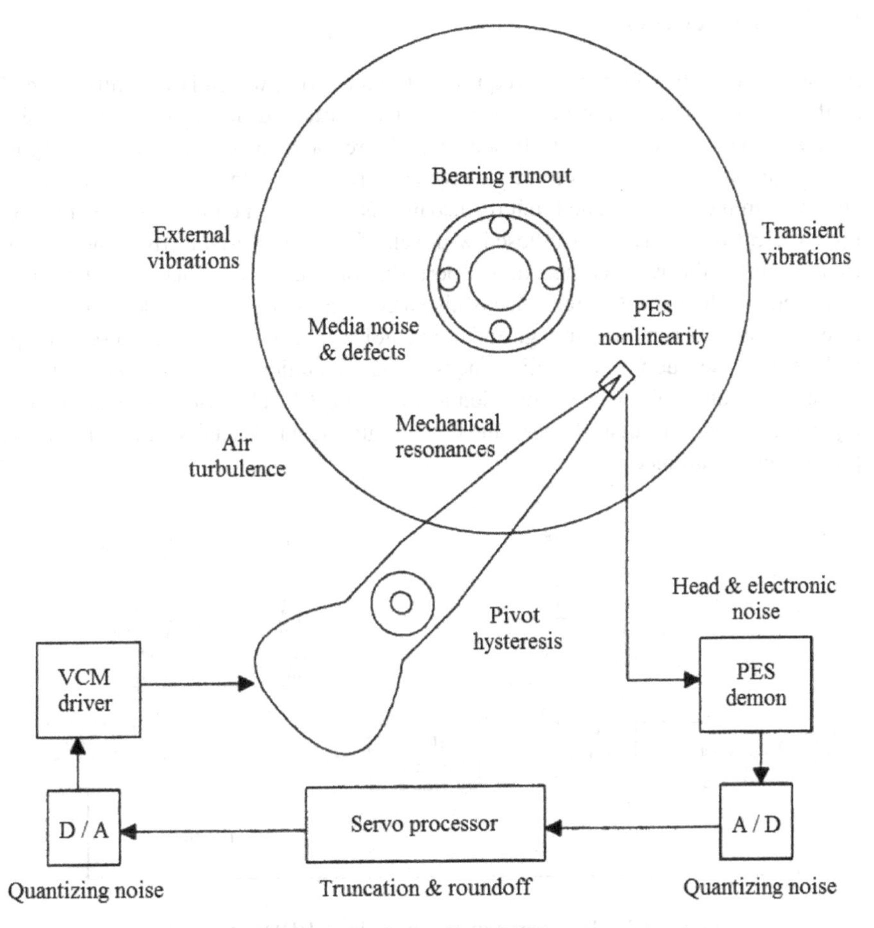

Figure 1.4. Sources of error in an HDD servo system

1. external shock and vibrations present in portable devices;
2. TMR caused by bearing hysteresis and poor velocity estimates during track-settling mode;
3. servo pattern nonlinearities and inaccuracies caused by head, media, and servo-writing effects;
4. mechanical resonances in suspension, actuator, disk, and housing;
5. electronic noise in recording channel entering the servo demodulator;
6. nonrepeatable spindle runout caused by bearings;
7. variations in RRO caused by thermal and other drifts.

A good HDD servo system has not only the desired track-seeking and following performance, but also the robustness to overcome all the above-listed disturbances and uncertainties. The following are robustness issues that one should consider in designing an HDD servo system:

Disturbance Rejection

As discussed earlier, higher TPI requires a tighter TMR, which is formally defined as three times the position error variance of the true position error signal (PES), *i.e.* $3\sigma_{\text{pes}}$. The sources of disturbances, which are the error sources contributing to $3\sigma_{\text{pes}}$, can be classified into three categories: input disturbance, output disturbance and measurement noise. The input disturbance is typically a color noise due to flexure, an electronic bias superimposed with selective energy arising from the natural frequencies of the various mechanical perturbations such as resonances, vibrations and friction. The output disturbance is also a color noise due to spindle rotation and its effects such as runout, windage and media noise. The measurement noise is a typical white noise due to the position-measurement techniques and/or sensors. These disturbances and noise can be modeled as in Figure 1.5. The objective would be to reject the effect of the disturbances and the measurement noise and achieve minimum position error variance.

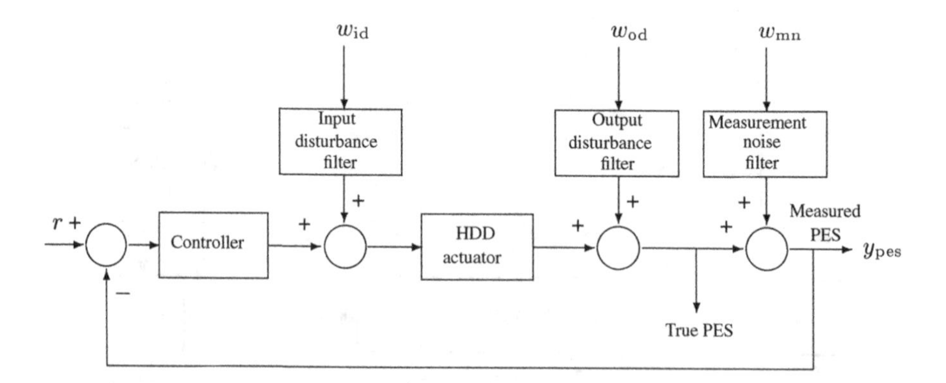

Figure 1.5. Modeling of disturbances in an HDD servo system

It has been shown [36] that, under some simplifications, σ_{pes} can be approximated as

$$\sigma_{\text{pes}} = \sqrt{\frac{1}{N-1}\sum_{i=1}^{N} y_{\text{pes}}^2(i)} \tag{1.1}$$

where N is the number of samples. Recently, Li *et al.* [37] gave a solution that minimizes σ_{pes} by converting the minimization of σ_{pes} into an equivalent H_2 optimal control problem for an auxiliary system, which contains the dynamics of the actuator, the input and output disturbances and the measurement noise, and which has a properly defined disturbance input and an output to be controlled. The problem to minimize σ_{pes} can then be solved using any appropriate H_2 optimal control methods (see, *e.g.*, Chapter 3).

Runout Compensation

Next, to understand runout disturbances, we recall the two main functions of HDD servo systems, *i.e.* track seeking and track following, which are usually achieved by two different controllers. The track-seeking controller moves the R/W head to the target track in minimum time; after this, when the control is switched to the track-following controller, it must make the R/W head follow the target track and keep the errors as small as possible. Thus, all HDDs must have a position-measurement mechanism. The position feedback signals in most HDD servomechanisms are derived through prerecorded position information recorded on one side of a disk surface at the time of manufacture using a servo writer. Ideally, servo tracks are perfect concentric circles. However, in the process of servo writing, the head that writes the signals cannot be kept perfectly still, due to, *e.g.*, the presence of vibration and NRRO effects, which result in tracks that are not perfect circles. This apparent track motion causes the R/W head to move in an attempt to minimize the position error, which results in positioning of the R/W head away from the real data track. Such an imperfection is termed a *runout*. This runout, depending upon its nature, can be classified as repeatable and nonrepeatable. In what follows we discuss these two types of runout and the methods available to compensate these disturbances collected from a literature survey.

Repeatable Runout. When the sampling frequency is equal to the spindle rotation frequency, or one of its multiples, the runout motion produced by the apparent track is repeated. This repeated runout, which is locked to the spindle rotation in both frequency and phase, is what we call a RRO. Thus, the major source of RROs is the eccentricity of the track. Other sources include the offset of the track center with respect to the spindle center, bearing geometry and wear, and motor geometry [38]. RROs caused by factors other than the eccentricity would cause a large amount of RROs at the rotational frequency of the spindle or its multiples, which is common to all tracks.

An RRO is a repetitive event in that both its amplitude and phase are locked to the rotation of the spindle. Therefore, this prior knowledge of an RRO can be used as a feedforward signal to compensate the tracking error. Roughly, there exist three approaches to reject RRO: 1) repetitive control, 2) feedback control based on the internal model principle, and 3) identification and feedforward control. Many versions of the above compensation techniques for RROs have been reported, and some include adaptive feed-forward cancellation and repetitive control (see, *e.g.*, [38–41], PID with repetitive control (see, *e.g.*, [42]) and recurrent neural networks (see, *e.g.*, [43]). To be a little more specific, assuming that the RRO, $d(t)$, is a time-varying unknown disturbance consisting of a sum of n sinusoids of known frequencies, *i.e.*

$$d(t) = \sum_{i=1}^{n} \Big[a_i(t)\, \cos(\omega_i t) + b_i(t)\, \sin(\omega_i t) \Big] \tag{1.2}$$

the adaptive feedforward compensation approach is to design a control

$$u(t) = \hat{d}(t) = \sum_{i=1}^{n} \Big[\hat{a}_i(t)\, \cos(\omega_i t) + \hat{b}_i(t)\, \sin(\omega_i t) + g_i k_i y(t) \Big] \tag{1.3}$$

Next, an appropriate adaptive algorithm is used to adjust the estimates $\hat{a}_i(t)$ and $\hat{b}_i(t)$ so that these estimates are made equal to the nominal values, *i.e.*

$$\hat{a}_i(t) = a_i(t), \quad \hat{b}_i(t) = b_i(t) \tag{1.4}$$

The RRO disturbance approximated as in Equation 1.2 can then be canceled by the reproduced signal. This technique is also demonstrated by using neural networks in [43].

Nonrepeatable Runout. NRROs are a product of disk drive vibration and electrical noise in the measurement channel (see, *e.g.*, [1]). More specifically, the causes of NRRO are spindle-bearing defects, windage-induced disk flutter, electronics noise in the measurement channel, *etc.*, present during servo-track writing. NRROs can be minimized via improved servo writing, use of better bearings, and improved design of electronics. Since RROs are the harmonics of the motor rotational frequency in the frequency domain, an NRRO is the subtraction of the harmonics from the total indicated runout (TIR), which can be defined as the distance difference between the R/W head and the previously written track in an HDD, and hence an NRRO in the time domain can be easily constructed by the inverse Fourier transform of an NRRO in the frequency domain [44]. Alternatively, for a better understanding of the time trends of an NRRO, specifically a motor NRRO, Ohmi [45] proposed subtracting the averaged RRO from the experimentally obtained radial vibration of a rotor, the TIR. That is, an NRRO can be derived as follows [45]:

$$\mathrm{RRO}\,(j) = \frac{1}{N} \sum_{i=1}^{N} \mathrm{TIR}\,(i,j), \quad j = 1, 2, \cdots, M \tag{1.5}$$

$$\mathrm{NRRO}\,(i,j) = \mathrm{TIR}\,(i,j) - \mathrm{RRO}\,(j), \quad j = 1, 2, \cdots, M \tag{1.6}$$

where N is the number of revolutions of the rotor, M is the number of samples per revolution, i represents the number of disk revolutions and j represents the number of phases from a fixed point of the slit. An NRRO can be taken care of by the servo controller through improved loop bandwidth. However, the increase in servo bandwidth required to reject an NRRO is mainly determined by three factors, *i.e.* the servo sampling rate, the spectrum of the measurement noise, and the existence of plant resonance modes. The effect of resonance modes and their compensation is discussed in the following section. Recent research suggests the use of improved mechanical design, with a damped disk substrate and fluid-bearing spindles [46], which imposes less stress on the servo loop, to reject NRROs.

Resonance Compensation

The actuator and HDD structures, of course, are not perfectly rigid and have hundreds of flexible modes. This flexibility gives rise to vibrations, which results in a longer time to settle at the target track and amounts to a significant component of the TMR. The servo bandwidth of modern disk drives is approaching 2 kHz, and it has been proved in the literature that resonance modes that exist within a decade away from the servo-crossover frequency degrade the system performance. In short, in modern disk drives, resonance or vibration modes are the major sources of NRROs. Each resonance mode can be modeled as a second-order transfer function. The VCM actuator transfer function displaying multiple resonance modes can be modeled as [35],

$$G(s) = \frac{k_{\mathrm{t}}}{Js^2} \prod_{i=1}^{n} \frac{\omega_i^2}{s^2 + 2\zeta_i\omega_i s + \omega_i^2} \tag{1.7}$$

where k_{t} is torque constant, J is inertia, ζ_i and ω_i are, respectively, the damping ratio and the natural frequency of the ith resonance mode. For simplicity, the frequency-response characteristics with a single resonance mode in actuator dynamics for a typical commercial drive are shown in Figure 1.6, in which the characteristics without resonance mode is shown by dashed lines. There is a significant difference in phase angle of the transfer function with the resonance mode. This tends to cause a loss of gain margin in the compensated loop and hence reduces its stability. Although there are hundreds of such resonances in an actual disk drive, many of the characteristics can be defined by considering only three or four modes, as other modes have an insignificant amplitude or are of too high a frequency to be of interest [1]. Some of the important resonance modes that must be considered in the design of high-density disk drives are the quasirigid body mode, the pivot bearing, the lateral elastic bending mode, and the vibrations of the individual disk platters.

Currently, the resonances of an HDD head actuator assembly caused by the pivot bearing have become a critical issue, since these resonances have been found to be the major design factor limiting the higher servo control bandwidth [47]. These resonances are excited during the track-seeking mode, and when the control is transferred to the track-following mode these vibrations result in an increased settling time. Recently, Mah *et al.* [48] have developed a novel moving-coil head actuator, which is designed deliberately to make sure that the force acting on the VCM is an orthogonal force so that there is no resulting force acting on the pivot bearing,

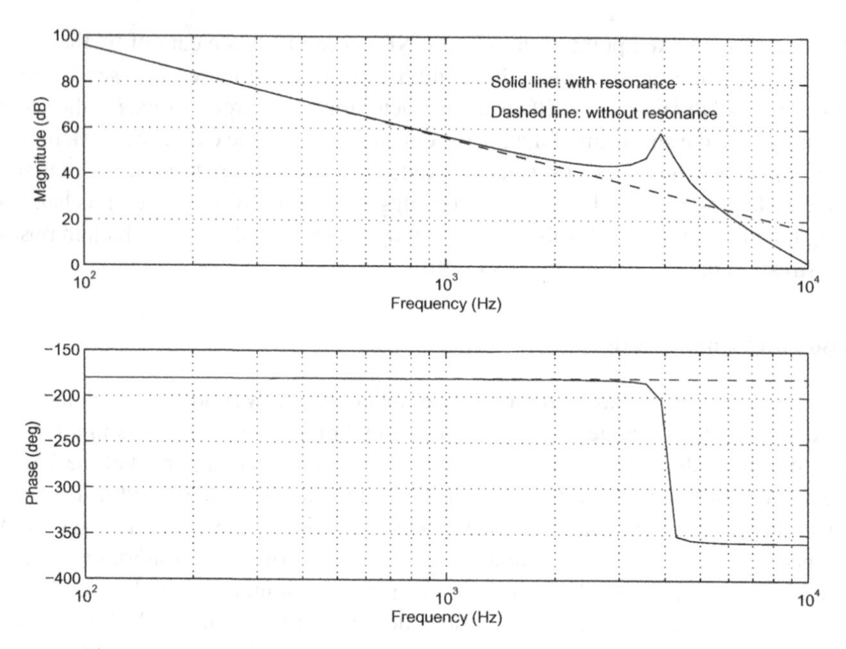

Figure 1.6. The ideal and "actual" frequency responses of HDD actuators

thereby minimizing residual vibrations. The rotational speeds of modern disk drives are progressively increasing, and hence the effect of the vibration of individual disk platters at their natural frequencies is a significant contributing factor to TMR in high-density disk drives. These resonances are driven primarily by internal windage excitation, and their behavior is dominated by the disk-material properties and geometry, and not by the spindle, encloser, or structural design (see, *e.g.*, [49]). Use of alternate disk substrate materials can control these effects. Structural resonance modes can be compensated by using a notch filter as a precompensator. Since almost all structural resonance modes have lightly damped poles, the idea is to cancel lightly damped poles and place a pair of well-damped poles instead by using a notch filter. Hanselmann and Mortix [50] proposed the use of three notch filters to suppress the plant model resonance modes. These filters are preferred instead of low-pass filters because the sharper the cutoff in the magnitude of the frequency response, the lower the phase introduced in the loop. The transfer function of an analog notch filter is commonly chosen as (see, *e.g.*, [51])

$$G_{\mathrm{n}}(s) = \frac{s^2 + (2\pi f_0)^2}{s^2 + \frac{2\pi f_0}{Q} s + (2\pi f_0)^2} \tag{1.8}$$

where f_0 is the center frequency and Q is the Q-factor. These notch filters can be realized by using switched capacitance filters. To use with digital control, digital notch filers can be realized using microprocessors or high-speed DSPs. Weaver and

Ehrlich [52] proposed the use of multirate filters to eliminate the resonance modes beyond the Nyquist frequency.

1.4 Implementation Setup

To make our work more complete, we have implemented almost all of our designs on actual HDDs with some highly advanced and accurate equipment. In what follows, we briefly summarize the key software and hardware tools used to obtain the simulation and implementation results.

1. MATLAB® AND Simulink®. All offline computation and simulation of the results in this book are done using the well-known products from Mathworks, Inc., MATLAB® with its simulation package Simulink®.
2. SOFTWARE TOOLKITS. Two software toolkits have been used in the development of results presented in this monograph. The linear systems toolkit [53], developed under a MATLAB® environment by the first author of the book and his coworkers, Zongli Lin and Xinmin Liu of the University of Virginia, collects a few tens of m-functions. These m-functions realize algorithms for computing linear system structures (such as the finite and infinite zero structures, invertibility structures, and many other properties) and algorithms for computing H_2 and H_∞ optimal controllers, as well as controllers that solve the H_∞ almost disturbance decoupling problem, and robust and perfect tracking problem. Another toolkit, the CNF control toolkit [54, 55], developed by our research team, has also been heavily used in designing track-seeking and track-following controllers in HDD servo systems presented in the coming chapters. Both toolkits are available for free at the website `http://hdd.ece.nus.edu.sg/~bmchen`.
3. dSPACE DSP SYSTEM. A dSpace DSP system is used in the actual implementation throughout the book. The system has the following main components:
   - *dSpace Add-on Card.* The main component of the dSpace DSP system is its add-on card, DS1102, which is built upon a Texas Instruments TMS320C31 floating-point DSP. The DSP has been supplemented by a set of analog-to-digital (A/D) and digital-to-analog (D/A) converters, a DSP microcontroller-based digital I/O subsystem and incremental sensor interfaces. Some major features of this add-on card are:
     a) a TMS320C31 floating-point DSP;
     b) two 16-bit 250 kHz and two 12-bit 800 kHz sampling A/D converters with input span of ± 10 V;
     c) a quad 12-bit D/A converter with programmable output voltages;
     d) a 16-bit fixed point digital I/O, a bit-selectable-parallel I/O port, four timers, six PWM circuits, and a serial interface.
   - *Real-time Interface (RTI) and Real-time Workshop (RTW).* The RTI acts as a link between Simulink® and the dSpace hardware. It has builtin hardware control functions and blocks for DS1102 add-on card based on Simulink®. This, together with the RTW, automatically generates real-time codes from

Simulink® offline models and implements these codes on the dSpace real-time hardware.
- *dSpace Control Desk*. This is a software platform that combines all the above tools of dSpace for controlling, monitoring, and automating the implementation process on the actual HDDs.

4. POLYTEC LASER DOPPLER VIBROMETER (LDV). The Polytec LDV is an optical instrument for accurately measuring velocity and displacement of vibrating surfaces completely without contact. The LDV system consists of two main components: 1) an optical sensor head or fiber optic unit (both are laser interferometers), which measures the dynamic Doppler shift from the vibrating object; and 2) a controller (processor), which provides power to the optics and demodulates the Doppler information using various types of Doppler signal decoder electronics, thereby producing an analog vibration signal (velocity and/or displacement) that can be viewed/measured by the customer using commercially available fast Fourier transform (FFT) analyzers and oscilloscopes. This instrument is used to measure the displacement and velocity of the R/W heads of HDDs.
5. DYNAMIC SIGNAL ANALYZER (DSA). The HP35670A dynamic signal analyzer is a dynamic monitoring and measuring instrument that can be used for characterizing the performance and stability of a control system. Performance parameters, such as rise time, overshoot, and settling time, are generally specified in the time domain. Stability criteria, gain/phase margins, are generally specified in the frequency domain. The HP35670A DSA is capable of measuring in both the time and frequency domains. The instrument can also be used for system identification.
6. VIBRATION-FREE TABLE. Since the success of the actual implementation depends largely on the accurate measurement of very small displacements of less than 1 μm, there is a need to isolate the HDD implementation setup from the external vibrations. A Vibraplane Model 9100/9200 series vibration-free workstation was used. These are designed and constructed to provide very effective isolation of vibrations at frequencies above 5 Hz and low amplification at low frequencies of 2–3 Hz. Hence, the use of this vibration-free table shows significant improvements in resolution and repeatability of the measurement.

The overall hardware setup in our laboratory at the National University of Singapore (NUS) is depicted in Figure 1.7.

1.5 Preview of Each Chapter

Since the publication of the first edition of this monograph [56], we have developed many new design methodologies and obtained many new theoretical and experimental results. We are to integrate all of these new developments and results in this second edition. This new edition can be naturally divided into two parts. The first part consists of Chapters 1 to 5, covering some background introductions to HDD servo

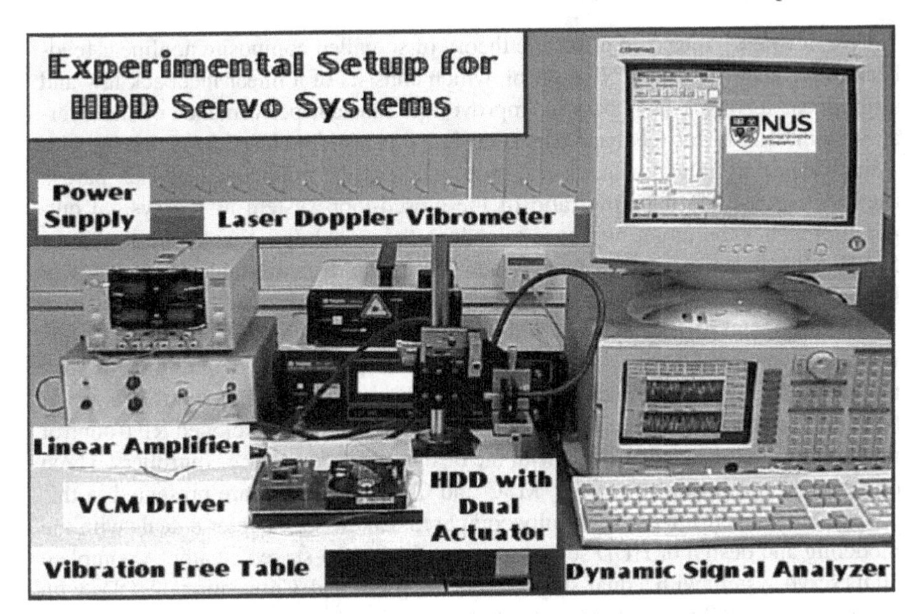

Figure 1.7. Implementation setup for HDD servo systems

systems, and some commonly used as well as newly developed linear and nonlinear control techniques. The second part, *i.e.* Chapters 6 to 10, deal with the design of specific single- and dual-stage actuated HDD servo systems.

More specifically, Chapter 2 recalls some commonly used system identification and modeling techniques, such as the prediction error identification and least squares estimation methods, applicable in the frequency domain, and the impulse response analysis and step response analysis in the time domain, as well as the physical effect approach together with Monte Carlo estimations. These techniques are employed to identify the models of VCM actuators and microactuators later in the book. Chapter 3 deals with linear systems and control techniques. In particular, we first recall a structural decomposition technique of linear systems, which has the distinct feature of displaying the finite and infinite zero structures as well as the invertibility structures of a given system, and plays a dominant role in the development of several linear control methods used in designing HDD servo systems. We also recall several commonly used linear control techniques, namely, the well-known classical PID control, H_2 optimal control, H_∞ control and almost disturbance decoupling, robust and perfect tracking (RPT) control, and the loop transfer recovery (LTR) technique. These methods are suitable for track-following control and have been used extensively in designing HDD servo systems in the literature. Chapter 4 focuses on some classical nonlinear control techniques such as the proximate time-optimal servomechanism (PTOS) and mode-switching control (MSC). PTOS is generally used to design a control law in the track-seeking stage of HDD servo systems, whereas the MSC design technique can be used to find a controller that is applicable for both track seeking and track following.

We devote Chapter 5 to cover the theory of so-called composite nonlinear feedback (CNF) control. The CNF control, which consists of a linear feedback law and a nonlinear feedback law, aims to improve the transient performance of the overall system. The linear feedback part is designed to yield a closed-loop system with a small damping ratio for quick response, whereas the nonlinear feedback part is used to increase the damping ratio of the closed-loop system as the system output approaches the target reference to reduce the overshoot. A software toolkit in MATLAB® is also presented for this new technique. It is available for free downloading at `http://hdd.ece.nus.edu.sg/~bmchen`.

In the second part, Chapters 6 and 7 focus on the design of HDD servo systems with a conventional single-stage VCM actuator. In particular, Chapter 6 deals with the modeling of the VCM actuator and design of track-following controllers using both the conventional PID control method and the recently developed RPT control method, whereas Chapter 7 deals with the design of track-seeking controllers. Three different designs using the PTOS, MSC and CNF approaches are presented in this chapter and their results are carefully compared. Likewise, Chapter 8 deals with the modeling and design of HDD servo systems with a dual-stage actuator. A complete HDD servo system with a dual-stage actuator is presented. Conventional HDDs with a single-stage VCM actuator usually have resonances in the positioning arm and low-frequency bearing effects. It is believed that the performances of such HDDs have been pushed almost to their limits. Dual-stage servo systems with high bandwidth and high accuracy control are a possible solution to overcome the problems associated with conventional HDDs.

Microdrives have become popular these days with high demand from many new applications. Unfortunately, many factors such as friction and nonlinearities, which can be safely neglected in normal drives, emerge as critical issues for microdrives. In Chapter 9, we present a comprehensive modeling and compensation of friction and nonlinearities as well as a complete servo system design of a typical microdrive. Chapter 10 considers a robust controller design for a piezoelectric bimorph nonlinear actuator using an H_∞ almost disturbance decoupling approach. Such an actuator has its own application domain including the use as a microactuator for the dual-stage actuated hard drives. Finally, we post in Chapter 11 a benchmark design problem for a typical HDD (Maxtor 51536U3) servo system, for which interested readers might try out to test their own techniques and designs.

2

System Modeling and Identification

2.1 Introduction

The purpose of this chapter is to revisit some basic theories and solutions of system identification, which are to be used later in the coming chapters to model various HDD systems. In general, the goal of system identification is to determine a mathematical model for a system or a process. Mathematical models may be developed either by use of "laws of nature", commonly known as *modeling* or based on experimentation, which is known as *system identification* [57]. In order to achieve a certain desirable performance for a given plant, it is necessary to derive a model for the plant that is adequate for controller design. The conventional design techniques in linear control systems require either parametric or nonparametric models. For example, design methods via root locus or robust control technique require a transfer function or a state-space description of the plant to be controlled. The plant model is either described by the coefficients of certain polynomials or by the elements of state-space matrices. In either case, we call these polynomial coefficients or matrix elements the *parameters* of the model. The category of such models is a *parametric description* of the plant model. On the other hand, design based on Nyquist, Bode and Nichols methods requires curves of amplitude and phase of transfer function from input to the output as functions of real frequency ω. If we have experimental data from a typical frequency response test, then we will be able to obtain certain functional curves for the plant. These curves are called nonparametric models of the plant, as there is no finite set of numbers that describes it exactly (see, *e.g.*, [1]).

Thus, for a given plant, the problem of system identification is to determine a system model from the relationship (either in the time or the frequency domain) between its input and output. The problem can be represented graphically as shown in Figure 2.1, in which $u(t)$ is the known input signal, $n(t)$ is the observation noise, and $y(t)$ is the measured output. A large variety of methods have been developed for solving such a problem (see, *e.g.*, [58] and references cited therein). These methods include classical identification techniques (such as the impulse response analysis, step response analysis, frequency response identification) and equation error approaches and model adjustment techniques (such as the least square estimation, maximum

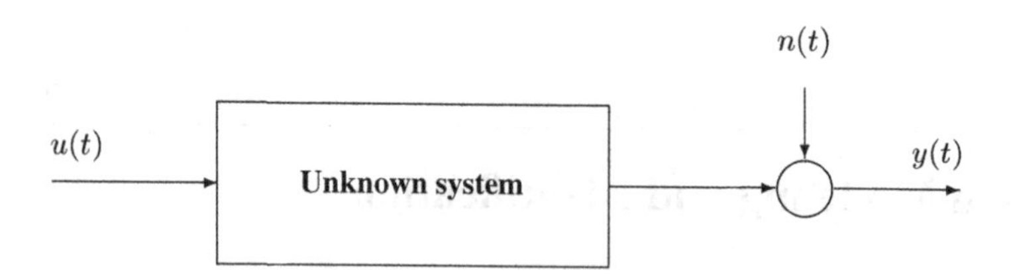

Figure 2.1. The unknown system to be identified

likelihood, and stochastic approximation, to name a few). The detailed derivations of these techniques can be found in a number of advanced texts devoted to system identification, *e.g.*, [57–61]. We note that the techniques are applicable only to linear systems. For systems with nonlinearities, other advanced modeling techniques have to be used. One of the key issues in modeling systems with nonlinearities is to determine the physical structures of the nonlinear components. Another issue is that there is generally no analytical solution for nonlinear differential equations, which causes trouble in identifying the model parameters. In the last section of this chapter, we present a so-called physical effect approach modeling method of [62] together with Monte Carlo estimations (see, *e.g.*, [63–65]). The technique will be utilized to identify a comprehensive model for a microdrive in Chapter 9, in which friction and nonlinearities associated with its VCM actuator are highly noticeable.

2.2 Time-domain Methods

In this section, we restrict our attention to identifying both parametric and nonparametric models through some commonly used time-domain techniques. Interested readers are referred to [57, 58, 66] for detailed materials on the identification through impulse and step response characteristics.

2.2.1 Impulse Response Analysis

Parametric Models. Parametric models are described by parameters of differential equations or transfer functions. From these analytic representations, plots or values of interest of frequency response can in general be generated without much difficulty, whereas the reverse process of deriving parameters from nonparametric model descriptions is much more difficult.

A fairly general parametric model of a single-input and single-output (SISO) system can be described by the following differential equation (see, *e.g.*, [66]),

$$\begin{aligned} a_n y^{(n)}(t) + \cdots + a_1 \dot{y}(t) + a_0 y(t) \\ = b_0 u(t-\tau_\mathrm{d}) + b_1 \dot{u}(t-\tau_\mathrm{d}) + \cdots + b_m u^{(m)}(t-\tau_\mathrm{d}) \end{aligned} \tag{2.1}$$

Solving the differential equation for the input signal

$$u(t) = \delta(t) \tag{2.2}$$

with $\delta(t)$ being the unit impulse function, gives the impulse transfer function $h(t)$,

$$y(t) = h(t) \tag{2.3}$$

as the corresponding output function. Note that it is hard to generate an impulse input in the continuous-time domain, and hence this method is impractical.

Nonparametric Models. Consider again a SISO system as in Figure 2.1 with a scalar input signal $u(t)$ and a scalar output signal $y(t)$. Assume that the system is linear, time invariant and causal.

It is well known that a linear, time-invariant causal system can be described by its impulse response $g(\tau)$ as follows:

$$y(t) = \int_0^{\infty} g(\tau)u(t-\tau)\,\mathrm{d}\tau + n(t) \tag{2.4}$$

Knowing $\{g(\tau)\}_{\tau=0}^{\infty}$ and knowing $u(s)$ for $s \leq t$, we can consequently compute the corresponding output $y(s)$, $s \leq t$ for any input. The impulse response is thus a complete characterization of the system.

The discrete equivalent of the output $y(t)$ can be written at the sampling instants $t_k = kT$, $k = 0, 1, 2, \cdots$, as

$$y(t_k) = \int_0^{\infty} g(\tau)u(t_k-\tau)\,\mathrm{d}\tau + n(t_k) \tag{2.5}$$

where T is the sampling period. Since, the input $u(t_k)$ is kept constant between the sampling instants:

$$u(t) = u(t_k) = u_k, \qquad kT \leq t < (k+1)T \tag{2.6}$$

we can derive that

$$y(t_k) = \sum_{i=0}^{\infty} g(i)u(t_k - i) + n(t_k), \quad t_k = kT,\ k = 0, 1, 2, \cdots \tag{2.7}$$

Now, let $G(z)$ be the transfer function of the system from input to output with z being the usual forward shift operator, *i.e.*

$$G(z) = \sum_{k=0}^{\infty} g(k)z^{-k} \tag{2.8}$$

Then, Equation 2.7 can be written as

$$y(t_k) = G(z)u(t_k) + n(t_k) \tag{2.9}$$

If the system in Equation 2.9 is subjected to a pulse input

$$u(t_k) = \begin{cases} \alpha, & t_k = 0 \\ 0, & t_k \neq 0 \end{cases} \tag{2.10}$$

then the output is

$$y(t_k) = \alpha g(t_k) + n(t_k) \tag{2.11}$$

If the noise level is low, then the estimates of the coefficients of the impulse response $\{g(t_k)\}$ from an experiment will be

$$\hat{g}(t_k) = \frac{y(t_k)}{\alpha} \tag{2.12}$$

and the errors $n(t_k)/\alpha$. This simple analysis is called impulse response analysis. Unfortunately, many physical processes do not allow the error $n(t_k)/\alpha$ to be insignificant compared with the impulse response coefficients. Moreover, such an input could induce nonlinear effects that would disturb the linearized behavior of the model. As such, identification methods depended on impulse inputs are rarely used in practical situations.

2.2.2 Step Response Analysis

Parametric Models. Parametric models usually are described by their frequency response function,

$$G(\omega) = \frac{b_0 + b_1 j\omega + ...b_m (j\omega)^m}{a_0 + a_1 j\omega + ...a_n (j\omega)^n} \cdot \exp(-j\omega\tau) \tag{2.13}$$

where a_i, b_i, m, n and τ are parameters to be identified.

Many researchers proposed methods for determining parameter values from time functions of the process output provided that the process input is a well-determined signal. Of primary importance are step functions as input signals and step responses as output signals.

Almost all methods for the evaluation of step response functions use a small number of characteristic values of the response function. A first-order lag model with time delay and a frequency response

$$\hat{G}(\omega) = \frac{K}{1 + j\omega\tau_1} \cdot \exp(-j\omega\tau) \tag{2.14}$$

is widely used in step response analysis. In fact, there are quite a number of systems, especially in process control, that can be approximated by a first-order model with an appropriate delay. The parameters K, τ_1 and τ can be derived from the step response shown in Figure 2.2 with u_0 being the amplitude of the input step signal (see, *e.g.*, [66]). In short, if the system model is of first order, one may need only obtain two pieces of information: (i) the steady-state response to the step input, and (ii) the time constant. The latter can be obtained either from the tangent with maximum slope of the step response or from the 10 to 90% rise time.

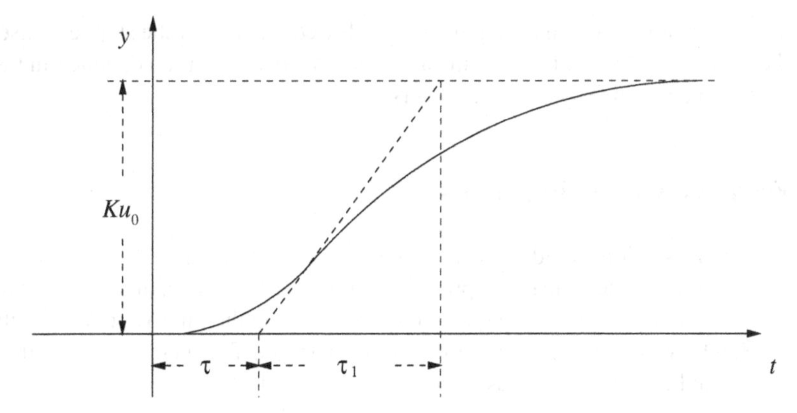

Figure 2.2. A typical step response

For a second-order system model (with two poles and no zero), there are two possible situations: 1) when the two poles are real and 2) when the poles are a complex conjugate pair. Formulae for finding these from measurements of the (a) steady-state response, (b) maximum overshoot, (c) time required to reach the first-peak, and (d) time required to reach 50% of the steady-state value (for overdamped systems) can be easily derived. For the general case of higher-order practical systems, it is perhaps best to use a gradient method to find the parameters of the model of a given order such that the integral of the square of the error is minimized (see, *e.g.*, [57]).

Nonparametric Models. Since the impulse response of a system is the derivative of the step response, the identification problem in this case may be regarded as the determination of the transfer function from the impulse response. Alternatively, a step function

$$u(t_k) = \begin{cases} \alpha, & t_k \geq 0 \\ 0, & t_k < 0 \end{cases} \tag{2.15}$$

when applied to Equation 2.9 gives the output

$$y(t_k) = \alpha \sum_{i=0}^{k} g(i) + n(t_k), \quad t_k = kT, \ k = 0, 1, \cdots \tag{2.16}$$

Then, the estimates can be obtained as

$$\hat{g}(t_k) = \frac{y(t_k) - y(t_{k-1})}{\alpha} \tag{2.17}$$

which has an error $[n(t_k) - n(t_{k-1})]/\alpha$. Hence, we would suffer again from large errors in most practical applications. But, if the goal is to determine some basic control-related characteristics, then the step responses from Equation 2.16 can very well furnish that information to a sufficient degree of accuracy. In fact, some well-known rules for tuning simple regulators, such as the Ziegler–Nichols rule, are based on model information hidden in step responses. Based on plots of the step responses, some key characteristics of the system can be graphically constructed, which in turn can be used to determine system parameters.

2.3 Frequency-domain Methods

We recall in this section two identification methods in the frequency domain, namely, the predication error identification approach and the least square estimation method, Both are particularly important to our studies in modeling the micro and VCM actuators in HDD servo systems in the forthcoming chapters. The theories behind these techniques can be found in various references (see, *e.g.*, [57, 67]).

2.3.1 Prediction Error Identification Approach

The prediction error approach is a black-box identification method. It includes the following three steps.

1. PARAMETER IDENTIFICATION. Suppose a system can be described as

$$y(k) = G(z^{-1})u(k) + H(z^{-1})n(k) \tag{2.18}$$

where $u(k)$ and $y(k)$ are its process input and output; $n(k)$ is noise input and supposed to be white; and

$$\left.\begin{aligned} G(z^{-1}) &= B(z^{-1})/A(z^{-1}) \\ H(z^{-1}) &= D(z^{-1})/A(z^{-1}) \\ A(z^{-1}) &= 1 + a_1 z^{-1} + a_2 z^{-2} + \cdots + a_{n_\mathrm{a}} z^{-n_\mathrm{a}} \\ B(z^{-1}) &= b_1 z^{-1} + b_2 z^{-2} + \cdots + b_{n_\mathrm{b}} z^{-n_\mathrm{b}} \\ D(z^{-1}) &= d_1 z^{-1} + d_2 z^{-2} + \cdots + d_{n_\mathrm{d}} z^{-n_\mathrm{d}} \end{aligned}\right\} \tag{2.19}$$

The predictor is:

$$\hat{y}(k|\theta) = [1 - H^{-1}(z^{-1},\theta)]y(k) + H^{-1}(z^{-1},\theta)G(z^{-1},\theta)u(k) \tag{2.20}$$

where

$$\theta = \begin{pmatrix} a_1 \\ \vdots \\ a_{n_\mathrm{a}} \\ b_1 \\ \vdots \\ b_{n_\mathrm{b}} \\ d_1 \\ \vdots \\ d_{n_\mathrm{d}} \end{pmatrix} \tag{2.21}$$

is the parameter vector of the system. Then, the prediction error given by a model is

$$e(k,\theta) = y(k) - \hat{y}(k|\theta) \tag{2.22}$$

Next, we define a loss function as

$$V_N(\theta, Z_N) = \frac{1}{N}\sum_{k=1}^{N} \ell(e(k,\theta)) \tag{2.23}$$

where $\ell(\cdot)$ is a scalar-valued positive function, and

$$Z_N = \Big[y(1), \cdots, y(N), u(1), \cdots, u(N)\Big] \tag{2.24}$$

is a set of input and output data from the experiment test. The desired system parameters can then be obtained by the minimization of this loss function, *i.e.*

$$\hat{\theta}_N = \arg\min V_N(\theta, Z_N) \tag{2.25}$$

2. DETERMINATION OF MODEL ORDER. The loss function $V_N(\theta, Z_N)$ can also be used to determine the order of a system. If the order of a model is lower than that of the system, then the value of the loss function will decrease significantly with the increase of the order of the model. However, when the order of the model is higher than that of the system, the increase of model order does not provide any more innovation for parameter identification, thus the value of $V_N(\theta, Z_N)$ will not decrease much. Therefore, the order of the system to be identified can be determined based on the decrease rate of $V_N(\theta, Z_N)$. Figure 2.3 shows a typical plot of the loss function versus identified model orders. It is clear from the plot that the order of the corresponding system to be identified is four.
3. MODEL VALIDATION. The last step of the prediction error identification method is to verify the correctness of the model obtained. It is clear that the residuals of the model can be obtained as

$$\begin{aligned} e(k,\theta) &= y(k) - \hat{y}(k|\theta) \\ &= \hat{H}(z^{-1},\theta)\Big[y(k) - \hat{G}(z^{-1},\theta)u(k)\Big] \end{aligned} \tag{2.26}$$

Obviously, if the model is correct, *i.e.*

$$\hat{G}(z^{-1},\theta) = G_0(z^{-1}) \quad \text{and} \quad \hat{H}(z^{-1},\theta) = H_0(z^{-1}) \tag{2.27}$$

the residual will tend to a white-noise sequence $n(k)$. However, the nonwhiteness of the residuals does not necessarily mean that the model is incorrect. In that case, the crosscorrelation of the input u and residuals e can be used to verify the model. If u and e are independent, this means that all information in the residuals is explained by the process model $\hat{G}$, and we can conclude that the estimate is correct. Otherwise the result is incorrect. The crosscorrelation of u and e is

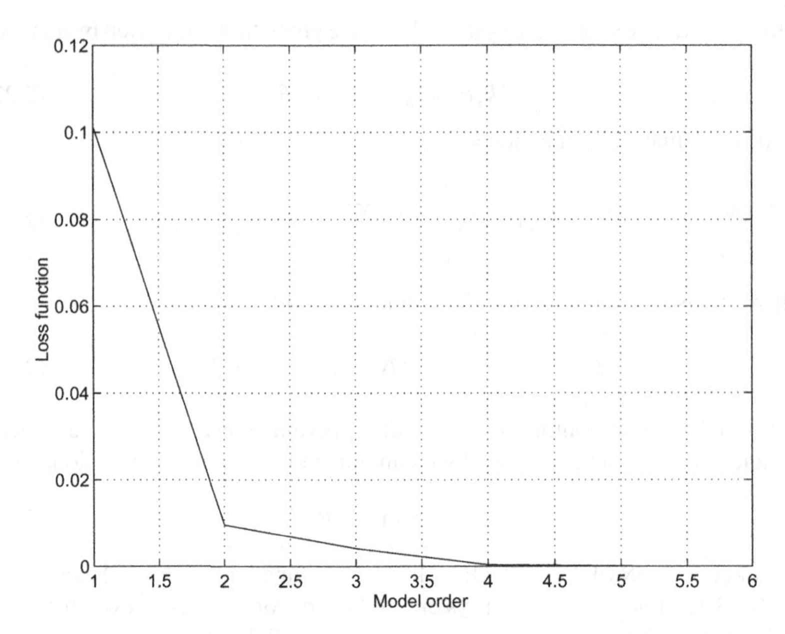

Figure 2.3. Values of loss function versus identified model orders

$$R_{eu}(\tau) = \boldsymbol{E}\{e(t+\tau)u(t)\} \tag{2.28}$$

where $\boldsymbol{E}\{\cdot\}$ is the expected value. If the residuals and input are independent, we have

$$\sqrt{N}R_{eu} \to \mathcal{N}(0,P) \quad \text{as } N \to \infty \tag{2.29}$$

where $P = \sum_{k=-\infty}^{\infty} R_e(k)R_u(k)$, and $\mathcal{N}(0,P)$ denotes the normal random distribution with zero mean and a variance P. Let N_α be the α-level of the $\mathcal{N}(0,P)$ such that

$$\boldsymbol{P}\left\{|R_{eu}| \leq \sqrt{\frac{P}{N}}N_\alpha\right\} = \alpha \tag{2.30}$$

where $\boldsymbol{P}\{\cdot\}$ is the probability. Define the following null hypothesis:

$$H_0 : |R_{eu}| \leq \sqrt{\frac{P}{N}}N_\alpha \tag{2.31}$$

If H_0 is accepted, then we can say that the model is acceptable with a probability of $1-\alpha$. Figure 2.4 shows a typical plot of the values of the crosscorrelation function between the input and the error residual. It can be seen that, for such a model, all the data are within the 95% confidence region. Hence, we can say that the corresponding identified model is acceptable with a probability of 95%.

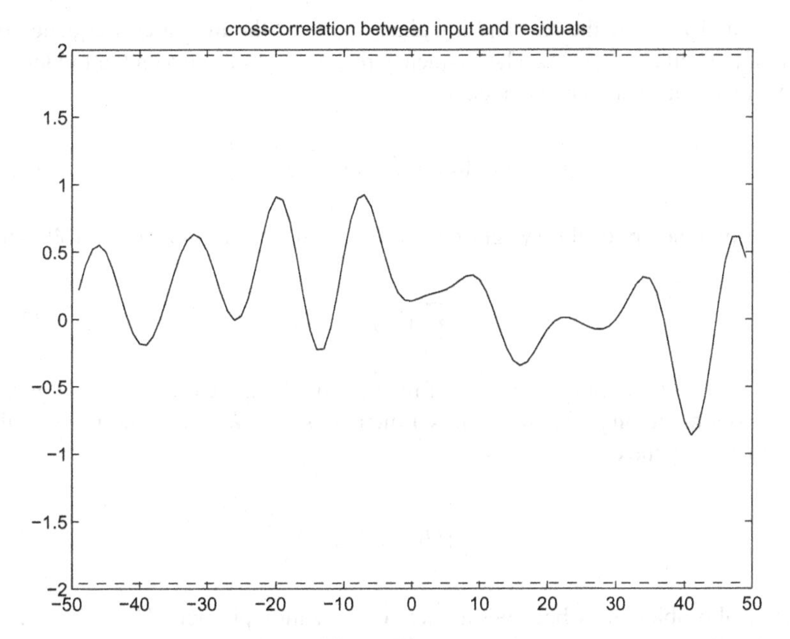

Figure 2.4. Model validation test

2.3.2 Least Square Estimation Method

We will utilize the frequency-response identification method (see, *e.g.*, [59]) to model our actuator. Such a method is applicable to minimum phase processes. We expect from the properties of the physical system that the VCM actuator is of minimum phase. The detailed procedure proceeds as follows: we first assume that the transfer function of a minimum phase plant is given by

$$G(s) = \frac{N(s)}{D(s)} = \frac{b_0 + b_1 s + b_2 s^2 + \cdots + b_m s^m}{1 + a_1 s + a_2 s^2 + \cdots + a_n s^n} \tag{2.32}$$

for some appropriate coefficients a_k, $k = 1, 2, \cdots, n$, and b_k, $k = 0, 1, \cdots, m$, with $n \geq m$. These parameters are to be identified. Then, its corresponding frequency response is given by

$$G(j\omega) = \frac{\alpha(\omega) + j\omega\beta(\omega)}{\sigma(\omega) + j\omega\tau(\omega)} = \frac{N(j\omega)}{D(j\omega)} \tag{2.33}$$

where

$$\left.\begin{aligned} \alpha(\omega_i) &= b_0 - b_2\omega_i^2 + b_4\omega_i^4 - \cdots \\ \beta(\omega_i) &= b_1 - b_3\omega_i^2 + b_5\omega_i^4 - \cdots \\ \sigma(\omega_i) &= 1 - a_2\omega_i^2 + a_4\omega_i^4 - \cdots \\ \tau(\omega_i) &= a_1 - a_3\omega_i^2 + a_5\omega_i^4 - \cdots \end{aligned}\right\} \tag{2.34}$$

Let $R(\omega)$ and $I(\omega)$ be the real and imaginary parts of the measured frequency response of the actuator system. The frequency response error between the model and the actual measurement data is given by

$$\mathcal{E}(j\omega) = [R(\omega) + jI(\omega)] - \frac{N(j\omega)}{D(j\omega)} \tag{2.35}$$

Thus, the parameters of the system can be obtained by minimizing the following index:

$$J = \sum_{i=1}^{L} |\mathcal{E}(j\omega_i)|^2 \tag{2.36}$$

where L is the total number of points of the measured data. Unfortunately, this is a nonlinear optimization problem, and it is difficult to solve. We then follow the results of [59] to modify the error norm as

$$J = \sum_{i=1}^{L} |D(j\omega_i)\mathcal{E}(j\omega_i)|^2 \tag{2.37}$$

The original problem now becomes a linear optimization problem. Using Equations 2.33 and 2.35, we can rewrite Equation 2.37 as follows

$$J = \sum_{i=1}^{L} \left\{ [X(\omega_i)]^2 + [Y(\omega_i)]^2 \right\} \tag{2.38}$$

where

$$X(\omega_i) = \sigma(\omega_i)R(\omega_i) - \omega_i\tau(\omega_i)I(\omega_i) - \alpha(\omega_i) \tag{2.39}$$

and

$$Y(\omega_i) = \omega_i\tau(\omega_i)R(\omega_i) + \sigma(\omega_i)I(\omega_i) - \omega_i\beta(\omega_i) \tag{2.40}$$

Therefore, J can be minimized by finding $\hat{b}_0, \hat{b}_1, \cdots, \hat{b}_m$ and $\hat{a}_1, \hat{a}_2, \cdots, \hat{a}_n$ such that

$$\left.\begin{aligned}
\left.\frac{\partial J}{\partial b_0}\right|_{b_0=\hat{b}_0} &= \sum_{i=1}^{L} \left\{2X(\omega_i)(-1)\right\}\Big|_{b_0=\hat{b}_0} = 0 \\
\left.\frac{\partial J}{\partial b_1}\right|_{b_1=\hat{b}_1} &= \sum_{i=1}^{L} \left\{2Y(\omega_i)(-\omega_i)\right\}\Big|_{b_1=\hat{b}_1} = 0 \\
&\vdots \\
\left.\frac{\partial J}{\partial a_1}\right|_{a_1=\hat{a}_1} &= \sum_{i=1}^{L} 2\omega_i \Big[Y(\omega_i)R(\omega_i) - X(\omega_i)I(\omega_i)\Big]\Big|_{a_1=\hat{a}_1} = 0 \\
\left.\frac{\partial J}{\partial a_2}\right|_{a_2=\hat{a}_2} &= \sum_{i=1}^{L} -2\omega_i^2 \Big[X(\omega_i)R(\omega_i) + Y(\omega_i)I(\omega_i)\Big]\Big|_{a_2=\hat{a}_2} = 0 \\
&\vdots
\end{aligned}\right\} \tag{2.41}$$

Rearranging the above equations, we obtain

$$\begin{pmatrix} \boldsymbol{A}_{11} & \boldsymbol{A}_{12} \\ \boldsymbol{A}_{21} & \boldsymbol{A}_{22} \end{pmatrix} \begin{pmatrix} \boldsymbol{b} \\ \boldsymbol{a} \end{pmatrix} = \begin{pmatrix} \boldsymbol{B}_1 \\ \boldsymbol{B}_2 \end{pmatrix} \tag{2.42}$$

where

$$\boldsymbol{A}_{11} = \begin{bmatrix} V_0 & 0 & -V_2 & 0 & V_4 & \cdots \\ 0 & V_2 & 0 & -V_4 & 0 & \cdots \\ V_2 & 0 & -V_4 & 0 & V_6 & \cdots \\ 0 & V_4 & 0 & -V_6 & 0 & \cdots \\ V_4 & 0 & -V_6 & 0 & V_8 & \cdots \\ \vdots & \vdots & \vdots & \vdots & \vdots & \ddots \end{bmatrix}, \quad \boldsymbol{b} = \begin{bmatrix} \hat{b}_0 \\ \hat{b}_1 \\ \hat{b}_2 \\ \hat{b}_3 \\ \hat{b}_4 \\ \vdots \end{bmatrix} \tag{2.43}$$

$$\boldsymbol{A}_{12} = \begin{bmatrix} T_1 & S_2 & -T_3 & -S_4 & T_5 & \cdots \\ -S_2 & T_3 & S_4 & -T_5 & -S_6 & \cdots \\ T_3 & S_4 & -T_5 & -S_6 & T_7 & \cdots \\ -S_4 & T_5 & S_6 & -T_7 & -S_8 & \cdots \\ T_5 & S_6 & -T_7 & -S_8 & T_9 & \cdots \\ \vdots & \vdots & \vdots & \vdots & \vdots & \ddots \end{bmatrix}, \quad \boldsymbol{a} = \begin{bmatrix} \hat{a}_1 \\ \hat{a}_2 \\ \hat{a}_3 \\ \hat{a}_4 \\ \hat{a}_5 \\ \vdots \end{bmatrix} \tag{2.44}$$

$$\boldsymbol{A}_{21} = \begin{bmatrix} T_1 & -S_2 & -T_3 & S_4 & T_5 & \cdots \\ S_2 & T_3 & -S_4 & -T_5 & S_6 & \cdots \\ T_3 & -S_4 & -T_5 & S_6 & T_7 & \cdots \\ S_4 & T_5 & -S_6 & -T_7 & S_8 & \cdots \\ T_5 & -S_6 & -T_7 & S_8 & T_9 & \cdots \\ \vdots & \vdots & \vdots & \vdots & \vdots & \ddots \end{bmatrix}, \quad \boldsymbol{B}_1 = \begin{bmatrix} S_0 \\ T_1 \\ S_2 \\ T_3 \\ S_4 \\ \vdots \end{bmatrix} \tag{2.45}$$

$$\boldsymbol{A}_{22} = \begin{bmatrix} U_2 & 0 & -U_4 & 0 & U_6 & \cdots \\ 0 & U_4 & 0 & -U_6 & 0 & \cdots \\ U_4 & 0 & -U_6 & 0 & U_8 & \cdots \\ 0 & U_6 & 0 & -U_8 & 0 & \cdots \\ U_6 & 0 & -U_8 & 0 & U_{10} & \cdots \\ \vdots & \vdots & \vdots & \vdots & \vdots & \ddots \end{bmatrix}, \quad \boldsymbol{B}_2 = \begin{bmatrix} 0 \\ U_2 \\ 0 \\ U_4 \\ 0 \\ \vdots \end{bmatrix} \tag{2.46}$$

and where

$$V_k = \sum_{i=1}^{L} \omega_i^k, \qquad S_k = \sum_{i=1}^{L} \omega_i^k R(\omega_i) \tag{2.47}$$

$$T_k = \sum_{i=1}^{L} \omega_i^k I(\omega_i), \qquad U_k = \sum_{i=1}^{L} \omega_i^k [R^2(\omega_i) + I^2(\omega_i)] \tag{2.48}$$

The desired parameters of the corresponding transfer function model can be obtained by solving the above equations.

We note that the methods recalled above are merely for the modeling and identification of the HDD VCM and microactuators in the forthcoming chapters. If systems to be identified are highly uncertain with disturbances, it would be more appropriate to use the methods reported in a recent monograph by Chen and Gu [68] to yield more accurate results.

2.4 Physical Effect Approach with Monte Carlo Estimations

Diverse methods have been proposed (see, *e.g.*, [13, 59, 69]) to identify the model of linear plants, in which the nonlinearities of the plants are assumed to be negligible. As such, these methods cannot be applied to identify the model of plants with significant nonlinearities. One of the key issues in identifying the model of a nonlinear plant is to configure its physical structure. Another issue is that there is generally no analytical solution for nonlinear differential equations, which causes problems in identifying the parameters in the model of nonlinear plants.

The so-called physical effect approach [62] with the Monte Carlo method (see, *e.g.*, [63–65]) can be applied to tackle these two problems efficiently. The approach is to examine and analyze physical effects that occur in or between physical parts of a nonlinear plant to configure a structured model and then to identify the parameters numerically in the obtained structured model with the Monte Carlo method. We proceed to summarize such an efficient parameter identification approach for nonlinear systems in the following.

2.4.1 Structural Analysis of Physical Effects

The structural analysis is to determine the physical structure of the nonlinear plants using analytical derivation based on some natural laws and theories, such as the well-known Newton's laws of motion, principles of circuits, electromagnetic effects, stochastic and probability theory, if applicable. It focuses on identifying the physical effects that occur in or between the components of the plants under consideration and on identifying their interconnection properties. Unfortunately, it is not always possible to capture all the properties of a physical plant. Instead, the structural analysis is more on the characterization of the major characteristics of physical plants based on some appropriate assumptions. For example, in HDD servo systems, Newton's laws can be applied to analyze the motion of the VCM actuator under the assumption that it is a rigid body. The permanent magnet associated with the coil is assumed to be constant for simplicity. Some other typical models of physical effects, such as the commonly used friction models and models for springs, can be applied to simplify the analysis. In principle, the structure of the nonlinearities has to be sufficient to characterize the main properties of the given system. One would generally obtain the structured model of a nonlinear plant in the following form:

$$\begin{cases} \dot{x} = f(x, u, P) \\ y = g(x, u, P) \end{cases} \tag{2.49}$$

where x, u and y are, respectively, the state, input and measurement output variables with appropriate dimensions. f and g are appropriate nonlinear functions. Lastly, P represents the set of all unknown parameters of the model, which is to be identified using some parameter-identification techniques.

2.4.2 Monte Carlo Estimations

In most situations, it is hard to find analytical solutions to the associated nonlinear differential equations. Thus, it is difficult to identify the parameters in nonlinear models. Alternatively, one has to search for solutions through numerical means. The Monte Carlo method (see, *e.g.*, [63–65]) is one such numerical technique, which can loosely be classified as a method purely based on numerical simulation. The essence of the Monte Carlo method is to employ certain stochastic techniques to identify the unknown parameters of a structured nonlinear model. The Monte Carlo method has been widely used from field to field even in the deterministic setting. It is very efficient in approximating solutions to various mathematical problems, for which their analytical forms are hard, if not impossible, to be determined.

Considering the structured model of Equation 2.49, the Monte Carlo estimation method is an optimization process in determining the set of the unknown parameters, P. It is to minimize the following performance index:

$$\min_{P} \left\{ \frac{1}{N} \sum_{k=1}^{N} [y_{\rm s}(t_k) - y_{\rm e}(t_k)]' W(t_k) [y_{\rm s}(t_k) - y_{\rm e}(t_k)] \right\} \tag{2.50}$$

subject to

$$\begin{cases} \dot{x} = f(x, u_{\rm e}, P) \\ y_{\rm s} = g(x, u_{\rm e}, P) \end{cases} \tag{2.51}$$

where N is the number of the sampled data, $W(t_k)$, $k = 1, 2, \ldots, N$, are appropriate weighting matrices, $y_{\rm s}$ and $y_{\rm e}$ are, respectively, the simulated and experimental measurement output responses corresponding to a set of preselected input signals $u_{\rm e}$. Note that $u_{\rm e}$ should be carefully chosen to ensure that the unknown parameters in the structured model of the plant can be properly identified.

Various numerical optimization algorithms, such as genetic algorithm, simulated annealing and gradient algorithm, neural network and their combinations, can be adopted to solve the above problem. As usual, all these numerical methods might yield a locally optimal solution. To ensure obtaining a meaningful solution, one should have certain background knowledge on the problem to be solved. For example, in HDD servo systems, the ranges of certain portions of the system dynamics are known to us. Such information is useful and should be incorporated in the above optimization process.

Similarly, an optimization process in the frequency domain can also be formulated to identify certain parts of the system dynamics. In fact, it is effective to combine the results identified using both the time- and frequency-domain methods. In general, it would yield a more accurate solution. The overall Monte Carlo estimation processes in the time domain and frequency domain are, respectively, depicted in Figures 2.5 and 2.6.

2.4.3 Verification and Validation

Once a model with all its parameters is identified, it is necessary to perform a series of model verification and validation processes. Model verification is a series of

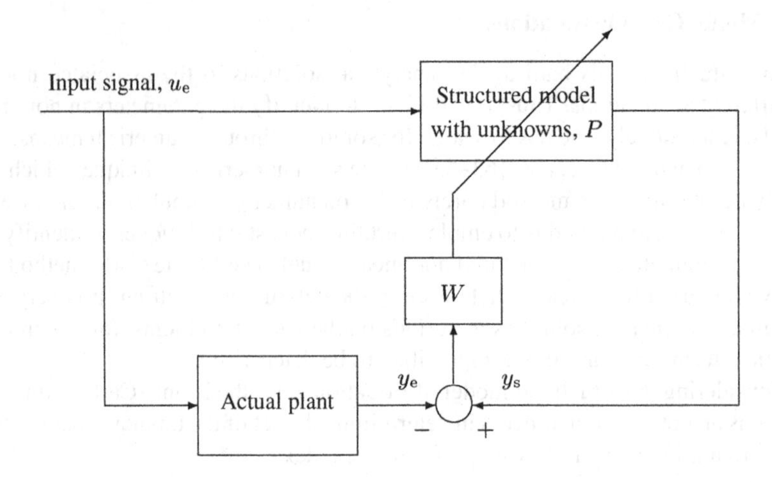

Figure 2.5. Monte Carlo estimation in the time-domain setting

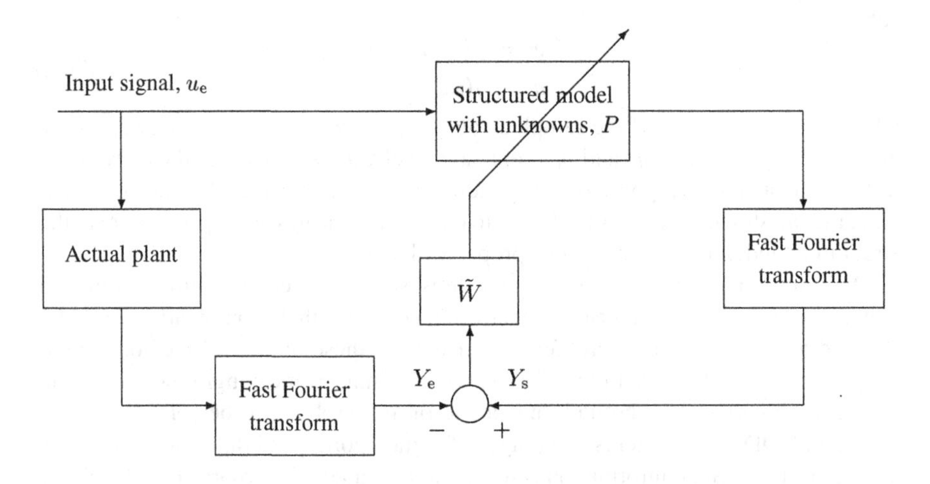

Figure 2.6. Monte Carlo estimation in the frequency-domain setting

quantitative examinations and comparisons between the actual experimental data and those generated from the identified model. It is to verify whether the identified model is a true representation of the real plants based on some intensive tests with various input-output responses other than those used in the identification process. On the other hand, validation is on qualitative examinations, which are to verify whether the features of the identified model are capable of displaying all of the essential characteristics of the actual plant. It is to recheck the process of the physical effect analysis, the correctness of the natural laws and theories used as well as the assumptions made.

In conclusion, verification and validation are two necessary steps that one needs to perform to ensure that the identified model is accurate and reliable. As mentioned earlier, the above technique will be utilized to identify the model of a commercial microdrive in Chapter 9.

3

Linear Systems and Control

3.1 Introduction

It is our belief that a good unambiguous understanding of linear system structures, *i.e.* the finite and infinite zero structures as well as the invertibility structures of linear systems, is essential for a meaningful control system design. As a matter of fact, the performance and limitation of an overall control system are primarily dependent on the structural properties of the given open-loop system. In our opinion, a control system engineer should thoroughly study the properties of a given plant before carrying out any meaningful design. Many of the difficulties one might face in the design stage may be avoided if the designer has fully understood the system properties or limitations. For example, it is well understood in the literature that a nonminimum phase zero would generally yield a poor overall performance no matter what design methodology is used. A good control engineer should try to avoid these kinds of problem at the initial stage by adding or adjusting sensors or actuators in the system. Sometimes, a simple rearrangement of existing sensors and/or actuators could totally change the system properties. We refer interested readers to the work by Liu *et al.* [70] and a recent monograph by Chen *et al.* [71] for details.

As such, we first recall in this chapter a structural decomposition technique of linear systems, namely the special coordinate basis of [72, 73], which has a unique feature of displaying the structural properties of linear systems. The detailed derivation and proof of such a technique can also be found in Chen *et al.* [71]. We then present some common linear control system design techniques, such as PID control, H_2 optimal control, H_∞ control, linear quadratic regulator (LQR) with loop transfer recovery design (LTR), together with some newly developed design techniques, such as the robust and perfect tracking (RPT) method. Most of these results will be intensively used later in the design of HDD servo systems, though some are presented here for the purpose of easy reference for general readers.

We have noticed that it is some kind of tradition or fashion in the HDD servo system research community in which researchers and practicing engineers prefer to carry out a control system design in the discrete-time setting. In this case, the designer would have to discretize the plant to be controlled (mostly using the ZOH technique) first and then use some discrete-time control system design technique to obtain a discrete-time control law. However, in our personal opinion, it is easier to design a controller directly in the continuous-time setting and then use some continuous-to-discrete transformations, such as the bilinear transformation, to discretize it when it is to be implemented in the real system. The advantage of such an approach follows from the following fact that the bilinear transformation does not introduce unstable invariant zeros to its discrete-time counterpart. On the other hand, it is well known in the literature that the ZOH approach almost always produces some additional nonminimum-phase invariant zeros for higher-order systems with faster sampling rates. These nonminimum phase zeros cause some additional limitations on the overall performance of the system to be controlled. Nevertheless, we present both continuous-time and discrete-time versions of these control techniques for completeness. It is up to the reader to choose the appropriate approach in designing their own servo systems.

Lastly, we would like to note that the results presented in this chapter are well studied in the literature. As such, all results are quoted without detailed proofs and derivations. Interested readers are referred to the related references for details.

3.2 Structural Decomposition of Linear Systems

Consider a general proper linear time-invariant system Σ, which could be of either continuous- or discrete-time, characterized by a matrix quadruple (A, B, C, D) or in the state-space form

$$\Sigma : \begin{cases} \delta(x) = A\,x + B\,u \\ \quad y \;\; = C\,x + D\,u \end{cases} \tag{3.1}$$

where $\delta(x) = \dot{x}(t)$ if Σ is a continuous-time system, or $\delta(x) = x(k+1)$ if Σ is a discrete-time system. Similarly, $x \in \mathbb{R}^n$, $u \in \mathbb{R}^m$ and $y \in \mathbb{R}^p$ are the state, input and output of Σ. They represent, respectively, $x(t)$, $u(t)$ and $y(t)$ if the given system is of continuous-time, or represent, respectively, $x(k)$, $u(k)$ and $y(k)$ if Σ is of discrete-time. Without loss of any generality, we assume throughout this section that both $[\,B' \;\; D'\,]$ and $[\,C \;\; D\,]$ are of full rank. The transfer function of Σ is then given by

$$H(\varsigma) = C(\varsigma I - A)^{-1}B + D \tag{3.2}$$

where $\varsigma = s$, the Laplace transform operator, if Σ is of continuous-time, or $\varsigma = z$, the z-transform operator, if Σ is of discrete-time. It is simple to verify that there exist nonsingular transformations U and V such that

$$UDV = \begin{bmatrix} I_{m_0} & 0 \\ 0 & 0 \end{bmatrix} \tag{3.3}$$

where m_0 is the rank of matrix D. In fact, U can be chosen as an orthogonal matrix. Hence, hereafter, without loss of generality, it is assumed that the matrix D has the form given on the right-hand side of Equation 3.3. One can now rewrite system Σ of Equation 3.1 as

$$\begin{cases} \delta(x) = A \quad x + [B_0 \quad B_1] \begin{pmatrix} u_0 \\ u_1 \end{pmatrix} \\ \begin{pmatrix} y_0 \\ y_1 \end{pmatrix} = \begin{bmatrix} C_0 \\ C_1 \end{bmatrix} x + \begin{bmatrix} I_{m_0} & 0 \\ 0 & 0 \end{bmatrix} \begin{pmatrix} u_0 \\ u_1 \end{pmatrix} \end{cases} \tag{3.4}$$

where the matrices B_0, B_1, C_0 and C_1 have appropriate dimensions. Theorem 3.1 below on the special coordinate basis (SCB) of linear systems is mainly due to the results of Sannuti and Saberi [72, 73]. The proofs of all its properties can be found in Chen *et al.* [71] and Chen [74].

Theorem 3.1. *Given the linear system Σ of Equation 3.1, there exist*

1. *coordinate-free non-negative integers $n_{\rm a}^-$, $n_{\rm a}^0$, $n_{\rm a}^+$, $n_{\rm b}$, $n_{\rm c}$, $n_{\rm d}$, $m_{\rm d} \le m - m_0$ and q_i, $i = 1, \cdots, m_{\rm d}$, and*
2. *nonsingular state, output and input transformations $\Gamma_{\rm s}$, $\Gamma_{\rm o}$ and $\Gamma_{\rm i}$ that take the given Σ into a special coordinate basis that displays explicitly both the finite and infinite zero structures of Σ.*

The special coordinate basis is described by the following set of equations:

$$x = \Gamma_{\rm s}\tilde{x}, \quad y = \Gamma_{\rm o}\tilde{y}, \quad u = \Gamma_{\rm i}\tilde{u} \tag{3.5}$$

$$\tilde{x} = \begin{pmatrix} x_{\rm a} \\ x_{\rm b} \\ x_{\rm c} \\ x_{\rm d} \end{pmatrix}, \quad x_{\rm a} = \begin{pmatrix} x_{\rm a}^- \\ x_{\rm a}^0 \\ x_{\rm a}^+ \end{pmatrix}, \quad x_{\rm d} = \begin{pmatrix} x_1 \\ x_2 \\ \vdots \\ x_{m_{\rm d}} \end{pmatrix} \tag{3.6}$$

$$\tilde{y} = \begin{pmatrix} y_0 \\ y_{\rm d} \\ y_{\rm b} \end{pmatrix}, \quad y_{\rm d} = \begin{pmatrix} y_1 \\ y_2 \\ \vdots \\ y_{m_{\rm d}} \end{pmatrix}, \quad \tilde{u} = \begin{pmatrix} u_0 \\ u_{\rm d} \\ u_{\rm c} \end{pmatrix}, \quad u_{\rm d} = \begin{pmatrix} u_1 \\ u_2 \\ \vdots \\ u_{m_{\rm d}} \end{pmatrix} \tag{3.7}$$

$$\delta(x_{\rm a}^-) = A_{\rm aa}^- x_{\rm a}^- + B_{\rm 0a}^- y_0 + L_{\rm ad}^- y_{\rm d} + L_{\rm ab}^- y_{\rm b} \tag{3.8}$$

$$\delta(x_{\rm a}^0) = A_{\rm aa}^0 x_{\rm a}^0 + B_{\rm 0a}^0 y_0 + L_{\rm ad}^0 y_{\rm d} + L_{\rm ab}^0 y_{\rm b} \tag{3.9}$$

$$\delta(x_{\rm a}^+) = A_{\rm aa}^+ x_{\rm a}^+ + B_{\rm 0a}^+ y_0 + L_{\rm ad}^+ y_{\rm d} + L_{\rm ab}^+ y_{\rm b} \tag{3.10}$$

$$\delta(x_{\rm b}) = A_{\rm bb} x_{\rm b} + B_{\rm 0b} y_0 + L_{\rm bd} y_{\rm d}, \quad y_{\rm b} = C_{\rm b} x_{\rm b} \tag{3.11}$$

$$\delta(x_{\rm c}) = A_{\rm cc} x_{\rm c} + B_{\rm 0c} y_0 + L_{\rm cb} y_{\rm b} + L_{\rm cd} y_{\rm d} + B_{\rm c}\left[E_{\rm ca}^- x_{\rm a}^- + E_{\rm ca}^0 x_{\rm a}^0 + E_{\rm ca}^+ x_{\rm a}^+\right] + B_{\rm c} u_{\rm c} \tag{3.12}$$

$$y_0 = C_{\rm 0c} x_{\rm c} + C_{\rm 0a}^- x_{\rm a}^- + C_{\rm 0a}^+ x_{\rm a}^0 + C_{\rm 0a}^+ x_{\rm a}^+ + C_{\rm 0d} x_{\rm d} + C_{\rm 0b} x_{\rm b} + u_0 \tag{3.13}$$

and for each $i = 1, \cdots, m_{\rm d}$,

$$\delta(x_i) = A_{q_i} x_i + L_{i0} y_0 + L_{i{\rm d}} y_{\rm d} + B_{q_i}\left[u_i + E_{i{\rm a}} x_{\rm a} + E_{i{\rm b}} x_{\rm b} + E_{i{\rm c}} x_{\rm c} + \sum_{j=1}^{m_{\rm d}} E_{ij} x_j\right] \tag{3.14}$$

$$y_i = C_{q_i} x_i, \quad y_d = C_d x_d \tag{3.15}$$

Here the states x_a^-, x_a^0, x_a^+, x_b, x_c *and* x_d *are, respectively, of dimensions* n_a^-, n_a^0, n_a^+, n_b, n_c *and* $n_d = \sum_{i=1}^{m_d} q_i$, *and* x_i *is of dimension* q_i *for each* $i = 1, \cdots, m_d$. *The control vectors* u_0, u_d *and* u_c *are, respectively, of dimensions* m_0, m_d *and* $m_c = m - m_0 - m_d$, *and the output vectors* y_0, y_d *and* y_b *are, respectively, of dimensions* $p_0 = m_0$, $p_d = m_d$ *and* $p_b = p - p_0 - p_d$. *The matrices* A_{q_i}, B_{q_i} *and* C_{q_i} *have the following form:*

$$A_{q_i} = \begin{bmatrix} 0 & I_{q_i-1} \\ 0 & 0 \end{bmatrix}, \quad B_{q_i} = \begin{bmatrix} 0 \\ 1 \end{bmatrix}, \quad C_{q_i} = [1, \ 0, \ \cdots, \ 0] \tag{3.16}$$

Assuming that x_i, $i = 1, 2, \cdots, m_d$, *are arranged such that* $q_i \le q_{i+1}$, *the matrix* L_{id} *has the particular form*

$$L_{id} = [\, L_{i1} \quad L_{i2} \quad \cdots \quad L_{ii-1} \quad 0 \quad \cdots \quad 0 \,] \tag{3.17}$$

The last row of each L_{id} *is identically zero. Moreover:*

1. *If* Σ *is a continuous-time system, then*

$$\lambda(A_{aa}^-) \subset \mathbb{C}^-, \quad \lambda(A_{aa}^0) \subset \mathbb{C}^0, \quad \lambda(A_{aa}^+) \subset \mathbb{C}^+ \tag{3.18}$$

2. *If* Σ *is a discrete-time system, then*

$$\lambda(A_{aa}^-) \subset \mathbb{C}^{\odot}, \quad \lambda(A_{aa}^0) \subset \mathbb{C}^{\circ}, \quad \lambda(A_{aa}^+) \subset \mathbb{C}^{\otimes} \tag{3.19}$$

Also, the pair (A_{cc}, B_c) *is controllable and the pair* (A_{bb}, C_b) *is observable.* ◊

Note that a detailed procedure of constructing the above structural decomposition can be found in Chen *et al.* [71]. Its software realization can be found in Lin *et al.* [53], which is free for downloading at `http://linearsystemskit.net`.

We can rewrite the special coordinate basis of the quadruple (A, B, C, D) given by Theorem 3.1 in a more compact form:

$$\begin{aligned} \tilde{A} &= \Gamma_s^{-1} A \Gamma_s = A_s + B_0 C_0 \\ &= \begin{bmatrix} A_{aa}^- & 0 & 0 & L_{ab}^- C_b & 0 & L_{ad}^- C_d \\ 0 & A_{aa}^0 & 0 & L_{ab}^0 C_b & 0 & L_{ad}^0 C_d \\ 0 & 0 & A_{aa}^+ & L_{ab}^+ C_b & 0 & L_{ad}^+ C_d \\ 0 & 0 & 0 & A_{bb} & 0 & L_{bd} C_d \\ B_c E_{ca}^- & B_c E_{ca}^0 & B_c E_{ca}^+ & L_{cb} C_b & A_{cc} & L_{cd} C_d \\ B_d E_{da}^- & B_d E_{da}^0 & B_d E_{da}^+ & B_d E_{db} & B_d E_{dc} & A_{dd} \end{bmatrix} \\ &\quad + \begin{bmatrix} B_{0a}^- \\ B_{0a}^0 \\ B_{0a}^+ \\ B_{0b} \\ B_{0c} \\ B_{0d} \end{bmatrix} [\, C_{0a}^- \quad C_{0a}^0 \quad C_{0a}^+ \quad C_{0b} \quad C_{0c} \quad C_{0d} \,] \end{aligned} \tag{3.20}$$

$$\tilde{B} = \Gamma_{\rm s}^{-1} B \Gamma_{\rm i} = [\, B_0 \quad B_{\rm s} \,] = \begin{bmatrix} B_{0\rm a}^- & 0 & 0 \\ B_{0\rm a}^0 & 0 & 0 \\ B_{0\rm a}^+ & 0 & 0 \\ B_{0\rm b} & 0 & 0 \\ B_{0\rm c} & 0 & B_{\rm c} \\ B_{0\rm d} & B_{\rm d} & 0 \end{bmatrix} \tag{3.21}$$

$$\tilde{C} = \Gamma_{\rm o}^{-1} C \Gamma_{\rm s} = \begin{bmatrix} C_0 \\ C_{\rm s} \end{bmatrix} = \begin{bmatrix} C_{0\rm a}^- & C_{0\rm a}^0 & C_{0\rm a}^+ & C_{0\rm b} & C_{0\rm c} & C_{0\rm d} \\ 0 & 0 & 0 & 0 & 0 & C_{\rm d} \\ 0 & 0 & 0 & C_{\rm b} & 0 & 0 \end{bmatrix} \tag{3.22}$$

$$\tilde{D} = \Gamma_{\rm o}^{-1} D \Gamma_{\rm i} = D_{\rm s} = \begin{bmatrix} I_{m_0} & 0 & 0 \\ 0 & 0 & 0 \\ 0 & 0 & 0 \end{bmatrix} \tag{3.23}$$

3.2.1 Interpretation

A block diagram of the structural decomposition of Theorem 3.1 is illustrated in Figure 3.1. In this figure, a signal given by a double-edged arrow is some linear combination of outputs y_i, $i = 0$ to $m_{\rm d}$, whereas a signal given by the double-edged arrow with a solid dot is some linear combination of all the states.

$$B_{0\rm a} := \begin{bmatrix} B_{0\rm a}^- \\ B_{0\rm a}^0 \\ B_{0\rm a}^+ \end{bmatrix}, \quad L_{\rm ab} := \begin{bmatrix} L_{\rm ab}^- \\ L_{\rm ab}^0 \\ L_{\rm ab}^+ \end{bmatrix}, \quad L_{\rm ad} := \begin{bmatrix} L_{\rm ad}^- \\ L_{\rm ad}^0 \\ L_{\rm ad}^+ \end{bmatrix} \tag{3.24}$$

and

$$A_{\rm aa} := \begin{bmatrix} A_{\rm aa}^- & 0 & 0 \\ 0 & A_{\rm aa}^0 & 0 \\ 0 & 0 & A_{\rm aa}^+ \end{bmatrix}, \quad E_{\rm ca} := [\, E_{\rm ca}^- \quad E_{\rm ca}^0 \quad E_{\rm ca}^+ \,] \tag{3.25}$$

Also, the block $\triangleright$ is either an integrator if Σ is of continuous-time or a backward-shifting operator if Σ is of discrete-time. We note the following intuitive points.

1. The input u_i controls the output y_i through a stack of q_i integrators (or backward-shifting operators), whereas x_i is the state associated with those integrators (or backward-shifting operators) between u_i and y_i. Moreover, (A_{q_i}, B_{q_i}) and (A_{q_i}, C_{q_i}), respectively, form controllable and observable pairs. This implies that all the states x_i are both controllable and observable.
2. The output $y_{\rm b}$ and the state $x_{\rm b}$ are not directly influenced by any inputs; however, they could be indirectly controlled through the output $y_{\rm d}$. Moreover, $(A_{\rm bb}, C_{\rm b})$ forms an observable pair. This implies that the state $x_{\rm b}$ is observable.
3. The state $x_{\rm c}$ is directly controlled by the input $u_{\rm c}$, but it does not directly affect any output. Moreover, $(A_{\rm cc}, B_{\rm c})$ forms a controllable pair. This implies that the state $x_{\rm c}$ is controllable.
4. The state $x_{\rm a}$ is neither directly controlled by any input nor does it directly affect any output.

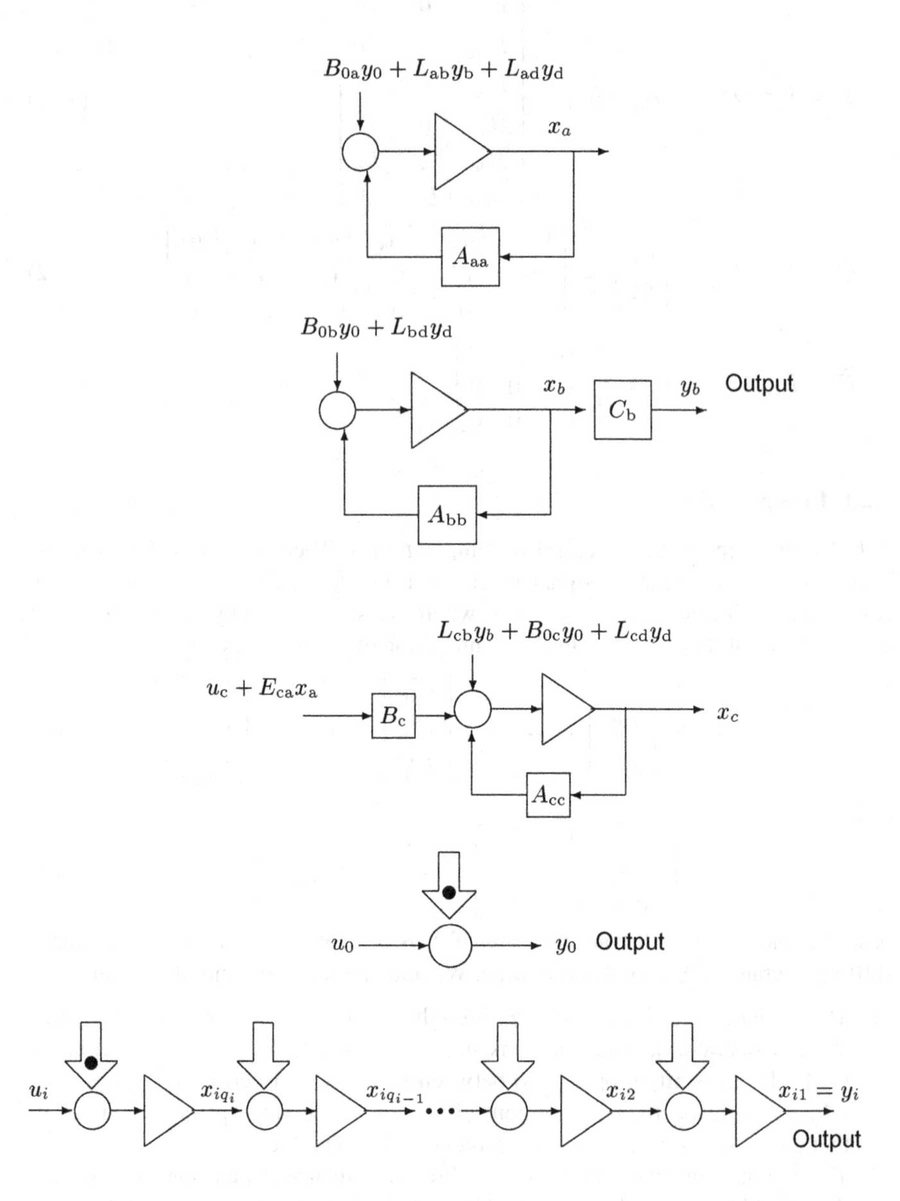

Figure 3.1. A block diagram representation of the special coordinate basis

3.2.2 Properties

In what follows, we state some important properties of the above special coordinate basis that are pertinent to our present work. As mentioned earlier, the proofs of these properties can be found in Chen *et al.* [71] and Chen [74].

Property 3.2. The given system Σ is observable (detectable) if and only if the pair $(A_{\mathrm{obs}}, C_{\mathrm{obs}})$ is observable (detectable), where

$$A_{\mathrm{obs}} := \begin{bmatrix} A_{\mathrm{aa}} & 0 \\ B_{\mathrm{c}}E_{\mathrm{ca}} & A_{\mathrm{cc}} \end{bmatrix}, \quad C_{\mathrm{obs}} := \begin{bmatrix} C_{0\mathrm{a}} & C_{0\mathrm{c}} \\ E_{\mathrm{da}} & E_{\mathrm{dc}} \end{bmatrix} \tag{3.26}$$

and where

$$E_{\mathrm{da}} := [\, E_{\mathrm{da}}^{-} \quad E_{\mathrm{da}}^{0} \quad E_{\mathrm{da}}^{+} \,], \quad C_{0\mathrm{a}} := [\, C_{0\mathrm{a}}^{-} \quad C_{0\mathrm{a}}^{0} \quad C_{0\mathrm{a}}^{+} \,] \tag{3.27}$$

Also, define

$$A_{\mathrm{con}} := \begin{bmatrix} A_{\mathrm{aa}} & L_{\mathrm{ab}}C_{\mathrm{b}} \\ 0 & A_{\mathrm{bb}} \end{bmatrix}, \quad B_{\mathrm{con}} := \begin{bmatrix} B_{0\mathrm{a}} & L_{\mathrm{ad}} \\ B_{0\mathrm{b}} & L_{\mathrm{bd}} \end{bmatrix} \tag{3.28}$$

Similarly, Σ is controllable (stabilizable) if and only if the pair $(A_{\mathrm{con}}, B_{\mathrm{con}})$ is controllable (stabilizable). ◊

The invariant zeros of a system Σ characterized by (A, B, C, D) can be defined via the Smith canonical form of the (Rosenbrock) system matrix [75] of Σ:

$$P_{\Sigma}(\varsigma) := \begin{bmatrix} \varsigma I - A & -B \\ C & D \end{bmatrix} \tag{3.29}$$

We have the following definition for the invariant zeros (see also [76]).

Definition 3.3. (Invariant Zeros). *A complex scalar $\alpha \in \mathbb{C}$ is said to be an invariant zero of Σ if*

$$\mathrm{rank}\,\{P_{\Sigma}(\alpha)\} < n + \mathrm{normrank}\,\{H(\varsigma)\} \tag{3.30}$$

where normrank$\{H(\varsigma)\}$ *denotes the normal rank of $H(\varsigma)$, which is defined as its rank over the field of rational functions of ς with real coefficients.* ◊

The special coordinate basis of Theorem 3.1 shows explicitly the invariant zeros and the normal rank of Σ. To be more specific, we have the following properties.

Property 3.4.

1. The normal rank of $H(\varsigma)$ is equal to $m_0 + m_{\mathrm{d}}$.
2. Invariant zeros of Σ are the eigenvalues of A_{aa}, which are the unions of the eigenvalues of A_{aa}^{-}, A_{aa}^{0} and A_{aa}^{+}. Moreover, the given system Σ is of minimum phase if and only if A_{aa} has only stable eigenvalues, marginal minimum phase if and only if A_{aa} has no unstable eigenvalue but has at least one marginally stable eigenvalue, and nonminimum phase if and only if A_{aa} has at least one unstable eigenvalue. ◊

The special coordinate basis can also reveal the infinite zero structure of Σ. We note that the infinite zero structure of Σ can be either defined in association with root-locus theory or as Smith–McMillan zeros of the transfer function at infinity. For the sake of simplicity, we only consider the infinite zeros from the point of view of Smith–McMillan theory here. To define the zero structure of $H(\varsigma)$ at infinity, one can use the familiar Smith–McMillan description of the zero structure at finite frequencies of a general not necessarily square but strictly proper transfer function matrix $H(\varsigma)$. Namely, a rational matrix $H(\varsigma)$ possesses an infinite zero of order k when $H(1/z)$ has a finite zero of precisely that order at $z = 0$ (see [75], [77–79]). The number of zeros at infinity, together with their orders, indeed defines an infinite zero structure. Owens [80] related the orders of the infinite zeros of the root-loci of a square system with a nonsingular transfer function matrix to the C^* structural invariant indices list $\mathcal{I}_4$ of Morse [81]. This connection reveals that, even for general not necessarily strictly proper systems, the *structure at infinity is in fact the topology of inherent integrations between the input and the output variables.* The special coordinate basis of Theorem 3.1 explicitly shows this topology of inherent integrations. The following property pinpoints this.

Property 3.5. Σ has $m_0 = \operatorname{rank}(D)$ infinite zeros of order 0. The infinite zero structure (of order greater than 0) of Σ is given by

$$S_\infty^\star(\Sigma) = \left\{ q_1, q_2, \cdots, q_{m_d} \right\} \tag{3.31}$$

That is, each q_i corresponds to an infinite zero of Σ of order q_i. Note that for an SISO system Σ, we have $S_\infty^\star(\Sigma) = \{q_1\}$, where q_1 is the *relative degree* of Σ. $\diamond$

The special coordinate basis can also exhibit the invertibility structure of a given system Σ. The formal definitions of right invertibility and left invertibility of a linear system can be found in [82]. Basically, for the usual case when $[\,B' \;\; D'\,]$ and $[\,C \;\; D\,]$ are of maximal rank, the system Σ, or equivalently $H(\varsigma)$, is said to be left invertible if there exists a rational matrix function, say $L(\varsigma)$, such that

$$L(\varsigma)H(\varsigma) = I_m \tag{3.32}$$

Σ or $H(\varsigma)$ is said to be right invertible if there exists a rational matrix function, say $R(\varsigma)$, such that

$$H(\varsigma)R(\varsigma) = I_p \tag{3.33}$$

Σ is invertible if it is both left and right invertible, and Σ is degenerate if it is neither left nor right invertible.

Property 3.6. The given system Σ is right invertible if and only if x_{b} (and hence y_{b}) are nonexistent, left invertible if and only if x_{c} (and hence u_{c}) are nonexistent, and invertible if and only if both x_{b} and x_{c} are nonexistent. Moreover, Σ is degenerate if and only if both x_{b} and x_{c} are present. $\diamond$

By now it is clear that the special coordinate basis decomposes the state space into several distinct parts. In fact, the state-space $\mathcal{X}$ is decomposed as

$$\mathcal{X} = \mathcal{X}_{\rm a}^- \oplus \mathcal{X}_{\rm a}^0 \oplus \mathcal{X}_{\rm a}^+ \oplus \mathcal{X}_{\rm b} \oplus \mathcal{X}_{\rm c} \oplus \mathcal{X}_{\rm d} \tag{3.34}$$

Here, $\mathcal{X}_{\rm a}^-$ is related to the stable invariant zeros, *i.e.* the eigenvalues of $A_{\rm aa}^-$ are the stable invariant zeros of Σ. Similarly, $\mathcal{X}_{\rm a}^0$ and $\mathcal{X}_{\rm a}^+$ are, respectively, related to the invariant zeros of Σ located in the marginally stable and unstable regions. On the other hand, $\mathcal{X}_{\rm b}$ is related to the right invertibility, *i.e.* the system is right invertible if and only if $\mathcal{X}_{\rm b} = \{0\}$, whereas $\mathcal{X}_{\rm c}$ is related to left invertibility, *i.e.* the system is left invertible if and only if $\mathcal{X}_{\rm c} = \{0\}$. Finally, $\mathcal{X}_{\rm d}$ is related to zeros of Σ at infinity.

There are interconnections between the special coordinate basis and various invariant geometric subspaces. To show these interconnections, we introduce the following geometric subspaces.

Definition 3.7. (Geometric Subspaces $\mathcal{V}^{\rm x}$ and $\mathcal{S}^{\rm x}$). *The weakly unobservable subspaces of Σ, $\mathcal{V}^{\rm x}$, and the strongly controllable subspaces of Σ, $\mathcal{S}^{\rm x}$, are defined as follows:*

1. *$\mathcal{V}^{\rm x}(\Sigma)$ is the maximal subspace of $\mathbb{R}^n$ that is $(A+BF)$-invariant and contained in* Ker $(C + DF)$ *such that the eigenvalues of $(A + BF)|\mathcal{V}^{\rm x}$ are contained in $\mathbb{C}^{\rm x} \subseteq \mathbb{C}$ for some constant matrix F.*
2. *$\mathcal{S}^{\rm x}(\Sigma)$ is the minimal $(A + KC)$-invariant subspace of $\mathbb{R}^n$ containing the subspace* Im $(B + KD)$ *such that the eigenvalues of the map that is induced by $(A + KC)$ on the factor space $\mathbb{R}^n/\mathcal{S}^{\rm x}$ are contained in $\mathbb{C}^{\rm x} \subseteq \mathbb{C}$ for some constant matrix K.*

Moreover, we let $\mathcal{V}^- = \mathcal{V}^{\rm x}$ and $\mathcal{S}^- = \mathcal{S}^{\rm x}$, if $\mathbb{C}^{\rm x} = \mathbb{C}^- \cup \mathbb{C}^0$; $\mathcal{V}^+ = \mathcal{V}^{\rm x}$ and $\mathcal{S}^+ = \mathcal{S}^{\rm x}$, if $\mathbb{C}^{\rm x} = \mathbb{C}^+$; $\mathcal{V}^{\odot} = \mathcal{V}^{\rm x}$ and $\mathcal{S}^{\odot} = \mathcal{S}^{\rm x}$, if $\mathbb{C}^{\rm x} = \mathbb{C}^{\odot} \cup \mathbb{C}^{\circ}$; $\mathcal{V}^{\otimes} = \mathcal{V}^{\rm x}$ and $\mathcal{S}^{\otimes} = \mathcal{S}^{\rm x}$, if $\mathbb{C}^{\rm x} = \mathbb{C}^{\otimes}$; and finally $\mathcal{V}^* = \mathcal{V}^{\rm x}$ and $\mathcal{S}^* = \mathcal{S}^{\rm x}$, if $\mathbb{C}^{\rm x} = \mathbb{C}$. ◊

We have the following property.

Property 3.8.

1. $\mathcal{X}_{\rm a}^- \oplus \mathcal{X}_{\rm a}^0 \oplus \mathcal{X}_{\rm c}$ spans $\begin{cases} \mathcal{V}^-(\Sigma), & \text{if } \Sigma \text{ is of continuous-time,} \\ \mathcal{V}^{\odot}(\Sigma), & \text{if } \Sigma \text{ is of discrete-time.} \end{cases}$
2. $\mathcal{X}_{\rm a}^+ \oplus \mathcal{X}_{\rm c}$ spans $\begin{cases} \mathcal{V}^+(\Sigma), & \text{if } \Sigma \text{ is of continuous-time,} \\ \mathcal{V}^{\otimes}(\Sigma), & \text{if } \Sigma \text{ is of discrete-time.} \end{cases}$
3. $\mathcal{X}_{\rm a}^- \oplus \mathcal{X}_{\rm a}^0 \oplus \mathcal{X}_{\rm a}^+ \oplus \mathcal{X}_{\rm c}$ spans $\mathcal{V}^*(\Sigma)$.
4. $\mathcal{X}_{\rm a}^+ \oplus \mathcal{X}_{\rm c} \oplus \mathcal{X}_{\rm d}$ spans $\begin{cases} \mathcal{S}^-(\Sigma), & \text{if } \Sigma \text{ is of continuous-time,} \\ \mathcal{S}^{\odot}(\Sigma), & \text{if } \Sigma \text{ is of discrete-time.} \end{cases}$
5. $\mathcal{X}_{\rm a}^- \oplus \mathcal{X}_{\rm a}^0 \oplus \mathcal{X}_{\rm c} \oplus \mathcal{X}_{\rm d}$ spans $\begin{cases} \mathcal{S}^+(\Sigma), & \text{if } \Sigma \text{ is of continuous-time,} \\ \mathcal{S}^{\otimes}(\Sigma), & \text{if } \Sigma \text{ is of discrete-time.} \end{cases}$
6. $\mathcal{X}_{\rm c} \oplus \mathcal{X}_{\rm d}$ spans $\mathcal{S}^*(\Sigma)$. ◊

Finally, for future development on deriving solvability conditions for H_∞ almost disturbance decoupling problems, we introduce two more subspaces of Σ. The original definitions of these subspaces were given by Scherer [83].

Definition 3.9. (Geometric Subspaces $\mathcal{V}_\lambda$ and $\mathcal{S}_\lambda$). *For any $\lambda \in \mathbb{C}$, we define*

$$\mathcal{V}_\lambda(\Sigma) := \left\{ \zeta \in \mathbb{C}^n \;\middle|\; \exists\, \omega \in \mathbb{C}^m : 0 = \begin{bmatrix} A - \lambda I & B \\ C & D \end{bmatrix} \begin{pmatrix} \zeta \\ \omega \end{pmatrix} \right\} \tag{3.35}$$

and

$$\mathcal{S}_\lambda(\Sigma) := \left\{ \zeta \in \mathbb{C}^n \;\middle|\; \exists\, \omega \in \mathbb{C}^{n+m} : \begin{pmatrix} \zeta \\ 0 \end{pmatrix} = \begin{bmatrix} A - \lambda I & B \\ C & D \end{bmatrix} \omega \right\} \tag{3.36}$$

$\mathcal{V}_\lambda(\Sigma)$ and $\mathcal{S}_\lambda(\Sigma)$ are associated with the so-called state zero directions of Σ if λ is an invariant zero of Σ. ◇

These subspaces $\mathcal{S}_\lambda(\Sigma)$ and $\mathcal{V}_\lambda(\Sigma)$ can also be easily obtained using the special coordinate basis. We have the following new property of the special coordinate basis.

Property 3.10.

$$\mathcal{S}_\lambda(\Sigma) = \mathrm{Im}\left\{ \Gamma_\mathrm{s} \begin{bmatrix} \lambda I - A_\mathrm{aa} & 0 & 0 & 0 \\ 0 & Y_{\mathrm{b}\lambda} & 0 & 0 \\ 0 & 0 & I_{n_\mathrm{c}} & 0 \\ 0 & 0 & 0 & I_{n_\mathrm{d}} \end{bmatrix} \right\} \tag{3.37}$$

where

$$\mathrm{Im}\,\{Y_{\mathrm{b}\lambda}\} = \mathrm{Ker}\,\left[C_\mathrm{b}(A_\mathrm{bb} + K_\mathrm{b}C_\mathrm{b} - \lambda I)^{-1}\right] \tag{3.38}$$

and where K_b is any appropriately dimensional matrix subject to the constraint that $A_\mathrm{bb} + K_\mathrm{b}C_\mathrm{b}$ has no eigenvalue at λ. We note that such a K_b always exists, as $(A_\mathrm{bb}, C_\mathrm{b})$ is completely observable.

$$\mathcal{V}_\lambda(\Sigma) = \mathrm{Im}\left\{ \Gamma_\mathrm{s} \begin{bmatrix} X_{\mathrm{a}\lambda} & 0 \\ 0 & 0 \\ 0 & X_{\mathrm{c}\lambda} \\ 0 & 0 \end{bmatrix} \right\} \tag{3.39}$$

where $X_{\mathrm{a}\lambda}$ is a matrix whose columns form a basis for the subspace,

$$\left\{ \zeta_\mathrm{a} \in \mathbb{C}^{n_\mathrm{a}} \;\middle|\; (\lambda I - A_\mathrm{aa})\zeta_\mathrm{a} = 0 \right\} \tag{3.40}$$

and

$$X_{\mathrm{c}\lambda} := \left(A_\mathrm{cc} + B_\mathrm{c}F_\mathrm{c} - \lambda I \right)^{-1} B_\mathrm{c} \tag{3.41}$$

with F_c being any appropriately dimensional matrix subject to the constraint that $A_\mathrm{cc} + B_\mathrm{c}F_\mathrm{c}$ has no eigenvalue at λ. Again, we note that the existence of such an F_c is guaranteed by the controllability of $(A_\mathrm{cc}, B_\mathrm{c})$. ◇

Clearly, if $\lambda \notin \lambda(A_\mathrm{aa})$, then we have $\mathcal{V}_\lambda(\Sigma) \subseteq \mathcal{V}^\times(\Sigma)$ and $\mathcal{S}_\lambda(\Sigma) \supseteq \mathcal{S}^\times(\Sigma)$. It is interesting to note that the subspaces $\mathcal{V}^\times(\Sigma)$ and $\mathcal{S}^\times(\Sigma)$ are dual in the sense that $\mathcal{V}^\times(\Sigma^\star) = \mathcal{S}^\times(\Sigma)^\perp$, where $\Sigma^\star$ is characterized by the quadruple (A', C', B', D'). Also, $\mathcal{S}_\lambda(\Sigma) = \mathcal{V}_{\bar{\lambda}}(\Sigma^\star)^\perp$.

3.3 PID Control

PID control is the most popular technique used in industry because it is relatively easy and simple to design and implement. Most importantly, it works in most practical situations, although its performance is somewhat limited owing to its restricted structure. Nevertheless, in what follows, we recall this well-known classical control system design methodology for ease of reference.

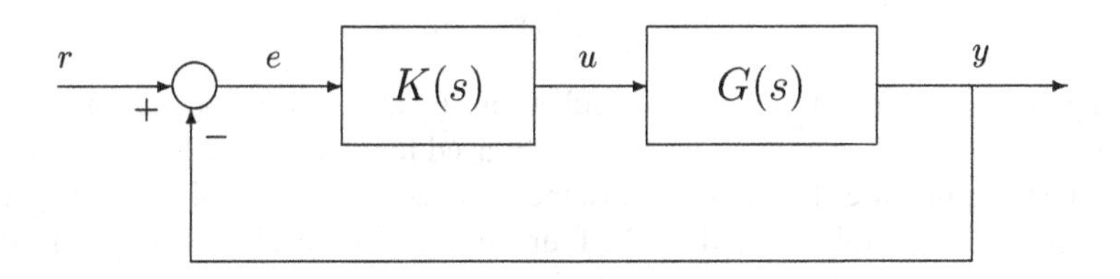

Figure 3.2. The typical PID control configuration

To be more specific, we consider the control system as depicted in Figure 3.2, in which $G(s)$ is the plant to be controlled and $K(s)$ is the PID controller characterized by the following transfer function

$$K(s) = K_{\mathrm{p}} \left(1 + \frac{1}{T_{\mathrm{i}} s} + T_{\mathrm{d}} s \right) \tag{3.42}$$

The control system design is then to determine the parameters K_{p}, T_{i} and T_{d} such that the resulting closed-loop system yields a certain desired performance, *i.e.* it meets certain prescribed design specifications.

3.3.1 Selection of Design Parameters

Ziegler–Nichols tuning is one of the most common techniques used in practical situations to design an appropriate PID controller for the class of systems that can be exactly modeled as, or approximated by, the following first-order system:

$$G(s) = \frac{Y(s)}{U(s)} = \frac{K}{\tau s + 1} \mathrm{e}^{-t_{\mathrm{d}} s} \tag{3.43}$$

One of the methods proposed by Ziegler and Nichols ([84, 85]) is first to replace the controller $K(s)$ in Figure 3.2 by a simple proportional gain. We then increase this proportional gain to a value, say K_{u}, for which we observe continuous oscillations in its step response, *i.e.* the system becomes marginally stable. Assume that the corresponding oscillating frequency is ω_{u}. The PID controller parameters are then given as follows:

$$K_{\mathrm{p}} = \frac{3K_{\mathrm{u}}}{5}, \quad T_{\mathrm{i}} = \frac{\pi}{\omega_{\mathrm{u}}}, \quad T_{\mathrm{d}} = \frac{\pi}{4\omega_{\mathrm{u}}} \tag{3.44}$$

Experience has shown that such controller settings provide a good closed-loop response for many systems. Unfortunately, it will be seen shortly in the coming chapters that the typical model of a VCM actuator is actually a double integrator and thus Ziegler–Nichols tuning cannot be directly applied to design a servo system for the VCM actuator.

Another common way to design a PID controller is the pole assignment method, in which the parameters K_p, T_i and T_d are chosen such that the dominant roots of the closed-loop characteristic equation, *i.e.*

$$1 + K(s)G(s) = 0 \tag{3.45}$$

are assigned to meet certain desired specifications (such as overshoot, rise time, settling time, *etc.*), while its remaining roots are placed far away to the left on the complex plane (roughly three to four times faster compared with the dominant roots). The detailed procedure of this method can be found in most classical control engineering texts (see, *e.g.*, [86]). For the PID control of discrete-time systems, interested readers are referred to [1] for more information.

3.3.2 Sensitivity Functions

System stability margins such as gain margin and phase margin are also very important factors in designing control systems. These stability margins can be obtained from either the well-known Bode plot or Nyquist plot of the open-loop system, *i.e.* $K(s)G(s)$. For an HDD servo system with a large number of resonance modes, its Bode plot might have more than one gain and/or phase crossover frequencies. Thus, it would be necessary to double check these margins using its Nyquist plot. Sensitivity function and complementary sensitivity function are two other measures for a good control system design. The sensitivity function is defined as the closed-loop transfer function from the reference signal, r, to the tracking error, e, and is given by

$$S(s) = \frac{1}{1 + K(s)G(s)} \tag{3.46}$$

The complementary sensitivity function is defined as the closed-loop transfer function between the reference, r, and the system output, y, *i.e.*

$$T(s) = \frac{K(s)G(s)}{1 + K(s)G(s)} \tag{3.47}$$

Clearly, we have $S(s) + T(s) \equiv 1$. A good design should have a sensitivity function that is small at low frequencies for good tracking performance and disturbance rejection and is equal to unity at high frequencies. On the other hand, the complementary sensitivity function should be made unity at low frequencies. It must roll off at high frequencies to possess good attenuation of high-frequency noise.

Note that for a two-degrees-of-freedom control system with a precompensator in the feedforward path right after the reference signal (see, for example, Figure

3.3), the sensitivity and complementary sensitivity functions still remain the same as those in Equations 3.46 and 3.47, which represent, respectively, the closed-loop transfer function from the disturbance at the system output point, if any, to the system output, and the closed-loop transfer function from the measurement noise, if any, to the system output. Thus, a feedforward precompensator does not cause changes in the sensitivity and complementary sensitivity functions. It does, however, help in improving the system tracking performance.

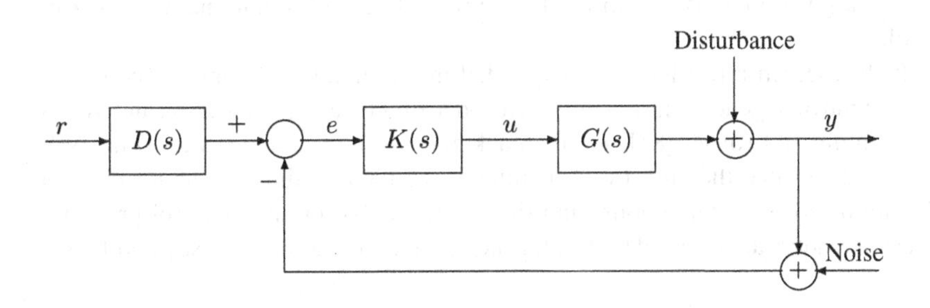

Figure 3.3. A two-degrees-of-freedom control system

3.4 H_2 Optimal Control

Most of the feedback design tools provided by the classical Nyquist–Bode frequency-domain theory are restricted to single-feedback-loop designs. Modern multivariable control theory based on state-space concepts has the capability to deal with multiple feedback-loop designs, and as such has emerged as an alternative to the classical Nyquist–Bode theory. Although it does have shortcomings of its own, a great asset of modern control theory utilizing the state-space description of systems is that the design methods derived from it are easily amenable to computer implementation. Owing to this, rapid progress has been made during the last two or three decades in developing a number of multivariable analysis and design tools using the state-space description of systems. One of the foremost and most powerful design tools developed in this connection is based on what is called linear quadratic Gaussian (LQG) control theory. Here, given a linear model of the plant in a state-space description, and assuming that the disturbance and measurement noise are Gaussian stochastic processes with known power spectral densities, the designer translates the design specifications into a quadratic performance criterion consisting of some state variables and control signal inputs. The object of design then is to minimize the performance criterion by using appropriate state or measurement feedback controllers while guaranteeing the closed-loop stability. A ubiquitous architecture for a measurement feedback controller has been observer based, wherein a state feedback control law is implemented by utilizing an estimate of the state. Thus, the design of a measurement feedback controller here is worked out in two stages. In the first stage, an optimal internally stabilizing static state feedback controller is designed, and in the second stage a state estimator is designed. The estimator, otherwise called an observer or filter, is traditionally designed to yield the least mean square error estimate of the state of the plant, utilizing only the measured output, which is often assumed to be corrupted by an additive white Gaussian noise. The LQG control problem as described above is posed in a stochastic setting. The same can be posed in a deterministic setting, known as an H_2 optimal control problem, in which the H_2 norm of a certain transfer function from an exogenous disturbance to a pertinent controlled output of a given plant is minimized by appropriate use of an internally stabilizing controller.

Much research effort has been expended in the area of H_2 optimal control or optimal control in general during the last few decades (see, *e.g.*, Anderson and Moore [87], Fleming and Rishel [88], Kwakernaak and Sivan [89], and Saberi *et al.* [90], and references cited therein). In what follows, we focus mainly on the formulation and solution to both continuous- and discrete-time H_2 optimal control problems. Interested readers are referred to [90] for more detailed treatments of such problems.

3.4.1 Continuous-time Systems

We consider a generalized system Σ with a state-space description,

$$\Sigma : \begin{cases} \dot{x} = A\ x + \ B\ \ u + \ E\ w \\ y = C_1\ x + D_{11}\ u + \ D_1\ w \\ h = C_2\ x + \ D_2\ \ u + D_{22}\ w \end{cases} \tag{3.48}$$

where $x \in \mathbb{R}^n$ is the state, $u \in \mathbb{R}^m$ is the control input, $w \in \mathbb{R}^q$ is the external disturbance input, $y \in \mathbb{R}^p$ is the measurement output, and $h \in \mathbb{R}^\ell$ is the controlled output of Σ. For the sake of simplicity in future development, throughout this chapter, we let Σ_{P} be the subsystem characterized by the matrix quadruple (A, B, C_2, D_2) and Σ_{Q} be the subsystem characterized by (A, E, C_1, D_1). Throughout this section, we assume that (A, B) is stabilizable and (A, C_1) is detectable.

Generally, we can assume that matrix D_{11} in Equation 3.48 is zero. This can be justified as follows: If $D_{11} \neq 0$, we define a new measurement output

$$y_{\mathrm{new}} = y - D_{11}u = C_1 x + D_1 w \tag{3.49}$$

that does not have a direct feedthrough term from u. Suppose we carry on our control system design using this new measurement output to obtain a proper control law, say, $u = K(s)y_{\mathrm{new}}$. Then, it is straightforward to verify that this control law is equivalent to the following one

$$u = [I + K(s)D_{11}]^{-1}K(s)y \tag{3.50}$$

provided that $[I + K(s)D_{11}]^{-1}$ is well posed, *i.e.* the inverse exists for almost all $s \in \mathbb{C}$. Thus, for simplicity, we assume that $D_{11} = 0$.

The standard H_2 optimal control problem is to find an internally stabilizing proper measurement feedback control law,

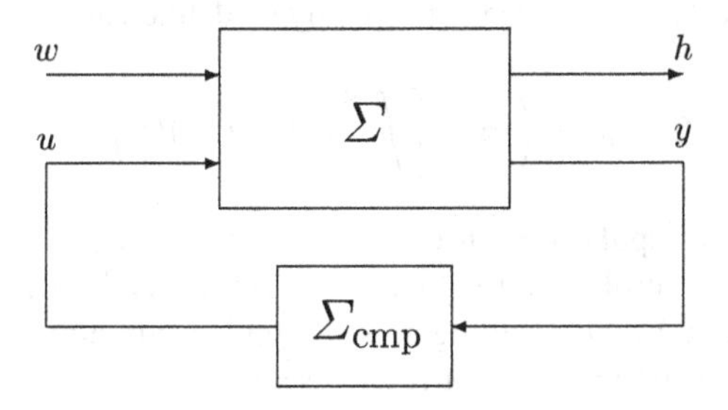

Figure 3.4. The typical control configuration in state-space setting

$$\Sigma_{\text{cmp}} : \begin{cases} \dot{v} = A_{\text{cmp}}\ v + B_{\text{cmp}}\ y \\ u = C_{\text{cmp}}\ v + D_{\text{cmp}}\ y \end{cases} \tag{3.51}$$

such that the H_2-norm of the overall closed-loop transfer matrix function from w to h is minimized (see also Figure 3.4). To be more specific, we will say that the control law Σ_{cmp} of Equation 3.51 is internally stabilizing when applied to the system Σ of Equation 3.48, if the following matrix is asymptotically stable:

$$A_{\text{cl}} := \begin{bmatrix} A + BD_{\text{cmp}}C_1 & BC_{\text{cmp}} \\ B_{\text{cmp}}C_1 & A_{\text{cmp}} \end{bmatrix} \tag{3.52}$$

i.e. all its eigenvalues lie in the open left-half complex plane. It is straightforward to verify that the closed-loop transfer matrix from the disturbance w to the controlled output h is given by

$$T_{hw}(s) = C_{\text{e}}(sI - A_{\text{e}})^{-1}B_{\text{e}} + D_{\text{e}} \tag{3.53}$$

where

$$\left.\begin{aligned} A_{\text{e}} &:= \begin{bmatrix} A + BD_{\text{cmp}}C_1 & BC_{\text{cmp}} \\ B_{\text{cmp}}C_1 & A_{\text{cmp}} \end{bmatrix} \\ B_{\text{e}} &:= \begin{bmatrix} E + BD_{\text{cmp}}D_1 \\ B_{\text{cmp}}D_1 \end{bmatrix} \\ C_{\text{e}} &:= \begin{bmatrix} C_2 + D_2D_{\text{cmp}}C_1 & D_2C_{\text{cmp}} \end{bmatrix} \\ D_{\text{e}} &:= D_2D_{\text{cmp}}D_1 + D_{22} \end{aligned}\right\} \tag{3.54}$$

It is simple to note that if Σ_{cmp} is a static state feedback law, *i.e.* $u = Fx$, then the closed-loop transfer matrix from w to h is given by

$$T_{hw}(s) = (C_2 + D_2F)(sI - A - BF)^{-1}E + D_{22} \tag{3.55}$$

The H_2-norm of a stable continuous-time transfer matrix, *e.g.*, $T_{hw}(s)$, is defined as follows:

$$\|T_{hw}\|_2 := \left(\frac{1}{2\pi}\text{trace}\left[\int_{-\infty}^{\infty} T_{hw}(j\omega)T_{hw}(j\omega)^{\text{H}}\ \mathrm{d}\omega\right]\right)^{1/2} \tag{3.56}$$

By Parseval's theorem, $\|T_{hw}\|_2$ can equivalently be defined as

$$\|T_{hw}\|_2 = \left(\text{trace} \left[\int_0^\infty g(t)g(t)' \, \mathrm{d}t \right] \right)^{1/2} \tag{3.57}$$

where $g(t)$ is the unit impulse response of $T_{hw}(s)$. Thus, $\|T_{hw}\|_2 = \|g\|_2$.

The H_2 optimal control is to design a proper controller Σ_{cmp} such that, when it is applied to the plant Σ, the resulting closed loop is asymptotically stable and the H_2-norm of $T_{hw}(s)$ is minimized. For future use, we define

$$\gamma_2^* := \inf \left\{ \|T_{hw}(\Sigma \times \Sigma_{\text{cmp}})\|_2 \mid \Sigma_{\text{cmp}} \text{ internally stabilizes } \Sigma \right\} \tag{3.58}$$

Furthermore, a control law Σ_{cmp} is said to be an H_2 optimal controller for Σ of Equation 3.48 if its resulting closed-loop transfer function from w to h has an H_2-norm equal to γ_2^*, *i.e.* $\|T_{hw}\|_2 = \gamma_2^*$.

It is clear to see from the definition of the H_2-norm that, in order to have a finite $\|T_{hw}\|_2$, the following must be satisfied:

$$D_e = D_2 D_{\text{cmp}} D_1 + D_{22} = 0 \tag{3.59}$$

which is equivalent to the existence of a static measurement prefeedback law $u = Sy + u_{\text{new}}$ to the system in Equation 3.48 such that $D_2 S D_1 + D_{22} = 0$. We note that the minimization of $\|T_{hw}\|_2$ is meaningful only when it is finite. As such, it is without loss of any generality to assume that the feedforward matrix $D_{22} = 0$ hereafter in this section. In fact, in this case, $\|T_{hw}\|_2$ can be easily obtained. Solving either one of the following Lyapunov equations:

$$A_{\text{e}}' P_{\text{e}} + P_{\text{e}} A_{\text{e}} + C_{\text{e}}' C_{\text{e}} = 0, \quad A_{\text{e}} Q_{\text{e}} + Q_{\text{e}} A_{\text{e}}' + B_{\text{e}} B_{\text{e}}' = 0 \tag{3.60}$$

for P_{e} or Q_{e}, then the H_2-norm of $T_{hw}(s)$ can be computed by

$$\|T_{hw}\|_2 = \sqrt{\text{trace}\left[B_{\text{e}}' P_{\text{e}} B_{\text{e}}\right]} = \sqrt{\text{trace}\left[C_{\text{e}} Q_{\text{e}} C_{\text{e}}'\right]} \tag{3.61}$$

In what follows, we present solutions to the problem without detailed proofs. We start first with the simplest case, when the given system Σ satisfies the following assumptions of the so-called *regular case*:

1. Σ_{P} has no invariant zeros on the imaginary axis and D_2 is of maximal column rank.
2. Σ_{Q} has no invariant zeros on the imaginary axis and D_1 is of maximal row rank.

The problem is called the *singular case* if Σ does not satisfy these conditions.

The solution to the regular case of the H_2 optimal control problem is very simple. The optimal controller is given by (see, *e.g.*, [91]),

$$\Sigma_{\text{cmp}} : \begin{cases} \dot{v} = (A + BF + KC_1)\, v - K\, y \\ u = F\, v \end{cases} \tag{3.62}$$

where

$$F = -(D_2'D_2)^{-1}(D_2'C_2 + B'P) \tag{3.63}$$

$$K = -(QC_1' + ED_1')(D_1D_1')^{-1} \tag{3.64}$$

and where $P = P' \geq 0$ and $Q = Q' \geq 0$ are, respectively, the stabilizing solutions of the following Riccati equations:

$$A'P+PA+C_2'C_2-(PB+C_2'D_2)(D_2'D_2)^{-1}(D_2'C_2+B'P) = 0 \tag{3.65}$$

$$QA'+AQ+EE'-(QC_1'+ED_1')(D_1D_1')^{-1}(D_1E'+C_1Q) = 0 \tag{3.66}$$

Moreover, the optimal value γ_2^* can be computed as follows:

$$\gamma_2^* = \left\{\text{trace}\Big[E'PE\Big] + \text{trace}\Big[(A'P + PA + C_2'C_2)\,Q\Big]\right\}^{1/2} \tag{3.67}$$

We note that if all the states of Σ are available for feedback, then the optimal controller is reduced to a static law $u = Fx$ with F being given as in Equation 3.63.

Next, we present two methods that solve the singular H_2 optimal control problem. As a matter of fact, in the singular case, it is in general infeasible to obtain an optimal controller, although it is possible under certain restricted conditions (see, *e.g.*, [90, 92]). The solutions to the singular case are generally suboptimal, and usually parameterized by a certain tuning parameter, say ε. A controller parameterized by ε is said to be suboptimal if there exists an $\varepsilon^* > 0$ such that for all $\varepsilon \leq \varepsilon^*$ the closed-loop system comprising the given plant and the controller is asymptotically stable, and the resulting closed-loop transfer function from w to h, which is obviously a function of ε, has an H_2-norm arbitrarily close to γ_2^* as ε tends to 0.

The following is a so-called *perturbation approach* (see, *e.g.*, [93]) that would yield a suboptimal controller for the general singular case. We note that such an approach is numerically unstable. The problem becomes very serious when the given system is ill-conditioned or has multiple time scales. In principle, the desired solution can be obtained by introducing some small perturbations to the matrices E, D_1, C_2 and D_2, *i.e.*

$$\tilde{E} := [\,E \quad \varepsilon I \quad 0\,], \quad \tilde{D}_1 := [\,D_1 \quad 0 \quad \varepsilon I\,] \tag{3.68}$$

and

$$\tilde{C}_2 := \begin{bmatrix} C_2 \\ \varepsilon I \\ 0 \end{bmatrix}, \quad \tilde{D}_2 := \begin{bmatrix} D_2 \\ 0 \\ \varepsilon I \end{bmatrix} \tag{3.69}$$

A full-order H_2 suboptimal output feedback controller is given by

$$\tilde{\Sigma}_{\text{cmp}}(\varepsilon) \; : \; \begin{cases} \dot{v} = A + B\tilde{F} + \tilde{K}C_1 \; v \; - \; \tilde{K} \; y \\ u = \qquad \tilde{F} \qquad\quad v \end{cases} \tag{3.70}$$

where

$$\tilde{F} = -(\tilde{D}_2'\tilde{D}_2)^{-1}(\tilde{D}_2'\tilde{C}_2 + B'\tilde{P}) \tag{3.71}$$

$$\tilde{K} = -(\tilde{Q}C_1' + \tilde{E}\tilde{D}_1')(\tilde{D}_1\tilde{D}_1')^{-1} \tag{3.72}$$

and where $\tilde{P} = \tilde{P}' > 0$ and $\tilde{Q} = \tilde{Q}' > 0$ are respectively the solutions of the following Riccati equations:

$$A'\tilde{P} + \tilde{P}A + \tilde{C}_2'\tilde{C}_2 - (\tilde{P}B + \tilde{C}_2'\tilde{D}_2)(\tilde{D}_2'\tilde{D}_2)^{-1}(\tilde{D}_2'\tilde{C}_2 + B'\tilde{P}) = 0 \tag{3.73}$$

$$\tilde{Q}A' + A\tilde{Q} + \tilde{E}\tilde{E}' - (\tilde{Q}C_1' + \tilde{E}\tilde{D}_1')(\tilde{D}_1\tilde{D}_1')^{-1}(\tilde{D}_1\tilde{E}' + C_1\tilde{Q}) = 0 \tag{3.74}$$

Alternatively, one could solve the singular case by using numerically stable algorithms (see, *e.g.*, [90]) that are based on a careful examination of the structural properties of the given system. We separate the problem into three distinct situations: 1) the state feedback case, 2) the full-order measurement feedback case, and 3) the reduced-order measurement feedback case. The software realization of these algorithms in MATLAB® can be found in [53]. For simplicity, we assume throughout the rest of this subsection that both subsystems Σ_{P} and Σ_{Q} *have no invariant zeros on the imaginary axis*. We believe that such a condition is always satisfied for most HDD servo systems. However, most servo systems can be represented as certain chains of integrators and thus could not be formulated as a regular problem without adding dummy terms. Nevertheless, interested readers are referred to the monograph [90] for the complete treatment of H_2 optimal control using the approach given below.

i. State Feedback Case. For the case when $y = x$ in the given system Σ of Equation 3.48, *i.e.* all the state variables of Σ are available for feedback, we have the following step-by-step algorithm that constructs an H_2 suboptimal static feedback control law $u = F(\varepsilon)x$ for Σ.

STEP 3.4.C.S.1: transform the system Σ_{P} into the special coordinate basis as given by Theorem 3.1. To all submatrices and transformations in the special coordinate basis of Σ_{P}, we append the subscript P to signify their relation to the system Σ_{P}. We also choose the output transformation Γ_{oP} to have the following form:

$$\Gamma_{\mathrm{oP}} = \begin{bmatrix} I_{m_{0\mathrm{P}}} & 0 \\ 0 & \Gamma_{\mathrm{orP}} \end{bmatrix} \tag{3.75}$$

where $m_{0\mathrm{P}} =$ rank (D_2). Next, define

$$A_{11\mathrm{P}} = \begin{bmatrix} A_{\mathrm{aaP}}^+ & L_{\mathrm{abP}}^+ C_{\mathrm{bP}} \\ 0 & A_{\mathrm{bbP}} \end{bmatrix}, \quad B_{11\mathrm{P}} = \begin{bmatrix} B_{0\mathrm{aP}}^+ \\ B_{0\mathrm{bP}} \end{bmatrix}, \quad A_{13\mathrm{P}} = \begin{bmatrix} L_{\mathrm{adP}}^+ \\ L_{\mathrm{bdP}} \end{bmatrix} \tag{3.76}$$

$$C_{21\mathrm{P}} = \Gamma_{\mathrm{orP}} \begin{bmatrix} 0 & 0 \\ 0 & C_{\mathrm{bP}} \end{bmatrix}, \quad C_{23\mathrm{P}} = \Gamma_{\mathrm{orP}} \begin{bmatrix} C_{\mathrm{dP}} C_{\mathrm{dP}}' \\ 0 \end{bmatrix} \tag{3.77}$$

$$A_{\mathrm{xP}} = A_{11\mathrm{P}} - A_{13\mathrm{P}}(C_{23\mathrm{P}}' C_{23\mathrm{P}})^{-1} C_{23\mathrm{P}}' C_{21\mathrm{P}} \tag{3.78}$$

$$B_{\mathrm{xP}} B_{\mathrm{xP}}' = B_{11\mathrm{P}} B_{11\mathrm{P}}' + A_{13\mathrm{P}}(C_{23\mathrm{P}}' C_{23\mathrm{P}})^{-1} A_{13\mathrm{P}}' \tag{3.79}$$

$$C_{\mathrm{xP}}' C_{\mathrm{xP}} = C_{21\mathrm{P}}' C_{21\mathrm{P}} - C_{21\mathrm{P}}' C_{21\mathrm{P}}(C_{23\mathrm{P}}' C_{23\mathrm{P}})^{-1} C_{23\mathrm{P}}' C_{21\mathrm{P}} \tag{3.80}$$

STEP 3.4.C.S.2: solve the following algebraic matrix Riccati equation:

$$P_x A_{xP} + A'_{xP} P_x - P_x B_{xP} B'_{xP} P_x + C'_{xP} C_{xP} = 0 \tag{3.81}$$

for $P_x > 0$ and define

$$F_{11} := \begin{bmatrix} F_{a0}^+ & F_{b0} \\ F_{a1}^+ & F_{b1} \end{bmatrix} = \begin{bmatrix} B'_{11P} P_x \\ (C'_{23P} C_{23P})^{-1} (A'_{13P} P_x + C'_{23P} C_{21P}) \end{bmatrix} \tag{3.82}$$

Then, partition $[\, F_{a1}^+ \quad F_{b1} \,]$ as

$$[\, F_{a1}^+ \quad F_{b1} \,] = \begin{bmatrix} F_{a11}^+ & F_{b11} \\ F_{a12}^+ & F_{b12} \\ \vdots & \vdots \\ F_{a1m_{dP}}^+ & F_{b1m_{dP}} \end{bmatrix} \tag{3.83}$$

where F_{a1i}^+ and F_{b1i} are of dimensions $1 \times n_{aP}^+$ and $1 \times n_{bP}$, respectively.

STEP 3.4.C.S.3: let Δ_{cP} be any arbitrary $m_{cP} \times n_{cP}$ matrix subject to the constraint that

$$A_{ccP}^{c} = A_{ccP} - B_{cP} \Delta_{cP} \tag{3.84}$$

is a stable matrix. Note that the existence of such a Δ_{cP} is guaranteed by the property that (A_{ccP}, B_{cP}) is controllable.

STEP 3.4.C.S.4: this step makes use of subsystems, $i = 1$ to m_{dP}, represented by Equation 3.14. Let $\Lambda_i = \{\, \lambda_{i1},\ \lambda_{i2},\ \cdots,\ \lambda_{iq_i} \,\}$, $i = 1$ to m_{dP}, be the sets of q_i elements all in $\mathbb{C}^-$, which are closed under complex conjugation, where q_i and m_{dP} are as defined in Theorem 3.1 but associated with the special coordinate basis of Σ_P. Let $\Lambda_{dP} := \Lambda_1 \cup \Lambda_2 \cup \cdots \cup \Lambda_{m_{dP}}$. For $i = 1$ to m_{dP}, we define

$$p_i(s) := \prod_{j=1}^{q_i} (s - \lambda_{ij}) = s^{q_i} + F_{i1} s^{q_i - 1} + \cdots + F_{iq_i - 1} s + F_{iq_i} \tag{3.85}$$

and

$$\tilde{F}_i(\varepsilon) := \frac{1}{\varepsilon^{q_i}} \left[F_{iq_i},\ \varepsilon F_{iq_i - 1},\ \cdots,\ \varepsilon^{q_i - 1} F_{i1} \right] \tag{3.86}$$

STEP 3.4.C.S.5: in this step, various gains calculated in Steps 3.4.C.S.2 to 3.4.C.S.4 are put together to form a composite state feedback gain for the given system Σ_P. Let

$$\tilde{F}_{a1}^+(\varepsilon) := \begin{bmatrix} F_{a11}^+ F_{1q_1} / \varepsilon^{q_1} \\ F_{a12}^+ F_{2q_2} / \varepsilon^{q_2} \\ \vdots \\ F_{a1m_{dP}}^+ F_{m_{dP} q_{m_{dP}}} / \varepsilon^{q_{m_{dP}}} \end{bmatrix} \tag{3.87}$$

and

$$\tilde{F}_{\rm b1}(\varepsilon) := \begin{bmatrix} F_{\rm b11}F_{1q_1}/\varepsilon^{q_1} \\ F_{\rm b12}F_{2q_2}/\varepsilon^{q_2} \\ \vdots \\ F_{{\rm b1}m_{\rm dP}}F_{m_{\rm dP}q_{m_{\rm dP}}}/\varepsilon^{q_{m_{\rm dP}}} \end{bmatrix} \tag{3.88}$$

Then, the H_2 suboptimal state feedback gain is given by

$$F(\varepsilon) = -\Gamma_{\rm iP}\left(\tilde{F}(\varepsilon) + \tilde{F}_0\right)\Gamma_{\rm SP}^{-1} \tag{3.89}$$

where

$$\tilde{F}(\varepsilon) = \begin{bmatrix} 0 & F_{\rm a0}^+ & F_{\rm b0} & 0 & 0 \\ 0 & \tilde{F}_{\rm a1}^+(\varepsilon) & \tilde{F}_{\rm b1}(\varepsilon) & 0 & \tilde{F}_{\rm d}(\varepsilon) \\ 0 & 0 & 0 & \Delta_{\rm cP} & 0 \end{bmatrix} \tag{3.90}$$

$$\tilde{F}_0 = \begin{bmatrix} C_{\rm 0aP}^- & C_{\rm 0aP}^+ & C_{\rm 0bP} & C_{\rm 0cP} & C_{\rm 0dP} \\ E_{\rm daP}^- & E_{\rm daP}^+ & E_{\rm dbP} & E_{\rm dcP} & E_{\rm dP} \\ E_{\rm caP}^- & E_{\rm caP}^+ & E_{\rm cbP} & 0 & 0 \end{bmatrix} \tag{3.91}$$

and where

$$E_{\rm dP} = \begin{bmatrix} E_{11} & \cdots & E_{1m_{\rm dP}} \\ \vdots & \ddots & \vdots \\ E_{m_{\rm dP}1} & \cdots & E_{m_{\rm dP}m_{\rm dP}} \end{bmatrix} \tag{3.92}$$

$$\tilde{F}_{\rm d}(\varepsilon) = {\rm diag}\left[\tilde{F}_1(\varepsilon),\ \tilde{F}_2(\varepsilon),\ \cdots,\ \tilde{F}_{m_{\rm dP}}(\varepsilon)\right] \tag{3.93}$$

This completes the algorithm. ◇

Theorem 3.11. *Consider the given system in Equation 3.48 with $D_{22} = 0$ and $y = x$, i.e. all states are measurable. Assume that $\Sigma_{\rm P}$ has no invariant zeros on the imaginary axis. Then, the closed-loop system comprising that of Equation 3.48 and $u = F(\varepsilon)x$ with $F(\varepsilon)$ being given as in Equation 3.89 has the following properties:*

1. *it is internally stable for sufficiently small ε;*
2. *the closed-loop transfer matrix from the disturbance w to the controlled output h, $T_{hw}(s,\varepsilon)$, possesses $\|T_{hw}(s,\varepsilon)\|_2 \to \gamma_2^*$ as $\varepsilon \to 0$.*

Clearly, $u = F(\varepsilon)x$ is an H_2 suboptimal controller for the system given in Equation 3.48. ◇

ii. Full-order Output Feedback Case. The following is a step-by-step algorithm for constructing a parameterized full-order output feedback controller that solves the general H_2 optimization problem.

STEP 3.4.C.F.1: (construction of the gain matrix $F_{\rm P}(\varepsilon)$). Define an auxiliary system

$$\begin{cases} \dot{x} = A\ x + B\ u + E\ w \\ y = \quad x \\ h = C_2\ x + D_2\ u \end{cases} \tag{3.94}$$

and then perform Steps 3.4.C.S.1 to 3.4.C.S.5 of the previous algorithm on the above system to obtain a parameterized gain matrix $F(\varepsilon)$. Let $F_P(\varepsilon) = F(\varepsilon)$.

STEP 3.4.C.F.2: (construction of the gain matrix $K_Q(\varepsilon)$). Define another auxiliary system

$$\begin{cases} \dot{x} = A'\,x + C_1'\,u + C_2'\,w \\ y = \quad x \\ h = E'\,x + D_1'\,u \end{cases} \tag{3.95}$$

and then perform Steps 3.4.C.S.1 to 3.4.C.S.5 on the above system to get the parameterized gain matrix $F(\varepsilon)$. We let $K_Q(\varepsilon) = F(\varepsilon)'$.

STEP 3.4.C.F.3: (construction of the full-order controller Σ_{FC}). Finally, the parameterized full-order output feedback controller is given by

$$\Sigma_{FC} : \begin{cases} \dot{v} = A_{FC}(\varepsilon)\,v + B_{FC}(\varepsilon)\,y \\ u = C_{FC}(\varepsilon)\,v + D_{FC}(\varepsilon)\,y \end{cases} \tag{3.96}$$

where

$$\left.\begin{aligned} A_{FC}(\varepsilon) &:= A + BF_P(\varepsilon) + K_Q(\varepsilon)C_1 \\ B_{FC}(\varepsilon) &:= -K_Q(\varepsilon) \\ C_{FC}(\varepsilon) &:= F_P(\varepsilon) \\ D_{FC}(\varepsilon) &:= 0 \end{aligned}\right\} \tag{3.97}$$

This concludes the algorithm for constructing the full-order measurement feedback controller. ◊

Theorem 3.12. *Consider the given system in Equation 3.48 with $D_{22} = 0$. Assume that Σ_P and Σ_Q have no invariant zeros on the imaginary axis. Then the closed-loop system comprising the given system and the full-order output feedback controller of Equation 3.96 has the following properties:*

1. *it is internally stable for sufficiently small ε;*
2. *the closed-loop transfer matrix from the disturbance w to the controlled output h, $T_{hw}(s,\varepsilon)$, possesses $\|T_{hw}(s,\varepsilon)\|_2 \to \gamma_2^*$ as $\varepsilon \to 0$.*

By definition, Equation 3.96 is an H_2 suboptimal controller for the system given in Equation 3.48. ◊

iii. Reduced-order Output Feedback Case. For the case when some measurement output channels are clean, *i.e.* they are not mixed with disturbances, then we can design an output feedback control law that has a dynamical order less than that of the given plant and yet has an identical performance compared with that of full-order control law. Such a control law is called the reduced-order output feedback controller. We note that the construction of a reduced-order controller was first reported by Chen *et al.* [94] for general linear systems, in which the direct feedthrough matrix from input is nonzero. It was shown in [94] that the reduced-order output feedback controller has the following advantages over the full-order counterpart:

1. the dynamical order of the reduced-order controller is generally smaller than that of the full-order counterpart;
2. the gain required for the same degree of performance for the reduced-order controller is smaller compared with that of the full-order counterpart.

We now proceed to design a reduced-order controller, which solves the general H_2 suboptimal problem. First, without loss of generality and for simplicity of presentation, we assume that the matrices C_1 and D_1 are already in the form

$$C_1 = \begin{bmatrix} 0 & C_{1,02} \\ I_k & 0 \end{bmatrix} \quad \text{and} \quad D_1 = \begin{bmatrix} D_{1,0} \\ 0 \end{bmatrix} \tag{3.98}$$

where $k = \ell - \operatorname{rank}(D_1)$ and $D_{1,0}$ is of full rank. Then the given system in Equation 3.48 can be written as

$$\begin{cases} \begin{pmatrix} \dot{x}_1 \\ \dot{x}_2 \end{pmatrix} = \begin{bmatrix} A_{11} & A_{12} \\ A_{21} & A_{22} \end{bmatrix} \begin{pmatrix} x_1 \\ x_2 \end{pmatrix} + \begin{bmatrix} B_1 \\ B_2 \end{bmatrix} u + \begin{bmatrix} E_1 \\ E_2 \end{bmatrix} w \\ \begin{pmatrix} y_0 \\ y_1 \end{pmatrix} = \begin{bmatrix} 0 & C_{1,02} \\ I_k & 0 \end{bmatrix} \begin{pmatrix} x_1 \\ x_2 \end{pmatrix} + \begin{bmatrix} D_{1,0} \\ 0 \end{bmatrix} w \\ \quad h = [C_{2,1} \quad C_{2,2}] \begin{pmatrix} x_1 \\ x_2 \end{pmatrix} + D_2 \ u + D_{22} \ w \end{cases} \tag{3.99}$$

where the original state x is partitioned into two parts, x_1 and x_2; and y is partitioned into y_0 and y_1 with $y_1 \equiv x_1$. Thus, one needs to estimate only the state x_2 in the reduced-order controller design. Next, define an auxiliary subsystem Σ_{QR} characterized by a matrix quadruple $(A_{\text{R}}, E_{\text{R}}, C_{\text{R}}, D_{\text{R}})$, where

$$(A_{\text{R}}, E_{\text{R}}, C_{\text{R}}, D_{\text{R}}) = \left(A_{22}, E_2, \begin{bmatrix} C_{1,02} \\ A_{12} \end{bmatrix}, \begin{bmatrix} D_{1,0} \\ E_1 \end{bmatrix} \right) \tag{3.100}$$

The following is a step-by-step algorithm that constructs the reduced-order output feedback controller for the general H_2 optimization.

STEP 3.4.C.R.1: (construction of the gain matrix $F_{\text{P}}(\varepsilon)$). Define an auxiliary system

$$\begin{cases} \dot{x} = A\ x + B\ u + E\ w \\ y = \quad x \\ h = C_2\ x + D_2\ u \end{cases} \tag{3.101}$$

and then perform Steps 3.4.C.S.1 to 3.4.C.S.5 on the above system to get the parameterized gain matrix $F(\varepsilon)$. We let $F_{\text{P}}(\varepsilon) = F(\varepsilon)$.

STEP 3.4.C.R.2: (construction of the gain matrix $K_{\text{R}}(\varepsilon)$). Define another auxiliary system

$$\begin{cases} \dot{x} = A_{\text{R}}'\ x + C_{\text{R}}'\ u + C_{2,2}'\ w \\ y = \quad x \\ h = E_{\text{R}}'\ x + D_{\text{R}}'\ u \end{cases} \tag{3.102}$$

and then perform Steps 3.4.C.S.1 to 3.4.C.S.5 on the above system to get the parameterized gain matrix $F(\varepsilon)$. We let $K_{\text{R}}(\varepsilon) = F(\varepsilon)'$.

STEP 3.4.C.R.3: (construction of the reduced-order controller Σ_{RC}). Let us partition $F_{\text{P}}(\varepsilon)$ and $K_{\text{R}}(\varepsilon)$ as

$$F_{\text{P}}(\varepsilon) = [\, F_{\text{P1}}(\varepsilon) \quad F_{\text{P2}}(\varepsilon)\,], \quad K_{\text{R}}(\varepsilon) = [\, K_{\text{R0}}(\varepsilon) \quad K_{\text{R1}}(\varepsilon)\,] \tag{3.103}$$

in conformity with the partitions of $x = \begin{pmatrix} x_1 \\ x_2 \end{pmatrix}$ and $y = \begin{pmatrix} y_0 \\ y_1 \end{pmatrix}$ of Equation 3.99, respectively. Then define

$$G_{\text{R}}(\varepsilon) = [\, -K_{\text{R0}}(\varepsilon), \;\; A_{21} + K_{\text{R1}}(\varepsilon)A_{11} - (A_{\text{R}} + K_{\text{R}}(\varepsilon)C_{\text{R}})K_{\text{R1}}(\varepsilon)\,]$$

Finally, the reduced-order output feedback controller is given by

$$\Sigma_{\text{RC}} \; : \; \begin{cases} \dot{v} = A_{\text{RC}}(\varepsilon)\; v + B_{\text{RC}}(\varepsilon)\; y \\ u = C_{\text{RC}}(\varepsilon)\; v + D_{\text{RC}}(\varepsilon)\; y \end{cases} \tag{3.104}$$

where

$$\left.\begin{aligned} A_{\text{RC}}(\varepsilon) &:= A_{\text{R}} + B_2 F_{\text{P2}}(\varepsilon) + K_{\text{R}}(\varepsilon)C_{\text{R}} + K_{\text{R1}}(\varepsilon)B_1 F_{\text{P2}}(\varepsilon) \\ B_{\text{RC}}(\varepsilon) &:= G_{\text{R}}(\varepsilon) + [B_2 + K_{\text{R1}}(\varepsilon)B_1] \\ &\qquad\qquad \times\, [\,0,\;\; F_{\text{P1}}(\varepsilon) - F_{\text{P2}}(\varepsilon)K_{\text{R1}}(\varepsilon)\,] \\ C_{\text{RC}}(\varepsilon) &:= F_{\text{P2}}(\varepsilon) \\ D_{\text{RC}}(\varepsilon) &:= [\,0,\;\; F_{\text{P1}}(\varepsilon) - F_{\text{P2}}(\varepsilon)K_{\text{R1}}(\varepsilon)\,] \end{aligned}\right\} \tag{3.105}$$

This concludes the algorithm for constructing the reduced-order output feedback controller. ◊

Theorem 3.13. *Consider the given system in Equation 3.48 with $D_{22} = 0$. Assume that Σ_{P} and Σ_{Q} have no invariant zeros on the imaginary axis. Then, the closed-loop system comprising the given system and the reduced-order output feedback controller in Equation 3.104 has the following properties:*

1. *it is internally stable for sufficiently small ε;*
2. *the closed-loop transfer matrix from the disturbance w to the controlled output h, $T_{hw}(s, \varepsilon)$, possesses $\|T_{hw}(s,\varepsilon)\|_2 \to \gamma_2^*$ as $\varepsilon \to 0$.*

By definition, Equation 3.104 is an H_2 suboptimal controller for the system given in Equation 3.48. ◊

3.4.2 Discrete-time Systems

We now consider a generalized discrete-time system Σ characterized by the following state-space equations

$$\Sigma \; : \; \begin{cases} x(k+1) = A\; x(k) + B\; u(k) + E\; w(k) \\ y(k) = C_1\; x(k) + D_{11}\; u(k) + D_1\; w(k) \\ h(k) = C_2\; x(k) + D_2\; u(k) + D_{22}\; w(k) \end{cases} \tag{3.106}$$

where $x \in \mathbb{R}^n$ is the state, $u \in \mathbb{R}^m$ is the control input, $w \in \mathbb{R}^q$ is the external disturbance input, $y \in \mathbb{R}^p$ is the measurement output, and $h \in \mathbb{R}^\ell$ is the controlled output of Σ. As usual, we let $\Sigma_{\rm P}$ be the subsystem characterized by the matrix quadruple (A, B, C_2, D_2) and $\Sigma_{\rm Q}$ be the subsystem characterized by (A, E, C_1, D_1). Without loss of any generality, we assume that $D_{11} = 0$, (A, B) is stabilizable and (A, C_1) is detectable.

The standard H_2 optimal control problem for a discrete-time system is to find an internally stabilizing proper measurement feedback control law,

$$\Sigma_{\rm cmp} : \begin{cases} v(k+1) = A_{\rm cmp}\, v(k) + B_{\rm cmp}\, y(k) \\ \quad u(k) \;\;\, = C_{\rm cmp}\, v(k) + D_{\rm cmp}\, y(k) \end{cases} \tag{3.107}$$

such that the H_2-norm of the overall closed-loop transfer matrix function from w to h is minimized. To be more specific, we will say that the control law $\Sigma_{\rm cmp}$ of Equation 3.107 is internally stabilizing when applied to the system Σ of Equation 3.106, if the following matrix is asymptotically stable:

$$A_{\rm cl} := \begin{bmatrix} A + BD_{\rm cmp}C_1 & BC_{\rm cmp} \\ B_{\rm cmp}C_1 & A_{\rm cmp} \end{bmatrix} \tag{3.108}$$

i.e. all its eigenvalues lie inside the open unit disc. The closed-loop transfer matrix from the disturbance w to the controlled output h is given by

$$T_{hw}(z) = C_{\rm e}(zI - A_{\rm e})^{-1}B_{\rm e} + D_{\rm e} \tag{3.109}$$

where

$$A_{\rm e} := \begin{bmatrix} A + BD_{\rm cmp}C_1 & BC_{\rm cmp} \\ B_{\rm cmp}C_1 & A_{\rm cmp} \end{bmatrix} \tag{3.110}$$

$$B_{\rm e} := \begin{bmatrix} E + BD_{\rm cmp}D_1 \\ B_{\rm cmp}D_1 \end{bmatrix} \tag{3.111}$$

$$C_{\rm e} := \begin{bmatrix} C_2 + D_2D_{\rm cmp}C_1 & D_2C_{\rm cmp} \end{bmatrix} \tag{3.112}$$

$$D_{\rm e} := D_2D_{\rm cmp}D_1 + D_{22} \tag{3.113}$$

The H_2-norm of a stable discrete-time transfer matrix, *e.g.*, $T_{hw}(z)$, is defined as follows:

$$||T_{hw}||_2 := \left(\frac{1}{2\pi} \text{trace} \left[\int_{-\pi}^{\pi} T_{hw}(\mathrm{e}^{j\omega}) T_{hw}(\mathrm{e}^{j\omega})^{\rm H} \, \mathrm{d}\omega \right] \right)^{1/2} \tag{3.114}$$

By Parseval's theorem, $||T_{hw}||_2$ can equivalently be defined as

$$||T_{hw}||_2 = \left(\text{trace} \left[\sum_{k=0}^{\infty} g(k) g(k)' \, \mathrm{d}t \right] \right)^{1/2} \tag{3.115}$$

where $g(k)$ is the impulse response of $T_{hw}(k)$. Thus, $||T_{hw}||_2 = ||g||_2$.

The H_2 optimal control for the discrete-time system of Equation 3.106 is to design a proper controller $\Sigma_{\rm cmp}$ such that, when it is applied to the plant Σ, the resulting closed loop is asymptotically stable and the H_2-norm of $T_{hw}(z)$ is minimized. For future use, we define

$$\gamma_2^* := \inf \left\{ \|T_{hw}(\Sigma \times \Sigma_{\rm cmp})\|_2 \mid \Sigma_{\rm cmp} \text{ internally stabilizes } \Sigma \right\} \tag{3.116}$$

Again, a control law $\Sigma_{\rm cmp}$ is said to be an H_2 optimal controller for Σ of Equation 3.106 if its resulting closed-loop transfer function from w to h has an H_2-norm equal to γ_2^*, *i.e.* $\|T_{hw}\|_2 = \gamma_2^*$.

For the case when $D_{\rm e} = 0$, $\|T_{hw}\|_2$ can be computed by

$$\|T_{hw}\|_2 = \sqrt{\operatorname{trace}\left[B_{\rm e}' P_{\rm e} B_{\rm e}\right]} = \sqrt{\operatorname{trace}\left[C_{\rm e} Q_{\rm e} C_{\rm e}'\right]} \tag{3.117}$$

where $P_{\rm e}$ and $Q_{\rm e}$ are, respectively, the solutions of the following Lyapunov equations:

$$A_{\rm e}' P_{\rm e} A - P_{\rm e} + C_{\rm e}' C_{\rm e} = 0, \quad A_{\rm e} Q_{\rm e} A_{\rm e}' - Q_{\rm e} + B_{\rm e} B_{\rm e}' = 0 \tag{3.118}$$

We now present solutions to the discrete-time H_2 optimal control problem without detailed proofs. As usual, we assume that $D_{22} = 0$ for convenience. We start first with the simplest case when the given system Σ satisfies the following assumptions of the so-called *regular case*:

1. $\Sigma_{\rm P}$ has no invariant zeros on the imaginary axis and is left invertible;
2. $\Sigma_{\rm Q}$ has no invariant zeros on the imaginary axis and is right invertible.

The problem is called the *singular case* if Σ does not satisfy these conditions.

Again, the solution to the regular case of the discrete-time H_2 optimal control problem is very simple as well. The optimal controller is given by $\Sigma_{\rm cmp}$:

$$\begin{cases} v(k+1) = (A+BF+KC_1-BNC_1)\, v(k) + (BN-K)\, y(k) \\ \quad u(k) \;\;\;\; = \quad (F-NC_1) \quad v(k) + \quad N \quad y(k) \end{cases} \tag{3.119}$$

where

$$F = -(B'PB + D_2'D_2)^{-1}(B'PA + D_2'C_2) \tag{3.120}$$

$$K = -(AQC_1' + ED_1')(D_1D_1' + C_1QC_1')^{-1} \tag{3.121}$$

and

$$N = -(B'PB + D_2'D_2)^{-1}\left[(B'PA + D_2'D_2)QC_1 + B'PED_1'\right] \times (D_1D_1' + C_1QC_1')^{-1} \tag{3.122}$$

and where $P = P' \geq 0$ and $Q = Q' \geq 0$ are, respectively, the stabilizing solutions of the following Riccati equations:

$$P = A'PA + C_2'C_2 - (C_2'D_2 + A'PB)(D_2'D_2 + B'PB)^{-1} \times (D_2'C_2 + B'PA) \quad (3.123)$$

$$Q = AQA' + EE' - (ED_1' + AQC_1')(D_1D_1' + C_1QC_1')^{-1} \times (D_1E' + C_1QA') \quad (3.124)$$

We note that the above discrete-time Riccati equations can be solved using the non-iterative method given in [74]. If all the states of Σ are available for feedback, then the optimal controller is reduced to a static law $u(k) = Fx(k)$ with F being given as in Equation 3.120.

Next, we present solutions to the singular H_2 optimal control problem. Similarly, the solutions to the singular case are generally suboptimal, and usually parameterized by a certain tuning parameter, say ε. Again, a discrete-time controller parameterized by ε is said to be suboptimal for the system in Equation 3.106 if there exists an $\varepsilon^* > 0$ such that for all $\varepsilon \leq \varepsilon^*$ the closed-loop system comprising the given plant and the controller is asymptotically stable, and the resulting closed-loop transfer function from w to h, which is obviously a function of ε, has an H_2-norm arbitrarily close to γ_2^* as ε tends to 0.

The following *perturbation approach* would yield a suboptimal controller for the general discrete-time singular case. Given any $\varepsilon > 0$, define

$$\tilde{E} := [\,E \quad \varepsilon I \quad 0\,], \quad \tilde{D}_1 := [\,D_1 \quad 0 \quad \varepsilon I\,] \quad (3.125)$$

and

$$\tilde{C}_2 := \begin{bmatrix} C_2 \\ \varepsilon I \\ 0 \end{bmatrix}, \quad \tilde{D}_2 := \begin{bmatrix} D_2 \\ 0 \\ \varepsilon I \end{bmatrix} \quad (3.126)$$

A full-order H_2 suboptimal output feedback controller is given by

$$\begin{cases} v(k+1) = (A+B\tilde{F}+\tilde{K}C_1-B\tilde{N}C_1)\, v(k) + (B\tilde{N}-\tilde{K})\, y(k) \\ \quad u(k) \;= \qquad (\tilde{F}-\tilde{N}C_1) \qquad v(k) + \quad \tilde{N} \quad y(k) \end{cases} \quad (3.127)$$

where

$$\tilde{F} = -(B'\tilde{P}B + \tilde{D}_2'\tilde{D}_2)^{-1}(B'\tilde{P}A + \tilde{D}_2'\tilde{C}_2) \quad (3.128)$$

$$\tilde{K} = -(A\tilde{Q}C_1' + \tilde{E}\tilde{D}_1')(\tilde{D}_1\tilde{D}_1' + C_1\tilde{Q}C_1')^{-1} \quad (3.129)$$

and

$$\tilde{N} = -(B'\tilde{P}B + \tilde{D}_2'\tilde{D}_2)^{-1}\left[(B'\tilde{P}A + \tilde{D}_2'\tilde{D}_2)\tilde{Q}C_1 + B'\tilde{P}\tilde{E}\tilde{D}_1'\right] \times (\tilde{D}_1\tilde{D}_1' + C_1\tilde{Q}C_1')^{-1} \quad (3.130)$$

and where $\tilde{P} = \tilde{P}' > 0$ and $\tilde{Q} = \tilde{Q}' > 0$ are, respectively, the stabilizing solutions of the following Riccati equations:

$$\tilde{P} = A'\tilde{P}A + \tilde{C}_2'\tilde{C}_2 - (\tilde{C}_2'\tilde{D}_2 + A'\tilde{P}B)(\tilde{D}_2'\tilde{D}_2 + B'\tilde{P}B)^{-1} \times (\tilde{D}_2'\tilde{C}_2 + B'\tilde{P}A) \quad (3.131)$$

$$\tilde{Q} = A\tilde{Q}A' + \tilde{E}\tilde{E}' - (\tilde{E}\tilde{D}_1' + A\tilde{Q}C_1')(\tilde{D}_1\tilde{D}_1' + C_1\tilde{Q}C_1')^{-1} \times (\tilde{D}_1\tilde{E}' + C_1\tilde{Q}A') \quad (3.132)$$

The following are alternative methods based on the structural decompositions of the given systems. Similarly, we separate the problem into three distinct situations: 1) the state feedback case, 2) the full-order measurement feedback case, and 3) the reduced-order measurement feedback case.

Similarly, for convenience, we assume throughout the rest of this subsection that both subsystems $\Sigma_{\rm P}$ and $\Sigma_{\rm Q}$ *have no invariant zeros on the unit circle.* The complete treatment of H_2 optimal control using the approach given below can be found in [90]. Interestingly, it turns out that for this case, although it is singular, we can always obtain a set of H_2 optimal controllers that need not be parameterized by any tuning scalar.

i. State Feedback Case. For the case when $y = x$ in the given system Σ of Equation 3.106, we have the following step-by-step algorithm that constructs an H_2 suboptimal static feedback control law $u = F(\varepsilon)x$ for Σ.

STEP 3.4.D.S.1: (decomposition of $\Sigma_{\rm P}$). Transform the subsystem $\Sigma_{\rm P}$, *i.e.* the matrix quadruple (A, B, C_2, D_2), into the special coordinate basis as given by Theorem 3.1. Denote the state, output and input transformation matrices as $\Gamma_{\rm sP}$, $\Gamma_{\rm oP}$ and $\Gamma_{\rm iP}$, respectively.

STEP 3.4.D.S.2: (gain matrix for the subsystem associated with $\mathcal{X}_{\rm c}$). Let $F_{\rm c}$ be any constant matrix subject to the constraint that

$$A_{\rm cc}^{\rm c} = A_{\rm cc} - B_{\rm c}F_{\rm c} \quad (3.133)$$

is a stable matrix. Note that the existence of such an $F_{\rm c}$ is guaranteed by the property of the special coordinate basis, *i.e.* $(A_{\rm cc}, B_{\rm c})$ is controllable.

STEP 3.4.D.S.3: (gain matrix for the subsystem associated with $\mathcal{X}_{\rm a}^+$, $\mathcal{X}_{\rm b}$ and $\mathcal{X}_{\rm d}$). Let

$$A_{\rm x} = \begin{bmatrix} A_{\rm aa}^+ & L_{\rm ad}^+C_{\rm d} & L_{\rm ab}^+C_{\rm b} \\ B_{\rm d}E_{\rm da}^+ & A_{\rm dd} & B_{\rm d}E_{\rm db} \\ 0 & L_{\rm bd}C_{\rm d} & A_{\rm bb} \end{bmatrix}, \quad E_{\rm x} = \begin{bmatrix} E_{\rm a}^+ \\ E_{\rm d} \\ E_{\rm b} \end{bmatrix} \quad (3.134)$$

$$B_{\rm x} = \begin{bmatrix} B_{\rm 0a}^+ & 0 \\ B_{\rm 0d} & B_{\rm d} \\ B_{\rm 0b} & 0 \end{bmatrix} \quad (3.135)$$

and

$$C_{\rm x} = \Gamma_{\rm oP} \begin{bmatrix} 0 & 0 & 0 \\ 0 & C_{\rm d} & 0 \\ 0 & 0 & C_{\rm b} \end{bmatrix}, \quad D_{\rm x} = \Gamma_{\rm oP} \begin{bmatrix} I & 0 \\ 0 & 0 \\ 0 & 0 \end{bmatrix} \quad (3.136)$$

Solve the following discrete-time Riccati equation

$$P_x = A_x' P_x A_x + C_x' C_x - (C_x' D_x + A_x' P_x B_x)(D_x' D_x + B_x' P_x B_x)^{-1} \times (D_x' C_x + B_x' P_x A_x) \quad (3.137)$$

for $P_x > 0$. Then partition

$$F_x = (B_x' P_x B_x + D_x' D_x)^{-1}(B_x' P_x A_x + D_x' C_x) = \begin{bmatrix} F_{a0}^+ & F_{d0} & F_{b0} \\ F_{ad}^+ & F_{dd} & F_{bd} \end{bmatrix} \quad (3.138)$$

STEP 3.4.D.S.4: (composition of gain matrix F). In this step, various gains calculated in the previous steps are put together to form a composite state feedback gain matrix F. It is given by

$$F = -\Gamma_{iP} \begin{bmatrix} C_{0a}^- & C_{0a}^+ + F_{a0}^+ & C_{0b} + F_{b0} & C_{0c} & C_{0d} + F_{d0} \\ E_{da}^- & F_{ad}^+ & F_{bd} & E_{dc} & F_{dd} \\ E_{ca}^- & E_{ca}^+ & 0 & F_c & 0 \end{bmatrix} \Gamma_{sP}^{-1} \quad (3.139)$$

This completes the algorithm. ◇

Theorem 3.14. *Consider the system given in Equation 3.106 with* $D_{22} = 0$ *and* $y = x$*, i.e. all states are measurable. Assume that* Σ_P *has no invariant zeros on the unit circle. Then, the closed-loop system comprising the given system and* $u(k) = Fx(k)$ *with* F *being given as in Equation 3.139 has the following properties:*

1. *it is internally stable;*
2. *the closed-loop transfer matrix from the disturbance* w *to the controlled output* h*,* $T_{hw}(z)$*, possesses* $\|T_{hw}\|_2 = \gamma_2^*$.

Thus, $u(k) = Fx(k)$ *is an* H_2 *optimal state feedback control law for the system in Equation 3.106.* ◇

ii. Full-order Output Feedback Case. The following is a step-by-step algorithm for constructing an H_2 optimal full-order output feedback controller.

STEP 3.4.D.F.1: (computation of N). Utilize the special coordinate basis properties to compute two constant matrices X and Y such that $\mathcal{V}^{\odot}(\Sigma_P) = \text{Ker}(X)$ and $\mathcal{S}^{\odot}(\Sigma_Q) = \text{Im}(Y)$. Then, compute

$$N = -(B'X'XB + D_2'D_2)^{\dagger} \begin{bmatrix} B'X' & D_2' \end{bmatrix} \begin{bmatrix} XAY & XE \\ C_2Y & D_{22} \end{bmatrix} \times \begin{bmatrix} Y'C_1' \\ D_1' \end{bmatrix} (C_1YY'C_1' + D_1D_1')^{\dagger} \quad (3.140)$$

STEP 3.4.D.F.2: (construction of the gain matrix $F_{\rm P}$). Define an auxiliary system

$$\begin{cases} x(k+1) = \tilde{A}\ x(k) + B\ u(k) + \tilde{E}\ w(k) \\ y(k) = x(k) \\ h(k) = \tilde{C}_2\ x(k) + D_2\ u(k) + 0\ w(k) \end{cases} \tag{3.141}$$

where

$$\tilde{A} = A + BNC_1, \quad \tilde{E} = E + BND_1, \quad \tilde{C}_2 = C_2 + D_2NC_1 \tag{3.142}$$

and then perform Steps 3.4.D.S.1 to 3.4.D.S.4 of the previous algorithm on the above system in Equation 3.141 to obtain a gain matrix F. We let $F_{\rm P} = F$.

STEP 3.4.D.F.3: (construction of the gain matrix $K_{\rm Q}$). Define another auxiliary system

$$\begin{cases} x(k+1) = \tilde{A}'\ x(k) + C_1'\ u(k) + \tilde{C}_2'\ w(k) \\ y(k) = x(k) \\ h(k) = \tilde{E}'\ x(k) + D_1'\ u(k) + 0\ w(k) \end{cases} \tag{3.143}$$

and then perform Steps 3.4.D.S.1 to 3.4.D.S.4 of the previous algorithm on the above system to get a gain matrix F. Similarly, we let $K_{\rm Q} = F'$.

STEP 3.4.D.F.4: (construction of the full-order controller $\Sigma_{\rm FC}$). Finally, the parameterized full-order output feedback controller is given by

$$\Sigma_{\rm FC} : \begin{cases} v(k+1) = A_{\rm FC}\ v(k) + B_{\rm FC}\ y(k) \\ u(k) = C_{\rm FC}\ v(k) + D_{\rm FC}\ y(k) \end{cases} \tag{3.144}$$

where

$$\left. \begin{aligned} A_{\rm FC} &:= A + BNC_1 + BF_{\rm P} + K_{\rm Q}C_1 \\ B_{\rm FC} &:= -K_{\rm Q} \\ C_{\rm FC} &:= F_{\rm P} \\ D_{\rm FC} &:= N \end{aligned} \right\} \tag{3.145}$$

This concludes the algorithm for constructing the full-order measurement feedback controller. ◊

Theorem 3.15. *Consider the system given in Equation 3.106 with $D_{22} = 0$. Assume that $\Sigma_{\rm P}$ and $\Sigma_{\rm Q}$ have no invariant zeros on the unit circle. Then the closed-loop system comprising the given system and the full-order output feedback controller of Equation 3.144 has the following properties:*

1. *it is internally stable;*
2. *the closed-loop transfer matrix from the disturbance w to the controlled output h, $T_{hw}(z)$, possesses $\|T_{hw}\|_2 = \gamma_2^*$.*

Hence, Equation 3.144 is an H_2 optimal control law for the system given in Equation 3.106. ◊

iii. Reduced-order Output Feedback Case. We now follow the procedure as in the continuous-time case to design a reduced-order output feedback controller. For simplicity of presentation, we assume that the matrices C_1 and D_1 are already in the form

$$C_1 = \begin{bmatrix} 0 & C_{1,02} \\ I_k & 0 \end{bmatrix} \quad \text{and} \quad D_1 = \begin{bmatrix} D_{1,0} \\ 0 \end{bmatrix} \tag{3.146}$$

where $k = \ell - \text{rank}(D_1)$ and $D_{1,0}$ is of full rank. Next, we follow Steps 3.4.D.F.1 and 3.4.D.F.2 of the previous subsection to compute the constant matrix N, and form the following system:

$$\begin{cases} x(k+1) = \tilde{A}\ x(k) + B\ u(k) + \tilde{E}\ w(k) \\ y(k) \quad = C_1\ x(k) \qquad\qquad\quad + D_1\ w(k) \\ h(k) \quad = \tilde{C}_2\ x(k) + D_2\ u(k) +\ 0\ \ w(k) \end{cases} \tag{3.147}$$

where $\tilde{A}$, $\tilde{E}$ and $\tilde{C}_2$ are defined as in Equation 3.142. Then, partition Equation 3.147 as follows:

$$\begin{cases} \begin{pmatrix} x_1(k+1) \\ x_2(k+1) \end{pmatrix} = \begin{bmatrix} A_{11} & A_{12} \\ A_{21} & A_{22} \end{bmatrix} \begin{pmatrix} x_1(k) \\ x_2(k) \end{pmatrix} + \begin{bmatrix} B_1 \\ B_2 \end{bmatrix} u(k) + \begin{bmatrix} E_1 \\ E_2 \end{bmatrix} w(k) \\ \begin{pmatrix} y_0(k) \\ y_1(k) \end{pmatrix} = \begin{bmatrix} 0 & C_{1,02} \\ I_k & 0 \end{bmatrix} \begin{pmatrix} x_1(k) \\ x_2(k) \end{pmatrix} + \begin{bmatrix} D_{1,0} \\ 0 \end{bmatrix} w(k) \\ h(k) = [C_{2,1} \quad C_{2,2}] \begin{pmatrix} x_1(k) \\ x_2(k) \end{pmatrix} + D_2\ u(k) + 0\ \ w(k) \end{cases}$$

where the state x of Equation 3.147 is partitioned into two parts, x_1 and x_2; and y is partitioned to y_0 and y_1 with $y_1 \equiv x_1$. Thus, one needs to estimate only the state x_2 in the reduced-order controller design. Next, define an auxiliary subsystem Σ_{QR} characterized by a matrix quadruple $(A_{\text{R}}, E_{\text{R}}, C_{\text{R}}, D_{\text{R}})$, where

$$(A_{\text{R}}, E_{\text{R}}, C_{\text{R}}, D_{\text{R}}) = \left(A_{22}, E_2, \begin{bmatrix} C_{1,02} \\ A_{12} \end{bmatrix}, \begin{bmatrix} D_{1,0} \\ E_1 \end{bmatrix} \right) \tag{3.148}$$

The following is a step-by-step algorithm that constructs the reduced-order output feedback controller for the general H_2 optimization.

STEP 3.4.D.R.1: (construction of the gain matrix F_{P}). Define an auxiliary system

$$\begin{cases} x(k+1) = \tilde{A}\ x(k) + B\ u(k) + \tilde{E}\ w(k) \\ y(k) \quad = \quad x(k) \\ h(k) \quad = \tilde{C}_2\ x(k) + D_2\ u(k) +\ 0\ \ w(k) \end{cases} \tag{3.149}$$

and then perform Steps 3.4.D.S.1 to 3.4.D.S.4 of the previous algorithm on the above system to obtain a parameterized gain matrix F. Moreover, let $F_{\text{P}} = F$.

STEP 3.4.D.R.2: (construction of the gain matrix K_{R}). Define another auxiliary system

$$\left\{\begin{array}{lcl} x(k+1) &=& A'_{\mathrm{R}}\, x(k) + C'_{\mathrm{R}}\, u(k) + C'_{2,2}\, w(k) \\ y(k) &=& x(k) \\ h(k) &=& E'_{\mathrm{R}}\, x(k) + D'_{\mathrm{R}}\, u(k) + \;0\; w(k) \end{array}\right. \tag{3.150}$$

and then perform Steps 3.4.D.S.1 to 3.4.D.S.4 of the previous algorithm on the above system to obtain a parameterized gain matrix F. Similarly, let $K_{\mathrm{R}} = F'$.

STEP 3.4.D.R.3: (construction of the reduced-order controller Σ_{RC}). Let us partition F_{P} and K_{R} as

$$F_{\mathrm{P}} = [\, F_{\mathrm{P1}} \quad F_{\mathrm{P2}} \,] \quad \text{and} \quad K_{\mathrm{R}} = [\, K_{\mathrm{R0}} \quad K_{\mathrm{R1}} \,] \tag{3.151}$$

in conformity with $x = \begin{pmatrix} x_1 \\ x_2 \end{pmatrix}$ and $y = \begin{pmatrix} y_0 \\ y_1 \end{pmatrix}$, respectively. Then define

$$G_{\mathrm{R}} = [\, -K_{\mathrm{R0}},\; A_{21} + K_{\mathrm{R1}} A_{11} - (A_{\mathrm{R}} + K_{\mathrm{R}} C_{\mathrm{R}}) K_{\mathrm{R1}} \,] \tag{3.152}$$

Finally, the parameterized reduced-order output feedback controller is given by

$$\Sigma_{\mathrm{RC}} : \left\{\begin{array}{lcl} v(k+1) &=& A_{\mathrm{RC}}\, v(k) + B_{\mathrm{RC}}\, y(k) \\ u(k) &=& C_{\mathrm{RC}}\, v(k) + D_{\mathrm{RC}}\, y(k) \end{array}\right. \tag{3.153}$$

where

$$\left.\begin{array}{l} A_{\mathrm{RC}} := A_{\mathrm{R}} + B_2 F_{\mathrm{P2}} + K_{\mathrm{R}} C_{\mathrm{R}} + K_{\mathrm{R1}} B_1 F_{\mathrm{P2}} \\ B_{\mathrm{RC}} := G_{\mathrm{R}} + [B_2 + K_{\mathrm{R1}} B_1]\, [\,0,\; F_{\mathrm{P1}} - F_{\mathrm{P2}} K_{\mathrm{R1}}\,] \\ C_{\mathrm{RC}} := F_{\mathrm{P2}} \\ D_{\mathrm{RC}} := [\,0,\; F_{\mathrm{P1}} - F_{\mathrm{P2}} K_{\mathrm{R1}}\,] + N \end{array}\right\} \tag{3.154}$$

This concludes the algorithm for constructing the reduced-order output feedback controller. ◊

Theorem 3.16. *Consider the system given in Equation 3.106 with* $D_{22} = 0$. *Assume that* Σ_{P} *and* Σ_{Q} *have no invariant zeros on the unit circle. Then, the closed-loop system comprising the given system and the reduced-order output feedback controller in Equation 3.153 has the following properties:*

1. *it is internally stable;*
2. *the closed-loop transfer matrix from the disturbance* w *to the controlled output* h, $T_{hw}(z)$, *possesses* $\|T_{hw}(z)\|_2 = \gamma_2^*$.

Thus, Equation 3.153 is an H_2 *optimal control law for the system given in Equation 3.106.* ◊

Lastly, we note that the result presented in this section, although it is not totally complete, is sufficient to obtain appropriate solutions for HDD servo systems and many engineering problems. We next move to H_∞ control and its related problems.

3.5 H_∞ Control and Disturbance Decoupling

The ultimate goal of a control system designer is to build a system that works in a real environment. Since the real environment may change and the operating conditions may vary from time to time, the control system must be able to withstand these variations. Even if the environment does not change, other factors of life are the model uncertainties, as well as noises. Any mathematical representation of a system often involves simplifying assumptions. Nonlinearities are either unknown, and hence unmodeled, or are modeled and later ignored in order to simplify analysis. High-frequency dynamics are often ignored at the design stage as well. In consequence, control systems designed based on simplified models may not work on real plants in real environments. The particular property that a control system must possess for it to operate properly in realistic situations is commonly called *robustness*. Mathematically, this means that the controller must perform satisfactorily not just for one plant, but for a family of plants. If a controller can be designed such that the whole system to be controlled remains stable when its parameters vary within certain expected limits, the system is said to possess robust stability. In addition, if it can satisfy performance specifications such as steady state tracking, disturbance rejection and speed of response requirements, it is said to possess robust performance. The problem of designing controllers that satisfy both robust stability and performance requirements is called robust control. H_∞ control theory is one of the cornerstones of modern control theory and was developed in an attempt to solve such a problem. Many robust control problems (such as the robust stability problem of unstructurally perturbed systems, the mixed-sensitivity problem, robust stabilization with additive and multiplicative perturbations, to name a few) can be cast into a standard H_∞ control problem (see, *e.g.*, [74]).

Since the original formulation of the H_∞ control problem by Zames [95], a great deal of work has been done on finding the solution to this problem. Practically all the research results of the early years involved a mixture of time-domain and frequency-domain techniques, including the following: *Interpolation approach* (see, *e.g.*, Limebeer and Anderson [96]); *Frequency domain approach* (see, *e.g.*, Doyle [97], Francis [98] and Glover [99]); *Polynomial approach* (see, *e.g.*, Kwakernaak [100]); and *J-spectral factorization approach* (see, *e.g.*, Kimura [101]). Recently, considerable attention has been focused on purely *time-domain methods* based on algebraic Riccati equations and/or singular perturbation approach (see, *e.g.* [74, 91, 102] and references cited therein). Along this line of research, connections are also made between H_∞ optimal control and differential games (see, *e.g.*, Başar and Bernhard [103]).

We also recall in this section the solutions to the problem of H_∞ almost disturbance decoupling with measurement feedback and internal stability. Although, in principle, it is a special case of the general H_∞ control problem, the problem of almost disturbance decoupling has a vast history behind it, occupying a central part of classical as well as modern control theory. Several important problems, such as robust control, decentralized control, non-interactive control, model reference or tracking control, H_2 and H_∞ optimal control problems can all be recast into an almost disturbance decoupling problem. Roughly speaking, the basic almost disturbance decoupling problem is to find an output feedback control law such that in the closed-loop system the disturbances are quenched, say in an L_p sense, up to any prespecified degree of accuracy while maintaining internal stability. Such a problem was originally formulated by Willems [104, 105] and termed almost disturbance decoupling problem with measurement feedback and internal stability (ADDPMS).

The formulation of H_∞ control is very similar to that of H_2 optimal control. In order to avoid unnecessary repetitions, we make use of some terms defined in the previous section, *e.g.*, the state-space equations of the given system and its subsystems $\Sigma_{\rm P}$ and $\Sigma_{\rm Q}$, the format of the control law and its corresponding closed-loop transfer matrix, as well as the definitions of the regular and singular problems.

3.5.1 Continuous-time Systems

We consider a continuous-time linear time-invariant system as given in Equation 3.48. For simplicity, we assume that (A, B) is stabilizable, (A, C_1) is detectable, $D_{11} = 0$ and $D_{22} = 0$. The standard H_∞ control problem for continuous-time systems is to find an internally stabilizing proper measurement feedback control law of the format in Equation 3.51 such that when it is applied to the given system the resulting closed-loop system is internally stable and the H_∞-norm of the overall closed-loop transfer matrix function from w to h, *i.e.* $T_{hw}(s)$, is minimized. The H_∞-norm of a stable continuous-time transfer matrix, *e.g.*, $T_{hw}(s)$, is defined as follows:

$$\|T_{hw}\|_\infty := \sup_{\omega\in[0,\infty)} \sigma_{\max}\left[T_{hw}(j\omega)\right] = \sup_{\|w\|_2=1} \frac{\|h\|_2}{\|w\|_2} \tag{3.155}$$

where w and h are, respectively, the input and output of $T_{hw}(s)$, and $\|\cdot\|_2$ is the l_2-norm of the corresponding signal. It is clear that the H_∞-norm of $T_{hw}(s)$ corresponds to the worst case gain from its input to its output. For future use, we define

$$\gamma_\infty^* := \inf\Big\{\|T_{hw}(\Sigma \times \Sigma_{\rm cmp})\|_\infty \mid \Sigma_{\rm cmp} \text{ internally stabilizes } \Sigma\Big\} \tag{3.156}$$

We note that the determination of this γ_∞^* is rather tedious. For a fairly large class of systems, γ_∞^* can be exactly computed using some numerically stable algorithms. In general, an iterative scheme is required to determine γ_∞^*. We refer interested readers to the work of Chen [74] for a detailed treatment of this particular issue. For simplicity, we assume throughout this section that γ_∞^* has been determined and hence it is known.

For the case when $\gamma_\infty^* = 0$, the corresponding H_∞ control problem is commonly known in the literature as the problem of H_∞ almost disturbance decoupling with internal stability. It can be shown that such a problem is solvable for Σ of Equation 3.48 if and only if the following conditions hold (see, *e.g.*, [74, 83]):

1. (A, B) is stabilizable;
2. (A, C_1) is detectable;
3. $D_{22} + D_2 S D_1 = 0$, where $S = -(D_2' D_2)^\dagger D_2' D_{22} D_1' (D_1 D_1')^\dagger$;
4. $\text{Im}\,(E + BSD_1) \subset \mathcal{S}^+(\Sigma_{\rm P}) \cap \{\cap_{\lambda\in\mathbb{C}^0} \mathcal{S}_\lambda(\Sigma_{\rm P})\}$;

5. $\text{Ker}\,(C_2 + D_2SC_1) \supset \mathcal{V}^+(\Sigma_Q) \cup \{\cup_{\lambda\in\mathbb{C}^0} \mathcal{V}_\lambda(\Sigma_Q)\}$; and
6. $\mathcal{V}^+(\Sigma_Q) \subset \mathcal{S}^+(\Sigma_P)$.

We note that if Σ_P is right invertible and of minimum phase, and Σ_Q is left invertible and of minimum phase, then conditions 4–6 are automatically satisfied.

It transpires that, for H_∞ control, it is almost impossible to find a control law with a finite gain to achieve the optimal performance, *i.e.* γ_∞^*. As such, we focus on designing H_∞ suboptimal controllers instead. To be more specific, given a scalar $\gamma > \gamma_\infty^*$, we focus on finding a control law that yields $\|T_{hw}\|_\infty < \gamma$, where $T_{hw}(s)$ is the corresponding closed-loop transfer matrix. Hereafter, we call a control law that possesses such a property an H_∞ γ-suboptimal controller.

Next, we proceed to construct a solution to the regular problem (for its definition see the previous section). Given a scalar $\gamma > \gamma_\infty^*$, we solve for positive semi-definite stabilizing solutions $P \geq 0$ and $Q \geq 0$ respectively to the following Riccati equations:

$$\begin{aligned} &A'P + PA + C_2'C_2 + \gamma^{-2}PEE'P \\ &\qquad -(PB + C_2'D_2)(D_2'D_2)^{-1}(B'P + D_2'C_2) = 0 \end{aligned} \tag{3.157}$$

and

$$\begin{aligned} &AQ + QA' + EE' + \gamma^{-2}QC_2'C_2Q \\ &\qquad -(QC_1' + ED_1')(D_1D_1')^{-1}(C_1Q + D_1E') = 0 \end{aligned} \tag{3.158}$$

The H_∞ γ-suboptimal control law is given by (see also [91]),

$$\Sigma_{\text{cmp}} : \begin{cases} \dot{v} = A_{\text{cmp}}\, v + B_{\text{cmp}}\, y \\ u = C_{\text{cmp}}\, v + \ \ 0 \ \ y \end{cases} \tag{3.159}$$

where

$$A_{\text{cmp}} = A + \gamma^{-2}EE'P + BF + (I - \gamma^{-2}QP)^{-1}K(C_1 + \gamma^{-2}D_1E'P) \tag{3.160}$$

$$B_{\text{cmp}} = -\left(I - \gamma^{-2}QP\right)^{-1}K \tag{3.161}$$

$$C_{\text{cmp}} = F \tag{3.162}$$

and where

$$F = -(D_2'D_2)^{-1}(D_2'C_2 + B'P), \quad K = -(QC_1' + ED_1')(D_1D_1')^{-1} \tag{3.163}$$

Note that, for the state feedback case, the H_∞ γ-suboptimal control law is given by $u = Fx$ with F being given as in Equation 3.163.

For the singular case, the following perturbation method can be utilized. For $\gamma > \gamma_\infty^*$ and a positive scalar $\varepsilon > 0$, define $\tilde{E}$, $\tilde{D}_1$, $\tilde{C}_2$ and $\tilde{D}_2$ as in Equations 3.68 and 3.69, and solve the following Riccati equations:

$$\begin{aligned} &A'\tilde{P} + \tilde{P}A + \tilde{C}_2'\tilde{C}_2 + \gamma^{-2}\tilde{P}\tilde{E}\tilde{E}'\tilde{P} \\ &\qquad -(\tilde{P}B + \tilde{C}_2'\tilde{D}_2)(\tilde{D}_2'\tilde{D}_2)^{-1}(B'\tilde{P} + \tilde{D}_2'\tilde{C}_2) = 0 \end{aligned} \tag{3.164}$$

and

$$A\tilde{Q} + \tilde{Q}A' + \tilde{E}\tilde{E}' + \gamma^{-2}\tilde{Q}\tilde{C}_2'\tilde{C}_2\tilde{Q} \\ -(\tilde{Q}C_1' + \tilde{E}\tilde{D}_1')(\tilde{D}_1\tilde{D}_1')^{-1}(C_1\tilde{Q} + \tilde{D}_1\tilde{E}') = 0 \tag{3.165}$$

for $\tilde{P} > 0$ and $\tilde{Q} > 0$. Then, it can be shown that there exists an $\varepsilon^* > 0$ such that for all $\varepsilon \in (0, \varepsilon^*]$ the following control law is an H_∞ γ-suboptimal for the given system:

$$\tilde{\Sigma}_{\rm cmp} \ : \ \begin{cases} \dot{v} = \tilde{A}_{\rm cmp}\ v + \tilde{B}_{\rm cmp}\ y \\ u = \tilde{C}_{\rm cmp}\ v + \ \ 0 \ \ \ y \end{cases} \tag{3.166}$$

where

$$\tilde{A}_{\rm cmp} = A + \gamma^{-2}EE'\tilde{P} + B\tilde{F} + (I - \gamma^{-2}\tilde{Q}\tilde{P})^{-1}\tilde{K}(C_1 + \gamma^{-2}D_1E'\tilde{P}) \tag{3.167}$$

$$\tilde{B}_{\rm cmp} = -\left(I - \gamma^{-2}\tilde{Q}\tilde{P}\right)^{-1}\tilde{K} \tag{3.168}$$

$$\tilde{C}_{\rm cmp} = \tilde{F} \tag{3.169}$$

and where

$$\tilde{F} = -(\tilde{D}_2'\tilde{D}_2)^{-1}(\tilde{D}_2'\tilde{C}_2 + B'\tilde{P}) \tag{3.170}$$

$$\tilde{K} = -(\tilde{Q}C_1' + \tilde{E}\tilde{D}_1')(\tilde{D}_1\tilde{D}_1')^{-1} \tag{3.171}$$

Note that for the state feedback case, the H_∞ γ-suboptimal control law is given by $u = \tilde{F}x$ with $\tilde{F}$ being given as in Equation 3.170.

Alternatively, the singular H_∞ control problem can also be solved using the singular perturbation approach as in the previous section for H_2 optimal control. In fact, we only need to modify the algorithms slightly for H_2 control to yield the required H_∞ γ-suboptimal controllers. We treat separately the state feedback case and the measurement feedback case. For simplicity, we assume that both subsystems $\Sigma_{\rm P}$ and $\Sigma_{\rm Q}$ do not have invariant zeros on the imaginary axis.

i. State Feedback Case. Given a scalar $\gamma > \gamma_\infty^*$, the following algorithm yields an H_∞ γ-suboptimal state feedback gain matrix for Σ of Equation 3.48 with $y = x$.

STEP 3.5.C.S.1: transform the system $\Sigma_{\rm P}$ into the special coordinate basis as given by Theorem 3.1. To all submatrices and transformations in the special coordinate basis of $\Sigma_{\rm P}$, we append the subscript P to signify their relation to the system $\Sigma_{\rm P}$. We also choose the output transformation $\Gamma_{\rm OP}$ to have the following form:

$$\Gamma_{\rm OP} = \begin{bmatrix} I_{m_{\rm OP}} & 0 \\ 0 & \Gamma_{\rm orP} \end{bmatrix} \tag{3.172}$$

where $m_{\rm OP} = \text{rank}\,(D_2)$. Next, compute

$$\Gamma_{\rm SP}^{-1} E = \begin{bmatrix} E_{\rm aP}^{-} \\ E_{\rm aP}^{+} \\ E_{\rm bP} \\ E_{\rm cP} \\ E_{\rm dP} \end{bmatrix} \tag{3.173}$$

and define $A_{11\rm P}$, $B_{11\rm P}$, $A_{13\rm P}$, $C_{21\rm P}$, $C_{23\rm P}$, $A_{\rm xP}$, $B_{\rm xP}B_{\rm xP}'$, and $C_{\rm xP}'C_{\rm xP}$ as in Equations 3.76–3.80. Finally, define

$$E_{\rm xP} = \begin{bmatrix} E_{\rm aP}^{+} \\ E_{\rm bP} \end{bmatrix} \tag{3.174}$$

STEP 3.5.C.S.2: solve the following algebraic matrix Riccati equation,

$$P_{\rm x} A_{\rm xP} + A_{\rm xP}' P_{\rm x} + P_{\rm x} E_{\rm xP} E_{\rm xP}' P_{\rm x}/\gamma^2 - P_{\rm x} B_{\rm xP} B_{\rm xP}' P_{\rm x} + C_{\rm xP}' C_{\rm xP} = 0 \tag{3.175}$$

for $P_{\rm x} > 0$ and define

$$F_{11} := \begin{bmatrix} F_{\rm a0}^{+} & F_{\rm b0} \\ F_{\rm a1}^{+} & F_{\rm b1} \end{bmatrix} = \begin{bmatrix} B_{11\rm P}' P_{\rm x} \\ (C_{23\rm P}' C_{23\rm P})^{-1} (A_{13\rm P}' P_{\rm x} + C_{23\rm P}' C_{21\rm P}) \end{bmatrix} \tag{3.176}$$

Then, partition $[\, F_{\rm a1}^{+} \;\; F_{\rm b1} \,]$ as

$$[\, F_{\rm a1}^{+} \;\; F_{\rm b1} \,] = \begin{bmatrix} F_{\rm a11}^{+} & F_{\rm b11} \\ F_{\rm a12}^{+} & F_{\rm b12} \\ \vdots & \vdots \\ F_{{\rm a1}m_{\rm dP}}^{+} & F_{{\rm b1}m_{\rm dP}} \end{bmatrix} \tag{3.177}$$

where $F_{{\rm a1}i}^{+}$ and $F_{{\rm b1}i}$ are of dimensions $1 \times n_{\rm aP}^{+}$ and $1 \times n_{\rm bP}$, respectively.

STEP 3.5.C.S.3: follow exactly Steps 3.4.C.S.3 to 3.4.C.S.5 of the previous section to construct a state feedback gain matrix $F(\varepsilon)$. ◊

We have the following result.

Theorem 3.17. *Consider the system given in Equation 3.48 with $D_{22} = 0$ and $y = x$, i.e. all states are measurable. Assume that $\Sigma_{\rm P}$ has no invariant zeros on the imaginary axis. Then, the closed-loop system comprising the given system and $u = F(\varepsilon)x$ with $F(\varepsilon)$ being given as in the above algorithm has the following properties:*

1. *it is internally stable for sufficiently small ε;*
2. *the closed-loop transfer matrix from the disturbance w to the controlled output h, $T_{hw}(s,\varepsilon)$, possesses $\|T_{hw}(s,\varepsilon)\|_\infty < \gamma$, for sufficiently small ε.*

Clearly, $u = F(\varepsilon)x$ is an H_∞ γ-suboptimal control law for the system given in Equation 3.48. ◊

ii. Measurement Feedback Case. Similarly, one can design either a full-order or a reduced-order output feedback control law that solves the H_∞ γ-suboptimal problem. Unfortunately, the reduced-order controller design for H_∞ control is quite different from its counterpart in H_2 control and is quite complicated. We only focus below on the design of a full-order H_∞ γ-suboptimal controller. Interested readers are referred to [74] for a more complete treatment. The following is a step-by-step algorithm to construct a full-order H_∞ γ-suboptimal control law for any $\gamma > \gamma_\infty^*$.

STEP 3.5.C.F.1: define an auxiliary full state feedback system

$$\begin{cases} \dot{x} = A\ x + \ B\ u + E\ w \\ y = \quad x \\ h = C_2\ x + D_2\ u \end{cases} \tag{3.178}$$

and perform Steps 3.5.C.S.1 to 3.5.C.S.3 of the previous algorithm to obtain a gain matrix $F(\varepsilon)$. Also, define

$$P := (\Gamma_{\rm sp}^{-1})' \begin{bmatrix} 0 & 0 & 0 & 0 \\ 0 & P_{\rm x} & 0 & 0 \\ 0 & 0 & 0 & 0 \\ 0 & 0 & 0 & 0 \end{bmatrix} \Gamma_{\rm sp}^{-1} \geq 0 \tag{3.179}$$

Note that $P_{\rm x} > 0$ is the solution of Equation 3.175.

STEP 3.5.C.F.2: define another auxiliary full state feedback system as follows:

$$\begin{cases} \dot{x} = A'\ x + C_1'\ u + C_2'\ w \\ y = \quad x \\ h = E'\ x + D_1'\ u \end{cases} \tag{3.180}$$

and perform Steps 3.5.C.S.1 to 3.5.C.S.3 of the previous algorithm, but for this auxiliary system, to obtain a gain matrix $F(\varepsilon)$. Let us define $K(\varepsilon) := F(\varepsilon)'$. Similarly, define a positive semi-definite matrix Q as in Equation 3.179, but for the current auxiliary system.

STEP 3.5.C.F.3: the full-order H_∞ γ-suboptimal controller is given by

$$\Sigma_{\rm cmp}: \begin{cases} \dot{v} = A_{\rm cmp}\ v + B_{\rm cmp}\ y \\ u = C_{\rm cmp}\ v + \quad 0 \quad y \end{cases} \tag{3.181}$$

where

$$\begin{aligned} A_{\rm cmp} = A &+ \gamma^{-2}EE'P + BF(\varepsilon) \\ &+ (I - \gamma^{-2}QP)^{-1}\Big[K(\varepsilon)(C_1 + \gamma^{-2}D_1E'P) \\ &+ \gamma^{-2}Q(A'P + PA + C_2'C_2 + \gamma^{-2}PEE'P) \\ &+ \gamma^{-2}Q(PB + C_2'D_2)F(\varepsilon)\Big] \end{aligned} \tag{3.182}$$

and

$$B_{\rm cmp} = -(I - \gamma^{-2}QP)^{-1}K(\varepsilon), \quad C_{\rm cmp} = F(\varepsilon) \tag{3.183}$$

This completes the algorithm. ◊

We have the following theorem.

Theorem 3.18. *Consider the system given in Equation 3.48 with $D_{22} = 0$. Assume that Σ_P and Σ_Q have no invariant zeros on the imaginary axis. Then, the closed-loop system comprising the given system and the measurement feedback controller in Equation 3.181 has the following properties:*

1. *it is internally stable for sufficiently small ε;*
2. *the closed-loop transfer matrix from the disturbance w to the controlled output h, $T_{hw}(s,\varepsilon)$, possesses $\|T_{hw}(s,\varepsilon)\|_\infty < \gamma$, for sufficiently small ε.*

Hence, Equation 3.181 is an H_∞ γ-suboptimal control law for the system given in Equation 3.48. ◊

Remark 3.19. For the case when $\gamma_\infty^* = 0$, *i.e.* the problem of H_∞ almost disturbance decoupling with internal stability for Σ of Equation 3.48 (with both subsystems Σ_P and Σ_Q having no invariant zeros on the imaginary axis) is solvable, then either the above algorithm or the algorithm in the previous section for H_2 optimal control would yield desired solutions. We note that, in general, suboptimal solutions are nonunique. In fact, one could utilize the algorithm for constructing the reduced-order output feedback H_2 suboptimal controller in the previous section to construct a reduced-order solution for the H_∞ almost disturbance decoupling problem. ◊

3.5.2 Discrete-time Systems

We now consider a discrete-time linear time-invariant system as given in Equation 3.106. Again, for simplicity, we assume that (A, B) is stabilizable, (A, C_1) is detectable, $D_{11} = 0$ and $D_{22} = 0$. The standard H_∞ control problem for discrete-time systems is to find an internally stabilizing proper measurement feedback control law of the format in Equation 3.107 such that, when it is applied to the system in Equation 3.106, the resulting closed-loop system is internally stable and the H_∞-norm of the overall closed-loop transfer matrix function from w to h, *i.e.* $T_{hw}(z)$, is minimized. The H_∞-norm of a stable discrete-time transfer matrix, *e.g.*, $T_{hw}(z)$, is defined as follows:

$$\|T_{hw}\|_\infty := \sup_{\omega\in[0,2\pi]} \sigma_{\max}\left[T_{hw}(\mathrm{e}^{j\omega})\right] = \sup_{\|w\|_2=1} \frac{\|h\|_2}{\|w\|_2} \tag{3.184}$$

where w and h are, respectively, the input and output of $T_{hw}(s)$, and $\|\cdot\|_2$ is the l_2-norm of the corresponding signal. Next, we define

$$\gamma_\infty^* := \inf\Big\{\|T_{hw}(\Sigma\times\Sigma_{\mathrm{cmp}})\|_\infty \mid \Sigma_{\mathrm{cmp}} \text{ internally stabilizes } \Sigma\Big\} \tag{3.185}$$

Again, we refer interested readers to the work of Chen [74] for the computation of γ_∞^*. For simplicity, we assume throughout this section that γ_∞^* has been determined and hence it is known.

It can be shown that the problem of H_∞ almost disturbance decoupling with internal stability is solvable for Σ of Equation 3.106, *i.e.* $\gamma_\infty^* = 0$, if and only if the following conditions hold (see, *e.g.*, Chen [74]):

1. (A, B) is stabilizable;
2. (A, C_1) is detectable;
3. $D_{22} + D_2 S D_1 = 0$, where $S = -(D_2' D_2)^\dagger D_2' D_{22} D_1' (D_1 D_1')^\dagger$;
4. $\text{Im}\,(E + BSD_1) \subset \left\{\mathcal{V}^\odot(\Sigma_\text{P}) + B\text{Ker}\,(D_2)\right\} \cap \left\{\bigcap_{|\lambda|=1} \mathcal{S}_\lambda(\Sigma_\text{P})\right\}$;
5. $\text{Ker}\,(C_2 + D_2 S C_1) \supset \left\{\mathcal{S}^\odot(\Sigma_\text{Q}) \cap C_1^{-1}\{\text{Im}\,(D_1)\}\right\} \cup \left\{\bigcup_{|\lambda|=1} \mathcal{V}_\lambda(\Sigma_\text{Q})\right\}$;
6. $\mathcal{S}^\odot(\Sigma_\text{Q}) \subset \mathcal{V}^\odot(\Sigma_\text{P})$.

We note that if Σ_P is right invertible and of minimum phase with no infinite zeros, and Σ_Q is left invertible and of minimum phase with no infinite zeros, then conditions 4–6 are automatically satisfied.

The problems of discrete-time H_∞ control and H_∞ almost disturbance decoupling can be solved explicitly in the discrete-time domain. Complete solutions to these problems have been reported by Chen [74]. However, by utilizing the bilinear transformations (see Chen *et al.* [71] for the mapping of structural properties under the bilinear transformations), we can convert these discrete-time problems into equivalent continuous-time problems, and thus all algorithms presented in the previous subsection can be readily applied to yield desired solutions. The procedure is fairly simple.

1. We first apply the well-known bilinear transformation to the given discrete-time system in Equation 3.106 to obtain an equivalent continuous-time counterpart, *i.e.*

$$\left.\begin{aligned} \dot{x} &= A_\text{d}\ x + B_\text{d}\ u + E_\text{d}\ w \\ y &= C_\text{1d}\ x + D_\text{11d}\ u + D_\text{1d}\ w \\ h &= C_\text{2d}\ x + D_\text{2d}\ u + D_\text{22d}\ w \end{aligned}\right\} \tag{3.186}$$

where

$$\left.\begin{aligned} A_\text{d} &= (A+I)^{-1}(A-I) \\ B_\text{d} &= \sqrt{2}(A+I)^{-1}B \\ E_\text{d} &= \sqrt{2}(A+I)^{-1}E \\ C_\text{1d} &= \sqrt{2}C_1(A+I)^{-1} \\ D_\text{11d} &= D_{11} - C_1(A+I)^{-1}B \\ D_\text{1d} &= D_1 - C_1(A+I)^{-1}E \\ C_\text{2d} &= \sqrt{2}C_2(A+I)^{-1} \\ D_\text{2d} &= D_2 - C_2(A+I)^{-1}B \\ D_\text{22d} &= D_{22} - C_2(A+I)^{-1}E \end{aligned}\right\} \tag{3.187}$$

Note that if A has eigenvalues at -1, then one can apply some prefeedback laws to remove them.

2. Next, utilize any method of the previous subsection to the system in Equation 3.186 to find an appropriate H_∞ γ-suboptimal controller, say,

$$\left.\begin{aligned} \dot{v} &= A_\text{c}\ v + B_\text{c}\ y \\ u &= C_\text{c}\ v + D_\text{c}\ y \end{aligned}\right\} \tag{3.188}$$

Note that some prefeedback might be necessary to wash out $D_{22\mathrm{d}}$.

3. Lastly, apply the inverse bilinear transformation to convert the controller in Equation 3.188 to a discrete-time equivalent system. It is known in the literature (see, *e.g.*, [99]) that the discrete-time controller obtained is H_∞ γ-suboptimal to the original discrete-time system in Equation 3.106. ◇

The above procedure is also applicable to finding discrete-time H_∞ almost disturbance decoupling controllers. As mentioned earlier, all results presented in the previous section on H_2 optimal control are applicable to solving the H_∞ almost disturbance decoupling problem when both Σ_{P} and Σ_{Q} have no invariant zeros on the unit circle.

3.6 Robust and Perfect Tracking Control

We present in this section a robust and perfect tracking (RPT) problem that was proposed and solved by Liu *et al.* [106] for continuous-time systems and Chen *et al.* [107] for discrete-time systems (see also Chen [74]). The RPT problem is to design a controller such that the resulting closed-loop system is asymptotically stable and the controlled output almost perfectly tracks a given reference signal in the presence of any initial conditions and external disturbances. By almost perfect tracking we mean the ability of a controller to track a given reference signal with an arbitrarily fast settling time in the face of external disturbances and initial conditions. Of course, in real life, a certain tradeoff has to be made in order to design a physically implementable control law. The results of this section will be utilized heavily in later chapters to solve track-following problems in hard disk drive servo systems.

3.6.1 Continuous-time Systems

Consider the following continuous-time system:

$$\Sigma : \begin{cases} \dot{x} = A\ x + B\ u + E\ w, & x(0) = x_0, \\ y = C_1\ x + D_1\ w \\ h = C_2\ x + D_2\ u + D_{22}\ w \end{cases} \tag{3.189}$$

where $x \in \mathbb{R}^n$ is the state, $u \in \mathbb{R}^m$ is the control input, $w \in \mathbb{R}^{\mathrm{q}}$ is the external disturbance, $y \in \mathbb{R}^{\mathrm{p}}$ is the measurement output, and $h \in \mathbb{R}^\ell$ is the output to be controlled. We also assume that the pair (A, B) is stabilizable and (A, C_1) is detectable. For future reference, we define Σ_{P} and Σ_{Q} to be the subsystems characterized by the matrix quadruples (A, B, C_2, D_2) and (A, E, C_1, D_1), respectively. Given the external disturbance $w \in L_p$, $p \in [1, \infty)$, and any reference signal vector $r \in \mathbb{R}^\ell$ with $r, \dot{r}, \cdots, r^{(\kappa-1)}$, $\kappa \geq 1$, being available, and $r^{(\kappa)}$ being either a vector of delta functions or in L_p, the RPT problem for the system in Equation 3.189 is to find a parameterized dynamic measurement control law of the following form

$$\begin{cases} \dot{v} = A_{\rm cmp}(\varepsilon)v + B_{\rm cmp}(\varepsilon)y + G_0(\varepsilon)r + \cdots + G_{\kappa-1}(\varepsilon)r^{(\kappa-1)} \\ u = C_{\rm cmp}(\varepsilon)v + D_{\rm cmp}(\varepsilon)y + H_0(\varepsilon)r + \cdots + H_{\kappa-1}(\varepsilon)r^{(\kappa-1)} \end{cases} \tag{3.190}$$

such that when the controller of Equation 3.190 is applied to the system of Equation 3.189, we have

1. there exists an $\varepsilon^* > 0$ such that the resulting closed-loop system with $r = 0$ and $w = 0$ is asymptotically stable for all $\varepsilon \in (0, \varepsilon^*]$; and
2. let $h(t, \varepsilon)$ be the closed-loop controlled output response and let $e(t, \varepsilon)$ be the resulting tracking error, *i.e.* $e(t, \varepsilon) := h(t, \varepsilon) - r(t)$. Then, for any initial condition of the state, $x_0 \in \mathbb{R}^n$,

$$\|e\|_p = \left(\int_0^\infty |e(t)|^p \, \mathrm{d}t \right)^{1/p} \to 0 \quad \text{as } \varepsilon \to 0 \tag{3.191}$$

We introduce in the above formulation some additional information besides the reference signal r, *i.e.* $\dot{r}, \ddot{r}, \cdots, r^{(\kappa-1)}$, as additional controller inputs. The idea of utilizing the reference signal r together with its derivatives in tracking control is similar to the command-generator system introduced in Lewis [108]. Note that, in general, these additional signals can easily be generated without any extra costs. For example, if $r(t) = t^2 \cdot 1(t)$, where $1(t)$ is a unit step function, then one can easily obtain its first-order derivative

$$\dot{r}(t) = 2t \cdot 1(t) + t^2 \cdot \delta(t) = 2t \cdot 1(t) \tag{3.192}$$

where $\delta(t)$ is a unit impulse function, and the second-order derivative

$$\ddot{r}(t) = 2 \cdot 1(t) \tag{3.193}$$

These $\dot{r}(t)$ and $\ddot{r}(t)$ can be used to improve the overall tracking performance, whereas $r^{(3)}(t) = 2 \cdot \delta(t)$ does not exist in the real world and hence cannot be used. We also note that our formulation covers all possible reference signals that have the form $r(t) = t^k$, $0 \le k < \infty$. Thus, our method could be applied to track approximately those reference signals that have a Taylor series expansion at $t = 0$. This can be done by truncating the higher-order terms of the Taylor series of the given signal. Also, it is simple to see that, when $r(t) \equiv 0$, the proposed problem reduces to the well-known perfect regulation problem with measurement feedback.

It is shown that the RPT problem for the system in Equation 3.189 is solvable if and only if the following conditions hold:

1. (A, B) is stabilizable and (A, C_1) is detectable;
2. $D_{22} + D_2 S D_1 = 0$, where $S = -(D_2' D_2)^\dagger D_2' D_{22} D_1' (D_1 D_1')^\dagger$;
3. $\Sigma_{\rm P}$, *i.e.* (A, B, C_2, D_2), is right invertible and of minimum phase;
4. $\mathrm{Ker}\,(C_2 + D_2 S C_1) \supset C_1^{-1}\{\mathrm{Im}\,(D_1)\}$.

We assume throughout the rest of this subsection that the above conditions are satisfied, and move on to construct solutions to the RPT problem. As usual, we focus on the following three cases: 1) the state feedback case; 2) the full-order measurement feedback case; and 3) the reduced-order measurement feedback case.

i. State Feedback Case. When all states of the plant are measured for feedback, the problem can be solved by a static control law. We construct in this subsection a parameterized state feedback control law,

$$u = F(\varepsilon)x + H_0(\varepsilon)r + \cdots + H_{\kappa-1}(\varepsilon)r^{(\kappa-1)} \tag{3.194}$$

that solves the RPT problem for the system in Equation 3.189. It is simple to note that we can rewrite the given reference in the following form:

$$\frac{\mathrm{d}}{\mathrm{d}t}\begin{pmatrix} r \\ \vdots \\ r^{(\kappa-2)} \\ r^{(\kappa-1)} \end{pmatrix} = \begin{bmatrix} 0 & I_\ell & \cdots & 0 \\ \vdots & \vdots & \ddots & \vdots \\ 0 & 0 & \cdots & I_\ell \\ 0 & 0 & \cdots & 0 \end{bmatrix}\begin{pmatrix} r \\ \vdots \\ r^{(\kappa-2)} \\ r^{(\kappa-1)} \end{pmatrix} + \begin{bmatrix} 0 \\ \vdots \\ 0 \\ I_\ell \end{bmatrix} r^{(\kappa)} \tag{3.195}$$

Combining Equation 3.195 with the given system, we obtain the following augmented system:

$$\Sigma_{\mathrm{AUG}} : \begin{cases} \dot{\boldsymbol{x}} = \boldsymbol{A}\ \boldsymbol{x} + \boldsymbol{B}\ u + \boldsymbol{E}\ \boldsymbol{w} \\ \boldsymbol{y} = \quad \boldsymbol{x} \\ e = \boldsymbol{C}_2\ \boldsymbol{x} + \boldsymbol{D}_2\ u \end{cases} \tag{3.196}$$

where

$$\boldsymbol{w} := \begin{pmatrix} w \\ r^{(\kappa)} \end{pmatrix}, \quad \boldsymbol{x} := \begin{pmatrix} r \\ \vdots \\ r^{(\kappa-2)} \\ r^{(\kappa-1)} \\ x \end{pmatrix} \tag{3.197}$$

$$\boldsymbol{A} = \begin{bmatrix} 0 & I_\ell & \cdots & 0 & 0 \\ \vdots & \vdots & \ddots & \vdots & \vdots \\ 0 & 0 & \cdots & I_\ell & 0 \\ 0 & 0 & \cdots & 0 & 0 \\ 0 & 0 & \cdots & 0 & A \end{bmatrix}, \quad \boldsymbol{B} = \begin{bmatrix} 0 \\ \vdots \\ 0 \\ 0 \\ B \end{bmatrix}, \quad \boldsymbol{E} = \begin{bmatrix} 0 & 0 \\ \vdots & \vdots \\ 0 & 0 \\ 0 & I_\ell \\ E & 0 \end{bmatrix} \tag{3.198}$$

and

$$\boldsymbol{C}_2 = [\,-I_\ell \ \ 0 \ \ 0 \ \ \cdots \ \ 0 \ \ C_2\,], \quad \boldsymbol{D}_2 = D_2 \tag{3.199}$$

It is then straightforward to show that the subsystem from u to e in the augmented system of Equation 3.196, *i.e.* the quadruple $(\boldsymbol{A}, \boldsymbol{B}, \boldsymbol{C}_2, \boldsymbol{D}_2)$, is right invertible and has the same infinite zero structure as that of Σ_{P}. Furthermore, its invariant zeros contain those of Σ_{P} and $\ell \times \kappa$ extra ones at $s = 0$. We are now ready to present a step-by-step algorithm to construct the required control law of the form in Equation 3.194.

STEP 3.6.C.S.1: this step transforms the subsystem from u to e of the augmented system in Equation 3.196 into the special coordinate basis of Theorem 3.1, *i.e.*

finds nonsingular state, input and output transformations $\Gamma_{\rm s}$, $\Gamma_{\rm i}$ and $\Gamma_{\rm o}$ to put the subsystem into the structural form of Theorem 3.1 and in a small variation of the compact form of Equations 3.20 to 3.23. It can be shown that the compact form of Equations 3.20 to 3.23 for the subsystem from u to e of Equation 3.196 can be written as

$$\tilde{A} = \begin{bmatrix} A_{\rm aa}^{0} & 0 & 0 & 0 \\ 0 & A_{\rm aa}^{-} & 0 & L_{\rm ad}^{-} C_{\rm d} \\ B_{\rm c} E_{\rm ca}^{0} & B_{\rm c} E_{\rm ca}^{-} & A_{\rm cc} & L_{\rm cd} C_{\rm d} \\ B_{\rm d} E_{\rm da}^{0} & B_{\rm d} E_{\rm da}^{-} & B_{\rm d} E_{\rm dc} & A_{\rm dd} \end{bmatrix} \tag{3.200}$$

$$A_{\rm aa}^{0} = \begin{bmatrix} 0 & I_\ell & \cdots & 0 \\ \vdots & \vdots & \ddots & \vdots \\ 0 & 0 & \cdots & I_\ell \\ 0 & 0 & \cdots & 0 \end{bmatrix}, \quad \tilde{B} = \begin{bmatrix} 0 & 0 & 0 \\ B_{\rm 0a}^{-} & 0 & 0 \\ B_{\rm 0c} & 0 & B_{\rm c} \\ B_{\rm 0d} & B_{\rm d} & 0 \end{bmatrix} \tag{3.201}$$

and

$$\tilde{C} = \begin{bmatrix} C_{\rm 0a}^{0} & C_{\rm 0a}^{-} & C_{\rm 0c} & C_{\rm 0d} \\ 0 & 0 & 0 & C_{\rm d} \end{bmatrix}, \quad \tilde{D} = \begin{bmatrix} I_{m_0} & 0 & 0 \\ 0 & 0 & 0 \end{bmatrix} \tag{3.202}$$

STEP 3.6.C.S.2: choose an appropriate dimensional matrix F_c such that

$$A_{\rm cc}^{\rm c} = A_{\rm cc} - B_{\rm c} F_{\rm c} \tag{3.203}$$

is asymptotically stable. The existence of such an $F_{\rm c}$ is guaranteed by the property that $(A_{\rm cc}, B_{\rm c})$ is completely controllable.

STEP 3.6.C.S.3: for each x_i of $x_{\rm d}$, which is associated with the infinite zero structure of $\Sigma_{\rm P}$ or the subsystem from u to e of Equation 3.196, we choose an F_i such that

$$p_i(s) = \prod_{j=1}^{q_i} (s - \lambda_{ij}) = s^{q_i} + F_{i1} s^{q_i - 1} + \cdots + F_{iq_{i-1}} s + F_{iq_i} \tag{3.204}$$

with all λ_{ij} being in $\mathbb{C}^-$. Let

$$F_i = [\, F_{iq_i} \quad F_{iq_{i-1}} \quad \cdots \quad F_{i1} \,], \quad i = 1, \cdots, m_{\rm d} \tag{3.205}$$

STEP 3.6.C.S.4: next, we construct

$$\boldsymbol{F}(\varepsilon) = -\Gamma_{\rm i} \begin{bmatrix} C_{\rm 0a}^{0} & C_{\rm 0a}^{-} & C_{\rm 0c} & C_{\rm 0d} \\ E_{\rm da}^{0} & E_{\rm da}^{-} & E_{\rm dc} & E_{\rm d} + F_{\rm d}(\varepsilon) \\ E_{\rm ca}^{0} & E_{\rm ca}^{-} & F_{\rm c} & 0 \end{bmatrix} \Gamma_{\rm s}^{-1} \tag{3.206}$$

where

$$E_{\rm d} = \begin{bmatrix} E_{11} & \cdots & E_{1m_{\rm d}} \\ \vdots & \ddots & \vdots \\ E_{m_{\rm d}1} & \cdots & E_{m_{\rm d}m_{\rm d}} \end{bmatrix} \tag{3.207}$$

$$F_{\mathrm{d}}(\varepsilon) = \text{blkdiag}\left\{\frac{F_1}{\varepsilon^{q_1}} S_1(\varepsilon), \frac{F_2}{\varepsilon^{q_2}} S_2(\varepsilon), \cdots, \frac{F_{m_\mathrm{d}}}{\varepsilon^{q_{m_\mathrm{d}}}} S_{m_\mathrm{d}}(\varepsilon)\right\} \tag{3.208}$$

and where

$$S_i(\varepsilon) = \text{diag}\left\{1, \varepsilon, \varepsilon^2, \cdots, \varepsilon^{q_i - 1}\right\} \tag{3.209}$$

STEP 3.6.C.S.5: finally, we partition

$$\boldsymbol{F}(\varepsilon) = [\, H_0(\varepsilon) \quad \cdots \quad H_{\kappa-1}(\varepsilon) \quad F(\varepsilon)\,] \tag{3.210}$$

where $H_i(\varepsilon) \in \mathbb{R}^{m \times \ell}$ and $F(\varepsilon) \in \mathbb{R}^{m \times n}$. ◇

We have the following result.

Theorem 3.20. *Consider the given system in Equation 3.189 with its external disturbance $w \in L_p$, $p \in [1, \infty)$, its initial condition $x(0) = x_0$. Assume that all its states are measured for feedback, i.e. $C_1 = I$ and $D_1 = 0$, and the RPT problem for the system in Equation 3.189 is solvable. Then, for any reference signal $r(t)$, which has all its ith-order derivatives, $i = 0, 1, \cdots, \kappa - 1$, $\kappa \geq 1$, being available and $r^{(\kappa)}(t)$ being either a vector of delta functions or in L_p, the RPT problem is solved by the control law of Equation 3.194 with $F(\varepsilon)$ and $H_i(\varepsilon)$, $i = 0, 1, \cdots, \kappa - 1$, as given in Equation 3.210.* ◇

The following remark gives an alternative approach for solving the proposed RPT problem via full state feedback. We leave the proof of this method to readers as an exercise.

Remark 3.21. Note that the required gain matrices for the state feedback RPT problem might be computed by solving the following Riccati equation:

$$\boldsymbol{P}\tilde{\boldsymbol{A}} + \tilde{\boldsymbol{A}}'\boldsymbol{P} + \tilde{\boldsymbol{C}}_2'\tilde{\boldsymbol{C}}_2 - \left(\boldsymbol{P}\boldsymbol{B} + \tilde{\boldsymbol{C}}_2'\tilde{\boldsymbol{D}}_2\right)\left(\tilde{\boldsymbol{D}}_2'\tilde{\boldsymbol{D}}_2\right)^{-1}\left(\boldsymbol{P}\boldsymbol{B} + \tilde{\boldsymbol{C}}_2'\tilde{\boldsymbol{D}}_2\right)' = 0$$

for a positive-definite solution $\boldsymbol{P} > 0$, where

$$\tilde{\boldsymbol{C}}_2 = \begin{bmatrix} \boldsymbol{C}_2 \\ \varepsilon I_{\kappa\ell+n} \\ 0 \end{bmatrix}, \quad \tilde{\boldsymbol{D}}_2 = \begin{bmatrix} \boldsymbol{D}_2 \\ 0 \\ \varepsilon I_m \end{bmatrix} \tag{3.211}$$

$$\tilde{\boldsymbol{A}} = \begin{bmatrix} \tilde{A}_0 & 0 \\ 0 & A \end{bmatrix}, \quad \tilde{A}_0 = -\varepsilon I_{\kappa\ell} + \begin{bmatrix} 0 & I_\ell & \cdots & 0 \\ \vdots & \vdots & \ddots & \vdots \\ 0 & 0 & \cdots & I_\ell \\ 0 & 0 & \cdots & 0 \end{bmatrix} \tag{3.212}$$

and where $\boldsymbol{B}$, $\boldsymbol{C}_2$ and $\boldsymbol{D}_2$ are as defined in Equations 3.198 and 3.199. The required gain matrix is then given by

$$\begin{aligned} \tilde{\boldsymbol{F}}(\varepsilon) &= -\left(\tilde{\boldsymbol{D}}_2'\tilde{\boldsymbol{D}}_2\right)^{-1}\left(\boldsymbol{P}\boldsymbol{B} + \tilde{\boldsymbol{C}}_2'\tilde{\boldsymbol{D}}_2\right)' \\ &= [\, H_0(\varepsilon) \quad \cdots \quad H_{\kappa-1}(\varepsilon) \quad F(\varepsilon)\,] \end{aligned} \tag{3.213}$$

where $H_i(\varepsilon) \in \mathbb{R}^{m \times \ell}$ and $F(\varepsilon) \in \mathbb{R}^{m \times n}$. ◇

Finally, we note that solutions to the Riccati equation in Remark 3.21 might have severe numerical problems as ε becomes smaller and smaller.

We consider two types of measurement feedback control laws: one is of full-order controllers whose dynamical order is equal to the order of the given system; the other is of reduced-order controllers with a dynamical order that is less than the order of the given system. Without loss of generality, we assume throughout this subsection that $D_{22} = 0$. If it is nonzero, it can always be washed out by the following preoutput feedback: $u = Sy$.

ii. Full-order Output Feedback Case. The following is a step-by-step algorithm for constructing a parameterized full-order measurement feedback controller, which solves the RPT problem.

STEP 3.6.C.F.1: for the given reference $r(t)$ and the given system in Equation 3.189, we first assume that all the state variables of Equation 3.189 are measurable and follow the procedures of the state feedback case to define an auxiliary system,

$$\begin{cases} \dot{\boldsymbol{x}} = \boldsymbol{A}\ \boldsymbol{x} + \boldsymbol{B}\ \boldsymbol{u} + \boldsymbol{E}\ \boldsymbol{w} \\ \boldsymbol{y} = \quad \boldsymbol{x} \\ \boldsymbol{e} = \boldsymbol{C}_2\ \boldsymbol{x} + \boldsymbol{D}_2\ \boldsymbol{u} \end{cases} \tag{3.214}$$

Then, we follow Steps 3.6.C.S.1 to 3.6.C.S.5 of the algorithm of the state feedback case to construct a state feedback gain matrix:

$$\boldsymbol{F}(\varepsilon) = [\, H_0(\varepsilon) \quad \cdots \quad H_{\kappa-1}(\varepsilon) \quad F(\varepsilon)\,] \tag{3.215}$$

STEP 3.6.C.F.2: let Σ_{Qaux} be characterized by a matrix quadruple

$$(A_{\mathrm{Qaux}}, E_{\mathrm{Qaux}}, C_{\mathrm{Qaux}}, D_{\mathrm{Qaux}}) := (A, [\, E \quad I_n\,], C_1, [\, D_1 \quad 0\,]) \tag{3.216}$$

This step is to transform this Σ_{Qaux} into the special coordinate basis of Theorem 3.1. Because of the special structure of the matrix E_{Qaux}, it is simple to show that Σ_{Qaux} is always right invertible and is free of invariant zeros. Utilize the results of Theorem 3.1 to find nonsingular state, input and output transformation Γ_{sQ}, Γ_{iQ} and Γ_{oQ} such that

$$\Gamma_{\mathrm{sQ}}^{-1} A \Gamma_{\mathrm{sQ}} = \begin{bmatrix} A_{\mathrm{ccQ}} & L_{\mathrm{cdQ}} \\ E_{\mathrm{dcQ}} & A_{\mathrm{ddQ}} \end{bmatrix} + \begin{bmatrix} B_{\mathrm{0cQ}} \\ B_{\mathrm{0dQ}} \end{bmatrix} [\, C_{\mathrm{0cQ}} \quad 0\,] \tag{3.217}$$

$$\Gamma_{\mathrm{sQ}}^{-1} E_{\mathrm{Qaux}} \Gamma_{\mathrm{iQ}} = \begin{bmatrix} B_{\mathrm{0cQ}} & 0 & I_{n-k} & 0 \\ B_{\mathrm{0dQ}} & I_k & 0 & 0 \end{bmatrix} \tag{3.218}$$

$$\Gamma_{\mathrm{oQ}}^{-1} C_1 \Gamma_{\mathrm{sQ}} = \begin{bmatrix} C_{\mathrm{0cQ}} & 0 \\ 0 & I_k \end{bmatrix} \tag{3.219}$$

$$\Gamma_{\mathrm{oQ}}^{-1} [\, D_1 \quad 0\,] \Gamma_{\mathrm{iQ}} = \begin{bmatrix} I_{\mathrm{p}-k} & 0 & 0 & 0 \\ 0 & 0 & 0 & 0 \end{bmatrix} \tag{3.220}$$

where $k = \mathrm{p} - \mathrm{rank}(D_1)$. It can be verified that the pair (A, C_1) is detectable if and only if the pair

$$\left(A_{\mathrm{ccQ}}, \begin{bmatrix} C_{0\mathrm{cQ}} \\ E_{\mathrm{dcQ}} \end{bmatrix}\right) \tag{3.221}$$

is detectable.

STEP 3.6.C.F.3: let K_{cQ} be an appropriate dimensional constant matrix such that the eigenvalues of the matrix

$$A^{\mathrm{c}}_{\mathrm{ccQ}} = A_{\mathrm{ccQ}} - K_{\mathrm{cQ}} \begin{bmatrix} C_{0\mathrm{cQ}} \\ E_{\mathrm{dcQ}} \end{bmatrix} = A_{\mathrm{ccQ}} - [\,K_{\mathrm{c0Q}} \quad K_{\mathrm{cdQ}}\,] \begin{bmatrix} C_{0\mathrm{cQ}} \\ E_{\mathrm{dcQ}} \end{bmatrix} \tag{3.222}$$

are all in $\mathbb{C}^-$. Next, we define a parameterized observer gain matrix,

$$K(\varepsilon) = -\Gamma_{\mathrm{sQ}} \begin{bmatrix} B_{0\mathrm{cQ}} + K_{\mathrm{c0Q}} & L_{\mathrm{cdQ}} + K_{\mathrm{cdQ}}/\varepsilon \\ B_{0\mathrm{dQ}} & A_{\mathrm{ddQ}} + I_k/\varepsilon \end{bmatrix} \Gamma_{\mathrm{oQ}}^{-1} \tag{3.223}$$

STEP 3.6.C.F.4: finally, we obtain the following full-order measurement feedback control law:

$$\begin{cases} \dot{v} = A_{\mathrm{cmp}}(\varepsilon)v - K(\varepsilon)y + BH_0(\varepsilon)r + \cdots + BH_{\kappa-1}(\varepsilon)r^{(\kappa-1)} \\ u = F(\varepsilon)\ v + H_0(\varepsilon)\,r + \cdots + H_{\kappa-1}(\varepsilon)\,r^{(\kappa-1)} \end{cases} \tag{3.224}$$

where $A_{\mathrm{cmp}}(\varepsilon) = A + BF(\varepsilon) + K(\varepsilon)C_1$. This completes the construction of the full-order measurement feedback controller. ◇

We have the following theorem.

Theorem 3.22. *Consider the given system in Equation 3.189 with its external disturbance $w \in L_p$, $p \in [1, \infty)$, its initial condition $x(0) = x_0$. Assume that the RPT problem is solvable for the system in Equation 3.189. Then, for any reference signal $r(t)$, which has all its ith-order derivatives, $i = 0, 1, \cdots, \kappa - 1$, $\kappa \geq 1$, being available and $r^{(\kappa)}(t)$ being either a vector of delta functions or in L_p, the RPT problem is solved by the parameterized full-order measurement feedback control laws as given in Equation 3.224.* ◇

The following remark yields an alternative way to compute the gain matrix $K(\varepsilon)$ in Step 3.6.C.F.3.

Remark 3.23. The gain matrix $K(\varepsilon)$ in Step 3.6.C.F.3 can also be computed by solving the following Riccati equation:

$$\begin{aligned} AQ + QA' + (EE' + I) \\ - (QC_1' + ED_1')(D_1D_1' + \varepsilon I)^{-1}(C_1Q + D_1E') = 0 \end{aligned} \tag{3.225}$$

for a positive-definite solution $Q > 0$. The required gain matrix $K(\varepsilon)$ is then given by

$$K(\varepsilon) = -(QC_1' + ED_1')(D_1D_1' + \varepsilon I)^{-1} \tag{3.226}$$

Again, this approach might have some numerical problems. ◇

iii. Reduced-order Output Feedback Case. We now present solutions to the RPT problem via reduced-order measurement feedback control laws. For simplicity of presentation, we assume that matrices C_1 and D_1 have already been transformed into the following forms:

$$C_1 = \begin{bmatrix} 0 & C_{1,02} \\ I_k & 0 \end{bmatrix} \quad \text{and} \quad D_1 = \begin{bmatrix} D_{1,0} \\ 0 \end{bmatrix} \tag{3.227}$$

where $D_{1,0}$ is of full row rank. Before we present a step-by-step algorithm to construct a parameterized reduced-order measurement feedback controller, we first partition the following system

$$\begin{cases} \dot{x} = A\ x + B\ u + [\,E \quad I_n\,]\ \tilde{w} \\ y = C_1\ x \qquad\qquad + [\,D_1 \quad 0\,]\ \tilde{w} \end{cases} \tag{3.228}$$

in conformity with the structures of C_1 and D_1 in Equation 3.227, *i.e.*

$$\begin{cases} \begin{pmatrix} \dot{x}_1 \\ \dot{x}_2 \end{pmatrix} = \begin{bmatrix} A_{11} & A_{12} \\ A_{21} & A_{22} \end{bmatrix} \begin{pmatrix} x_1 \\ x_2 \end{pmatrix} + \begin{bmatrix} B_1 \\ B_2 \end{bmatrix} u + \begin{bmatrix} E_1 & I_k & 0 \\ E_2 & 0 & I_{n-k} \end{bmatrix} \tilde{w} \\ \begin{pmatrix} y_0 \\ y_1 \end{pmatrix} = \begin{bmatrix} 0 & C_{1,02} \\ I_k & 0 \end{bmatrix} \begin{pmatrix} x_1 \\ x_2 \end{pmatrix} \qquad\qquad + \begin{bmatrix} D_{1,0} & 0 & 0 \\ 0 & 0 & 0 \end{bmatrix} \tilde{w} \end{cases} \tag{3.229}$$

where

$$\tilde{w} = \begin{pmatrix} w \\ x_0 \cdot \delta(t) \end{pmatrix} \tag{3.230}$$

Obviously, $y_1 = x_1$ is directly available and hence need not be estimated. Next, we define $\Sigma_{\rm QR}$ to be characterized by

$$(A_{\rm R}, E_{\rm R}, C_{\rm R}, D_{\rm R}) = \left(A_{22}, [\,E_2 \quad 0 \quad I_{n-k}\,], \begin{bmatrix} C_{1,02} \\ A_{12} \end{bmatrix}, \begin{bmatrix} D_{1,0} & 0 & 0 \\ E_1 & I_k & 0 \end{bmatrix} \right)$$

It is again straightforward to verify that $\Sigma_{\rm QR}$ is right invertible with no finite and infinite zeros. Moreover, $(A_{\rm R}, C_{\rm R})$ is detectable if and only if (A, C_1) is detectable. We are ready to present the following algorithm.

STEP 3.6.C.R.1: for the given reference $r(t)$ and the given system in Equation 3.189, we again assume that all the state variables of the given system are measurable and follow the procedures of the state feedback case to define an auxiliary system,

$$\begin{cases} \dot{\boldsymbol{x}} = \boldsymbol{A}\ \boldsymbol{x} + \boldsymbol{B}\ u + \boldsymbol{E}\ \boldsymbol{w} \\ \boldsymbol{y} = \quad \boldsymbol{x} \\ e = \boldsymbol{C}_2\ \boldsymbol{x} + \boldsymbol{D}_2\ u \end{cases} \tag{3.231}$$

Then, we follow Steps 3.6.C.S.1 to 3.6.C.S.5 of the algorithm of the state feedback case to construct a state feedback gain matrix

$$\boldsymbol{F}(\varepsilon) = [\,H_0(\varepsilon) \quad \cdots \quad H_{\kappa-1}(\varepsilon) \quad F(\varepsilon)\,] \tag{3.232}$$

Let us partition $F(\varepsilon)$ in conformity with x_1 and x_2 of Equation 3.229 as follows,

$$F(\varepsilon) = [\,F_1(\varepsilon) \quad F_2(\varepsilon)\,] \tag{3.233}$$

STEP 3.6.C.R.2: let $K_{\rm R}$ be an appropriate dimensional constant matrix such that the eigenvalues of

$$A_{\rm R} + K_{\rm R} C_{\rm R} = A_{22} + [\, K_{\rm R0} \quad K_{\rm R1} \,] \begin{bmatrix} C_{1,02} \\ A_{12} \end{bmatrix} \tag{3.234}$$

are all in $\mathbb{C}^-$. This can be done because $(A_{\rm R}, C_{\rm R})$ is detectable.

STEP 3.6.C.R.3: let

$$G_{\rm R} = [\, -K_{\rm R0}, \quad A_{21} + K_{\rm R1} A_{11} - (A_{\rm R} + K_{\rm R} C_{\rm R}) K_{\rm R1} \,] \tag{3.235}$$

and

$$\left.\begin{aligned} A_{\rm cmp}(\varepsilon) &= A_{\rm R} + B_2 F_2(\varepsilon) + K_{\rm R} C_{\rm R} + K_{\rm R1} B_1 F_2(\varepsilon) \\ B_{\rm cmp}(\varepsilon) &= G_{\rm R} + (B_2 + K_{\rm R1} B_1)\, [\, 0, \quad F_1(\varepsilon) - F_2(\varepsilon) K_{\rm R1} \,] \\ C_{\rm cmp}(\varepsilon) &= F_2(\varepsilon) \\ D_{\rm cmp}(\varepsilon) &= [\, 0, \quad F_1(\varepsilon) - F_2(\varepsilon) K_{\rm R1} \,] \end{aligned}\right\} \tag{3.236}$$

STEP 3.6.C.R.4: finally, we obtain the following reduced-order measurement feedback control law:

$$\begin{cases} \dot{v} = A_{\rm cmp}(\varepsilon) v + B_{\rm cmp}(\varepsilon) y + G_0(\varepsilon) r + \cdots + G_{\kappa-1}(\varepsilon) r^{(\kappa-1)} \\ u = C_{\rm cmp}(\varepsilon) v + D_{\rm cmp}(\varepsilon) y + H_0(\varepsilon) r + \cdots + H_{\kappa-1}(\varepsilon) r^{(\kappa-1)} \end{cases} \tag{3.237}$$

where for $i = 0, 1, \cdots, \kappa - 1$, $G_i(\varepsilon) = (B_2 + K_{\rm R1} B_1) H_i(\varepsilon)$. This completes the constructing procedure. ◊

Theorem 3.24. *Consider the given system in Equation 3.189 with its external disturbance $w \in L_p$, $p \in [1, \infty)$, its initial condition $x(0) = x_0$. Assume that the RPT problem is solvable for the system in Equation 3.189. Then, for any reference signal $r(t)$, which has all its ith-order derivatives, $i = 0, 1, \cdots, \kappa - 1$, $\kappa \geq 1$, being available and $r^{(\kappa)}(t)$ being either a vector of delta functions or in L_p, the RPT problem is solved by the parameterized reduced-order measurement feedback control laws of Equation 3.237.* ◊

3.6.2 Discrete-time Systems

We present in this subsection the RPT problem for the following discrete-time system:

$$\Sigma : \begin{cases} x(k+1) = A\, x(k) + B\, u(k) + E\, w(k), \quad x(0) = x_0 \\ y(k) \;\;= C_1\, x(k) \qquad\qquad\quad + D_1\, w(k) \\ h(k) \;\;= C_2\, x(k) + D_2\, u(k) + D_{22}\, w(k) \end{cases} \tag{3.238}$$

where $x \in \mathbb{R}^n$ is the state, $u \in \mathbb{R}^m$ is the control input, $w \in \mathbb{R}^q$ is the external disturbance, $y \in \mathbb{R}^p$ is the measurement output, and $h \in \mathbb{R}^\ell$ is the output to be controlled.

We also assume that the pair (A, B) is stabilizable and (A, C_1) is detectable. For future reference, we define $\Sigma_{\rm P}$ and $\Sigma_{\rm Q}$ to be the subsystems characterized by the matrix quadruples (A, B, C_2, D_2) and (A, E, C_1, D_1) respectively. Given the external disturbance $w \in L_p$, $p \in [1, \infty]$, and any reference signal vector $r \in \mathbb{R}^\ell$, the RPT problem for the discrete-time system in Equation 3.238 is to find a parameterized dynamic measurement feedback control law of the following form:

$$\left\{ \begin{array}{ll} v(k+1) &= A_{\rm cmp}(\varepsilon)v(k) + B_{\rm cmp}(\varepsilon)y(k) + G(\varepsilon)r(k) \\ u(k) &= C_{\rm cmp}(\varepsilon)v(k) + D_{\rm cmp}(\varepsilon)y(k) + H(\varepsilon)r(k) \end{array} \right. \tag{3.239}$$

such that, when the controller in Equation 3.239 is applied to the system in Equation 3.238,

1. there exists an $\varepsilon^* > 0$ such that the resulting closed-loop system with $r = 0$ and $w = 0$ is asymptotically stable for all $\varepsilon \in (0, \varepsilon^*]$; and
2. let $h(k, \varepsilon)$ be the closed-loop controlled output response and let $e(k, \varepsilon)$ be the resulting tracking error, *i.e.* $e(k, \varepsilon) := h(k, \varepsilon) - r(k)$. Then, for any initial condition of the state, $x_0 \in \mathbb{R}^n$, $\|e\|_p \to 0$ as $\varepsilon \to 0$.

It has been shown by Chen [74] that the above RPT problem is solvable for the system in Equation 3.238 if and only if the following conditions hold:

1. (A, B) is stabilizable and (A, C_1) is detectable;
2. $D_{22} + D_2 S D_1 = 0$, where $S = -(D_2' D_2)^\dagger D_2' D_{22} D_1' (D_1 D_1')^\dagger$;
3. $\Sigma_{\rm P}$ is right invertible and of minimum phase with no infinite zeros;
4. $\operatorname{Ker}(C_2 + D_2 S C_1) \supset C_1^{-1}\{\operatorname{Im}(D_1)\}$.

It turns out that the control laws, which solve the RPT for the given plant in Equation 3.238 under the solvability conditions, need not be parameterized by any tuning parameter. Thus, Equation 3.239 can be replaced by

$$\left\{ \begin{array}{ll} v(k+1) &= A_{\rm cmp}v(k) + B_{\rm cmp}y(k) + Gr(k) \\ u(k) &= C_{\rm cmp}v(k) + D_{\rm cmp}y(k) + Hr(k) \end{array} \right. \tag{3.240}$$

and, furthermore, the resulting tracking error $e(k)$ can be made identically zero for all $k \geq 0$.

Assume that all the solvability conditions are satisfied. We present in the following solutions to the discrete-time RPT problem.

i. State Feedback Case. When all states of the plant are measured for feedback, the problem can be solved by a static control law. We construct in this subsection a state feedback control law,

$$u = Fx + Hr \tag{3.241}$$

that solves the RPT problem for the system in Equation 3.238. We have the following algorithm.

STEP 3.6.D.S.1: this step transforms the subsystem from u to h of the given system in Equation 3.238 into the special coordinate basis of Theorem 3.1, *i.e.* finds nonsingular state, input and output transformations Γ_s, Γ_i and Γ_o to put it into the structural form of Theorem 3.1 as well as in the compact form of Equations 3.20 to 3.23, *i.e.*

$$\tilde{A} = \Gamma_s^{-1} A \Gamma_s = \begin{bmatrix} A_{aa}^- & 0 \\ B_c E_{ca}^- & A_{cc} \end{bmatrix} + B_0 C_{2,0} \tag{3.242}$$

$$\tilde{B} = \Gamma_s^{-1} B \Gamma_i = \Gamma_s^{-1} [\, B_0 \quad B_1 \,] \Gamma_i = \begin{bmatrix} B_{0a}^- & 0 \\ B_{0c} & B_c \end{bmatrix} \tag{3.243}$$

$$\tilde{D}_2 = \Gamma_o^{-1} D_2 \Gamma_i = [\, I_{m_0} \quad 0 \,] \tag{3.244}$$

$$\tilde{C}_2 = \Gamma_o^{-1} C_2 \Gamma_s = \Gamma_o^{-1} C_{2,0} \Gamma_s = [\, C_{2,0a}^- \quad C_{2,0c} \,] \tag{3.245}$$

STEP 3.6.D.S.2: choose an appropriate dimensional matrix F_c such that

$$A_{cc}^c = A_{cc} - B_c F_c \tag{3.246}$$

is asymptotically stable. The existence of such an F_c is guaranteed by the property that (A_{cc}, B_c) is completely controllable.

STEP 3.6.D.S.3: finally, we let

$$F = -\Gamma_i \begin{bmatrix} C_{2,0a}^- & C_{2,0c} \\ E_{ca}^- & F_c \end{bmatrix} \Gamma_s^{-1} \quad \text{and} \quad H = \Gamma_i \begin{bmatrix} I \\ 0 \end{bmatrix} \Gamma_o^{-1} \tag{3.247}$$

This ends the constructive algorithm. ◊

We have the following result.

Theorem 3.25. *Consider the given discrete-time system in Equation 3.238 with any external disturbance $w(k)$ and any initial condition $x(0)$. Assume that all its states are measured for feedback, i.e. $C_1 = I$ and $D_1 = 0$, and the solvability conditions for the RPT problem hold. Then, for any reference signal $r(k)$, the proposed RPT problem is solved by the control law of Equation 3.241 with F and H as given in Equation 3.247.* ◊

ii. Measurement Feedback Case. Without loss of generality, we assume throughout this subsection that matrix $D_{22} = 0$. If it is nonzero, it can always be washed out by the following preoutput feedback $u(k) = Sy(k)$. It turns out that, for discrete-time systems, the full-order observer-based control law is not capable of achieving the RPT performance, because there is a delay of one step in the observer itself. Thus, we focus on the construction of a reduced-order measurement feedback control law to solve the RPT problem. For simplicity of presentation, we assume that matrices C_1 and D_1 have already been transformed into the following forms,

$$C_1 = \begin{bmatrix} 0 & C_{1,02} \\ I_\kappa & 0 \end{bmatrix} \quad \text{and} \quad D_1 = \begin{bmatrix} D_{1,0} \\ 0 \end{bmatrix} \tag{3.248}$$

where $D_{1,0}$ is of full row rank. Before we present a step-by-step algorithm to construct a reduced-order measurement feedback controller, we first partition the following system

$$\begin{cases} x(k+1) = A\ x(k) + B\ u(k) + [\,E \quad I_n\,]\ \tilde{w}(k) \\ \ \ y(k) \ \ = C_1\ x(k) \qquad\qquad\quad + [\,D_1 \quad 0\,]\ \tilde{w}(k) \end{cases} \tag{3.249}$$

in conformity with the structures of C_1 and D_1 in Equation 3.248, *i.e.*

$$\begin{cases} \begin{pmatrix} \delta(x_1) \\ \delta(x_2) \end{pmatrix} = \begin{bmatrix} A_{11} & A_{12} \\ A_{21} & A_{22} \end{bmatrix} \begin{pmatrix} x_1 \\ x_2 \end{pmatrix} + \begin{bmatrix} B_1 \\ B_2 \end{bmatrix} u + \begin{bmatrix} E_1 & I_\kappa & 0 \\ E_2 & 0 & I_{n-\kappa} \end{bmatrix} \tilde{w} \\ \begin{pmatrix} y_0 \\ y_1 \end{pmatrix} = \begin{bmatrix} 0 & C_{1,02} \\ I_\kappa & 0 \end{bmatrix} \begin{pmatrix} x_1 \\ x_2 \end{pmatrix} + \begin{bmatrix} D_{1,0} & 0 & 0 \\ 0 & 0 & 0 \end{bmatrix} \tilde{w} \end{cases}$$

where $\delta(x_1) = x_1(k+1)$ and $\delta(x_2) = x_2(k+1)$. Obviously, $y_1 = x_1$ is directly available and hence need not be estimated. Next, let $\Sigma_{\rm QR}$ be characterized by

$$(A_{\rm R}, E_{\rm R}, C_{\rm R}, D_{\rm R}) = \left(A_{22}, [\,E_2 \quad 0 \quad I_{n-\kappa}\,], \begin{bmatrix} C_{1,02} \\ A_{12} \end{bmatrix}, \begin{bmatrix} D_{1,0} & 0 & 0 \\ E_1 & I_\kappa & 0 \end{bmatrix} \right)$$

It is straightforward to verify that $\Sigma_{\rm QR}$ is right invertible with no finite and infinite zeros. Moreover, $(A_{\rm R}, C_{\rm R})$ is detectable if and only if (A, C_1) is detectable. We are ready to present the following algorithm.

STEP 3.6.D.R.1: for the given system in Equation 3.238, we again assume that all the state variables of the given system are measurable and then follow Steps 3.6.D.S.1 to 3.6.D.S.3 of the algorithm of the previous subsection to construct gain matrices F and H. We also partition F in conformity with x_1 and x_2 as follows:

$$F = [\,F_1 \quad F_2\,] \tag{3.250}$$

STEP 3.6.D.R.2: let $K_{\rm R}$ be an appropriate dimensional constant matrix such that the eigenvalues of

$$A_{\rm R} + K_{\rm R} C_{\rm R} = A_{22} + [\,K_{\rm R0} \quad K_{\rm R1}\,] \begin{bmatrix} C_{1,02} \\ A_{12} \end{bmatrix} \tag{3.251}$$

are all in $\mathbb{C}^{\odot}$. This can be done because $(A_{\rm R}, C_{\rm R})$ is detectable.

STEP 3.6.D.R.3: let

$$G_{\rm R} = [\,-K_{\rm R0}, \quad A_{21} + K_{\rm R1} A_{11} - (A_{\rm R} + K_{\rm R} C_{\rm R}) K_{\rm R1}\,] \tag{3.252}$$

$$\left. \begin{aligned} A_{\rm cmp} &= A_{\rm R} + B_2 F_2 + K_{\rm R} C_{\rm R} + K_{\rm R1} B_1 F_2 \\ B_{\rm cmp} &= G_{\rm R} + (B_2 + K_{\rm R1} B_1)\,[\,0, \quad F_1 - F_2 K_{\rm R1}\,] \\ C_{\rm cmp} &= F_2 \\ D_{\rm cmp} &= [\,0, \quad F_1 - F_2 K_{\rm R1}\,] \end{aligned} \right\} \tag{3.253}$$

and

$$G = (B_2 + K_{\rm R1} B_1) H \tag{3.254}$$

STEP 3.6.D.R.4: finally, we obtain the following reduced-order measurement feedback control law:

$$\begin{cases} v(k+1) = A_{\text{cmp}}\ v(k) + B_{\text{cmp}}\ y(k) + G\ r(k) \\ \ \ u(k) \ \ = C_{\text{cmp}}\ v(k) + D_{\text{cmp}}\ y(k) + H\ r(k) \end{cases} \tag{3.255}$$

This completes the algorithm. ◇

Theorem 3.26. *Consider the given system in Equation 3.238 with any external disturbance $w(k)$ and any initial condition $x(0)$. Assume that the solvability conditions for the RPT problem hold. Then, for any reference signal $r(k)$, the proposed RPT problem is solved by the reduced-order measurement feedback control laws of Equation 3.255.* ◇

3.7 Loop Transfer Recovery Technique

Another popular design methodology for multivariable systems, which is based on the 'loop shaping' concept, is linear quadratic Gaussian (LQG) with loop transfer recovery (LTR). It involves two separate designs of a state feedback controller and an observer or an estimator. The exact design procedure depends on the point where the unstructured uncertainties are modeled and where the loop is broken to evaluate the open-loop transfer matrices. Commonly, either the input point or the output point of the plant is taken as such a point. We focus on the case when the loop is broken at the input point of the plant. The required results for the output point can be easily obtained by appropriate dualization. Thus, in the two-step procedure of LQG/LTR, the first step of design involves loop shaping by a state feedback design to obtain an appropriate loop transfer function, called the target loop transfer function. Such a loop shaping is an engineering art and often involves the use of linear quadratic regulator (LQR) design, in which the cost matrices are used as free design parameters to generate the target loop transfer function, and thus the desired sensitivity and complementary sensitivity functions. However, when such a feedback design is implemented via an observer-based controller (or Kalman filter) that uses only the measurement feedback, the loop transfer function obtained, in general, is not the same as the target loop transfer function, unless proper care is taken in designing the observers. This is when the second step of LQG/LTR design philosophy comes into the picture. In this step, the required observer design is attempted so as to recover the loop transfer function of the full state feedback controller. This second step is known as LTR.

The topic of LTR was heavily studied in the 1980s. Major contributions came from [109–119]. We present in the following the methods of LTR design at both the input point and output point of the given plant.

3.7.1 LTR at Input Point

It turns out that it is very simple to formulate the LTR design technique for both continuous- and discrete-time systems into a single framework. Thus, we do it in one shot. Let us consider a linear time-invariant multivariable system characterized by

$$\Sigma : \begin{cases} \delta(x) = A\,x + B\,u \\ \;\;y \;\;\;= C\,x + D\,u \end{cases} \tag{3.256}$$

where $\delta(x) = \dot{x}(t)$, if Σ is a continuous-time system, or $\delta(x) = x(k+1)$, if Σ is a discrete-time system. Similarly, $x \in \mathbb{R}^n$, $u \in \mathbb{R}^m$ and $y \in \mathbb{R}^p$ are the state, input and output of Σ. They represent, respectively, $x(t)$, $u(t)$ and $y(t)$ if the given system is of continuous-time, or represent, respectively, $x(k)$, $u(k)$ and $y(k)$ if Σ is of discrete-time. Without loss of any generality, we assume throughout this section that both $[\,B' \;\; D'\,]$ and $[\,C \;\; D\,]$ are of full rank. The transfer function of Σ is then given by

$$P(\varsigma) = C(\varsigma I - A)^{-1}B + D \tag{3.257}$$

where $\varsigma = s$, the Laplace transform operator, if Σ is of continuous-time, or $\varsigma = z$, the z-transform operator, if Σ is of discrete-time.

As mentioned earlier, there are two steps involved in LQG/LTR design. In the first step, we assume that all state variables of the system in Equation 3.256 are available and design a full state feedback control law

$$u = -Fx \tag{3.258}$$

such that

1. the closed-loop system is asymptotically stable, and
2. the open-loop transfer function when the loop is broken at the input point of the given system, *i.e.*
$$L_{\rm t}(\varsigma) = F(\varsigma I - A)^{-1}B \tag{3.259}$$
meets some frequency-dependent specifications.

Arriving at an appropriate value for F is concerned with the issue of loop shaping, which often includes the use of LQR design in which the cost matrices are used as free design parameters to generate $L_{\rm t}(\varsigma)$ that satisfies the given specifications.

To be more specific, if Σ is a continuous-time system, the target loop transfer function $L_{\rm t}(s)$ can be generated by minimizing the following cost function:

$$J_{\rm c} = \int_0^\infty (x'Qx + u'Ru)\;{\rm d}t \tag{3.260}$$

where $Q \geq 0$ and $R > 0$ are free design parameters provided that $(A, Q^{1/2})$ has no unobservable modes on the imaginary axis. The solution to the above problem is given by

$$F = R^{-1}B'P \tag{3.261}$$

where $P \geq 0$ is the stabilizing solution of the following algebraic Riccati equation (ARE):

$$PA + A'P - PBR^{-1}B'P + Q = 0 \tag{3.262}$$

It is known in the literature that a target loop transfer function $L_{\mathrm{t}}(s)$ with F given as in Equation 3.261 has a phase margin greater than $60°$ and an infinite gain margin.

Similarly, if Σ is a discrete-time system, we can generate a target loop transfer function $L_{\mathrm{t}}(z)$ by minimizing

$$J_{\mathrm{D}} = \sum_{k=0}^{\infty}\Big(x'(k)Qx(k) + u'(k)Ru(k)\Big) \tag{3.263}$$

where $Q \geq 0$ and $R > 0$ are free design parameters provided that $(A, Q^{1/2})$ has no unobservable modes on the unit circle.

$$F = (R + B'PB)^{-1}B'PA \tag{3.264}$$

where $P \geq 0$ is the stabilizing solution of the following ARE:

$$P = A'PA - A'PB(R + B'PB)^{-1}B'PA + Q \tag{3.265}$$

Unfortunately, there are no guaranteed phase and gain margins for the target loop transfer function $L_{\mathrm{t}}(z)$ resulting from the discrete-time linear quadratic regulator.

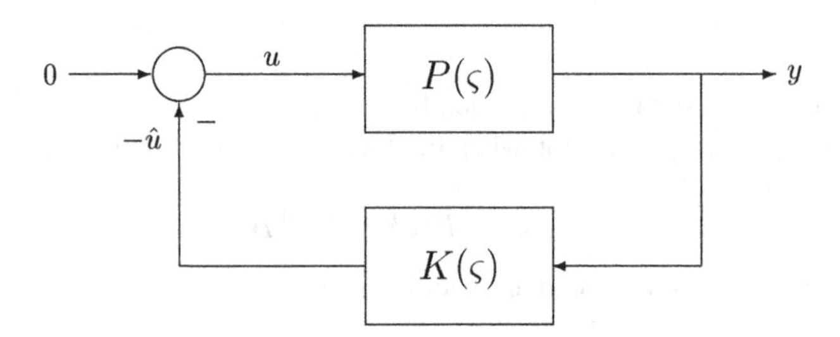

Figure 3.5. Plant-controller closed-loop configuration

Generally, it is unreasonable to assume that all the state variables of a given system can be measured. Thus, we have to implement the control law obtained in the first step by a measurement feedback controller. The technique of LTR is to design an appropriate measurement feedback control (see Figure 3.5) such that the resulting system is asymptotically stable and the achieved open-loop transfer function $L_{\mathrm{a}}(\varsigma)$ from u to $-\hat{u}$ is either exactly or approximately matched with the target loop transfer function $L_{\mathrm{t}}(\varsigma)$ obtained in the first step. In this way, all the nice properties associated with the target loop transfer function can be recovered by the measurement feedback controller. This is the so-called LTR design.

It is simple to observe that the achieved open-loop transfer function in the configuration of Figure 3.5 is given by

$$L_{\mathrm{a}}(\varsigma) = K(\varsigma)P(\varsigma) \tag{3.266}$$

Let us define *recovery error* as

$$E(\varsigma) := L_{\rm t}(\varsigma) - L_{\rm a}(\varsigma) = F(\varsigma I - A)^{-1}B - K(\varsigma)P(\varsigma) \tag{3.267}$$

The LTR technique is to design an appropriate stabilizing $K(\varsigma)$ such that the recovery error $E(\varsigma)$ is either identically zero or small in a certain sense. As usual, two commonly used structures for $K(\varsigma)$ are: 1) the full-order observer-based controller, and 2) the reduced-order observer-based controller.

i. Full-order Observer-based Controller. The dynamic equations of a full-order observer-based controller are well known and are given by

$$\begin{cases} \delta(\hat{x}) = A\hat{x} + Bu + K_{\rm f}(y - C\hat{x} - Du) \\ u \quad = -F\hat{x} \end{cases} \tag{3.268}$$

where $K_{\rm f}$ is the full-order observer gain matrix and is the only free design parameter. It is chosen so that $A - K_{\rm f}C$ is asymptotically stable. The transfer function of the full-order observer-based control is given by

$$K(\varsigma) = F(\varsigma I - A + BF + K_{\rm f}C - K_{\rm f}DF)^{-1}K_{\rm f} \tag{3.269}$$

It has been shown [110, 117] that the recovery error resulting from the full-order observer-based controller can be expressed as

$$E(\varsigma) = M_{\rm f}(\varsigma)[I + M_{\rm f}(\varsigma)]^{-1}[I + F(\varsigma I - A)^{-1}B] \tag{3.270}$$

where

$$M_{\rm f}(\varsigma) = F(\varsigma I - A + K_{\rm f}C)^{-1}(B - K_{\rm f}D) \tag{3.271}$$

Obviously, in order to render $E(\varsigma)$ to be zero or small, one has to design an observer gain $K_{\rm f}$ such that $M_{\rm f}(\varsigma)$, or equivalently $M_{\rm f}'(\varsigma)$, is zero or small (in a certain sense). Defining an auxiliary system,

$$\begin{cases} \dot{\tilde{x}} = A'\,\tilde{x} + C'\,\tilde{u} + F'\,\tilde{w} \\ \tilde{y} = \quad \tilde{x} \\ \tilde{h} = B'\,\tilde{x} + D'\,\tilde{u} \end{cases} \tag{3.272}$$

with a state feedback control law,

$$\tilde{u} = -K_{\rm f}'\,\tilde{x} \tag{3.273}$$

It is straightforward to verify that the closed-loop transfer matrix from $\tilde{w}$ to $\tilde{h}$ of the above system is equivalent to $M_{\rm f}'(\varsigma)$. As such, any of the methods presented in Sections 3.4 and 3.5 for H_2 and H_∞ optimal control can be utilized to find $K_{\rm f}$ to minimize either the H_2-norm or H_∞-norm of $M_{\rm f}'(\varsigma)$. In particular,

1. if the given plant Σ is a continuous-time system and if Σ is left invertible and of minimum phase,
2. if the given plant Σ is a discrete-time system and if Σ is left invertible and of minimum phase with no infinite zeros,

then either the H_2-norm or H_∞-norm of $M_{\mathrm{f}}(\varsigma)$ can be made arbitrarily small, and hence LTR can be achieved. If these conditions are not satisfied, the target loop transfer function $L_{\mathrm{t}}(\varsigma)$, in general, cannot be fully recovered!

For the case when the target loop transfer function can be approximately recovered, the following full-order Chen–Saberi–Sannuti (CSS) architecture-based control law (see [111, 117]),

$$\begin{cases} \delta(v) = Av + K_{\mathrm{f}}(y - Cv) \\ u \;\;\; = -Fv \end{cases} \tag{3.274}$$

which has a resulting recovery error,

$$E(\varsigma) = M_{\mathrm{f}}(\varsigma) = F(\varsigma I - A + K_{\mathrm{f}}C)^{-1}(B - K_{\mathrm{f}}D) \tag{3.275}$$

can be utilized to recover the target loop transfer function as well. In fact, when the same gain matrix K_{f} is used, the full-order CSS architecture-based controller would yield a much better recovery compared to that of the full order observer-based controller.

ii. Reduced-order Observer-based Controller. For simplicity, we assume that C and D have already been transformed into the form

$$C = \begin{bmatrix} 0 & C_{02} \\ I & 0 \end{bmatrix} \quad \text{and} \quad D = \begin{bmatrix} D_0 \\ 0 \end{bmatrix} \tag{3.276}$$

where D_0 is of full row rank. Then, the dynamic equations of Σ can be partitioned as follows:

$$\begin{cases} \begin{pmatrix} \delta(x_1) \\ \delta(x_2) \end{pmatrix} = \begin{bmatrix} A_{11} & A_{12} \\ A_{21} & A_{22} \end{bmatrix} \begin{pmatrix} x_1 \\ x_2 \end{pmatrix} + \begin{bmatrix} B_1 \\ B_2 \end{bmatrix} u \\ \begin{pmatrix} y_0 \\ y_1 \end{pmatrix} = \begin{bmatrix} 0 & C_{02} \\ I & 0 \end{bmatrix} \begin{pmatrix} x_1 \\ x_2 \end{pmatrix} + \begin{bmatrix} D_0 \\ 0 \end{bmatrix} u \end{cases} \tag{3.277}$$

where x_1 is readily accessible. Let

$$A_{\mathrm{r}} = A_{22}, \quad B_{\mathrm{r}} = B_2, \quad C_{\mathrm{r}} = \begin{bmatrix} C_{02} \\ A_{12} \end{bmatrix}, \quad D_{\mathrm{r}} = \begin{bmatrix} D_0 \\ B_1 \end{bmatrix} \tag{3.278}$$

and the reduced-order observer gain matrix K_{r} be such that $A_{\mathrm{r}} - K_{\mathrm{r}}C_{\mathrm{r}}$ is asymptotically stable. Next, we partition

$$F = [\, F_1 \quad F_2 \,], \quad K_{\mathrm{r}} = [\, K_{\mathrm{r}0} \quad K_{\mathrm{r}1} \,] \tag{3.279}$$

in conformity with the partitions of $x = \begin{pmatrix} x_1 \\ x_2 \end{pmatrix}$ and $y = \begin{pmatrix} y_0 \\ y_1 \end{pmatrix}$, respectively. Then, define

$$G_{\mathrm{r}} = [\, K_{\mathrm{r}0}, \quad A_{21} - K_{\mathrm{r}1}A_{11} + (A_{\mathrm{r}} - K_{\mathrm{r}}C_{\mathrm{r}})K_{\mathrm{r}1} \,] \tag{3.280}$$

The reduced-order observer-based controller is given by

$$\begin{cases} \delta(v) = (A_{\mathrm{r}} - K_{\mathrm{r}}C_{\mathrm{r}})v + (B_{\mathrm{r}} - K_{\mathrm{r}}D_{\mathrm{r}})u + G_{\mathrm{r}}y \\ u \;\;\; = -F_2 v - [\, 0, \quad F_1 + F_2K_{\mathrm{r}1} \,]\, y \end{cases} \tag{3.281}$$

It is again reported in [110, 117] that the recovery error resulting from the reduced-order observer-based controller can be expressed as

$$E(\varsigma) = M_r(\varsigma)[I + M_r(\varsigma)]^{-1}[I + F(\varsigma I - A)^{-1}B] \tag{3.282}$$

where

$$M_r(\varsigma) = F_2(\varsigma I - A_r + K_r C_r)^{-1}(B_r - K_r D_r) \tag{3.283}$$

Thus, making $E(\varsigma)$ zero or small is equivalent to designing a reduced-order observer gain K_r such that $M_r(\varsigma)$, or equivalently $M_r'(\varsigma)$, is zero or small. Following the same idea as in the full-order case, we define an auxiliary system

$$\begin{cases} \dot{\tilde{x}} = A_r' \, \tilde{x} + C_r' \, \tilde{u} + F_2' \, \tilde{w} \\ \tilde{y} = \tilde{x} \\ \tilde{h} = B_r' \, \tilde{x} + D_r' \, \tilde{u} \end{cases} \tag{3.284}$$

with a state feedback control law,

$$\tilde{u} = -K_r' \, \tilde{x} \tag{3.285}$$

Obviously, the closed-loop transfer matrix from $\tilde{w}$ to $\tilde{h}$ of the above system is equivalent to $M_r'(\varsigma)$. Hence, the methods of Sections 3.4 and 3.5 for H_2 and H_∞ optimal control again can be used to find K_r to minimize either the H_2-norm or H_∞-norm of $M_r'(\varsigma)$. In particular, for the case when Σ satisfies Condition 1 (for continuous-time systems) or Condition 2 (for discrete-time systems) stated in the full-order case, the target loop can be either exactly or approximately recovered. In fact, in this case, the following reduced-order CSS architecture-based controller

$$\begin{cases} \delta(v) = (A_r - K_r C_r) \, v + G_r \, y \\ u = -F_2 \, v - [\,0, \; F_1 + F_2 K_{r1}\,] \, y \end{cases} \tag{3.286}$$

which has a resulting recovery error,

$$E(\varsigma) = M_r(\varsigma) = F_2(\varsigma I - A_r + K_r C_r)^{-1}(B_r - K_r D_r) \tag{3.287}$$

can also be used to recover the given target loop transfer function. Again, when the same K_r is used, the reduced-order CSS architecture-based controller would yield a better recovery compared to that of the reduced-order observer-based controller (see [111, 117]).

3.7.2 LTR at Output Point

For the case when uncertainties of the given plant are modeled at the output point, the following dualization procedure can be used to find appropriate solutions. The basic idea is to convert the LTR design at the output point of the given plant into an equivalent LTR problem at the input point of an auxiliary system so that all the methods studied in the previous subsection can be readily applied.

1. Consider a plant Σ characterized by the quadruple (A, B, C, D). Let us design a Kalman filter or an observer first with a Kalman filter or observer gain matrix K_f such that $A - K_\mathrm{f}C$ is asymptotically stable and the resulting target loop

$$L_\mathrm{t}(\varsigma) = C(\varsigma I - A)^{-1}K_\mathrm{f} \tag{3.288}$$

meets all the design requirements specified at the output point. We are now seeking to design a measurement feedback controller $K(\varsigma)$ such that all the properties of $L_\mathrm{t}(\varsigma)$ can be recovered.

2. Define a dual system Σ_du characterized by $(A_\mathrm{du}, B_\mathrm{du}, C_\mathrm{du}, D_\mathrm{du})$ where

$$A_\mathrm{du} := A', \quad B_\mathrm{du} := C', \quad C_\mathrm{du} := B', \quad D_\mathrm{du} := D' \tag{3.289}$$

Let $F_\mathrm{du} := K_\mathrm{f}'$ and let $L_\mathrm{du}(\varsigma)$ be defined as

$$L_\mathrm{du}(\varsigma) := L_\mathrm{t}'(\varsigma) = F_\mathrm{du}(\varsigma I - A_\mathrm{du})^{-1}B_\mathrm{du} \tag{3.290}$$

Let $L_\mathrm{du}(\varsigma)$ be considered as a target loop transfer function for Σ_du when the loop is broken at the input point of Σ_du. Let a measurement feedback controller $K_\mathrm{du}(\varsigma)$ be used for Σ_du. Here, the controller $K_\mathrm{du}(\varsigma)$ could be based either on a full- or a reduced-order observer or CSS architecture depending upon what $K(\varsigma)$ is based on. Following the results given earlier for LTR at the input point to design an appropriate controller $K_\mathrm{du}(\varsigma)$, then the required controller for LTR at the output point of the original plant Σ is given by

$$K(\varsigma) = K_\mathrm{du}'(\varsigma) \tag{3.291}$$

This concludes the LTR design for the case when the loop is broken at the output point of the plant.

Finally, we note that there are another type of loop transfer recovery techniques that have been proposed in the literature, *i.e.* in Chen *et al.* [120–122], in which the focus is to recover a closed-loop transfer function instead of an open-loop one as in the conventional LTR design studied in this section. Interested readers are referred to [120–122] for details.

4

Classical Nonlinear Control

4.1 Introduction

Every physical system in real life has nonlinearities and very little can be done to overcome them. Many practical systems are sufficiently nonlinear so that important features of their performance may be completely overlooked if they are analyzed and designed through linear techniques. In HDD servo systems, major nonlinearities are frictions, high-frequency mechanical resonances and actuator saturation nonlinearities. Among all these, the actuator saturation could be the most significant nonlinearity in designing an HDD servo system. When the actuator saturates, the performance of the control system designed will seriously deteriorate. Interested readers are referred to a recent monograph by Hu and Lin [123] for a fairly complete coverage of many newly developed results on control systems with actuator nonlinearities.

The actuator saturation in the HDD has seriously limited the performance of its overall servo system, especially in the track-seeking stage, in which the HDD R/W head is required to move over a wide range of tracks. It will be obvious in the forthcoming chapters that it is impossible to design a pure linear controller that would achieve a desired performance in the track-seeking stage. Instead, we have no choice but to utilize some sophisticated nonlinear control techniques in the design. The most popular nonlinear control technique used in the design of HDD servo systems is the so-called proximate time-optimal servomechanism (PTOS) proposed by Workman [30], which achieves near time-optimal performance for a large class of motion control systems characterized by a double integrator. The PTOS was actually modified from the well-known time-optimal control. However, it is made to yield a minimum variance with smooth switching from the track-seeking to track-following modes. We also introduce another nonlinear control technique, namely a mode-switching control (MSC). The MSC we present in this chapter is actually a combination of the PTOS and the robust and perfect tracking (RPT) control of Chapter 3. In particular, in the MSC scheme for HDD servo systems, the track-seeking mode is controlled by a PTOS and the track-following mode is controlled by a RPT controller. The MSC is a type of variable-structure control systems, but its switching is in only one direction.

4.2 Time-optimal Control

We recall the technique of the time-optimal control (TOC) in this section. Given a dynamic system characterized by

$$\dot{x} = h(x, u, t) \tag{4.1}$$

where x is the state variable and u is the control input, the objective of optimal control is to determine a control input u that causes a controlled process to satisfy the physical constraints and at the same time optimize a certain performance criterion,

$$J = \int_{t_0}^{t_f} f(x, u, t)\, \mathrm{d}t \tag{4.2}$$

where t_0 and t_f are, respectively, initial time and final time of operation, and f is a scalar function. The TOC is a special class of optimization problems and is defined as the transfer of the system from an arbitrary initial state $x(t_0)$ to a specified target set point in minimum time. For simplicity, we let $t_0 = 0$. Hence, the performance criterion for the time-optimal problem becomes one of minimizing the following cost function with $f(x, u, t) = 1$, *i.e.*

$$J = \int_0^{t_f} \mathrm{d}t = t_f \tag{4.3}$$

Let us now derive the TOC law using Pontryagin's principle and the calculus of variation (see, *e.g.*, [124]) for a simple dynamic system obeying Newton's law, *i.e.* for a double-integrator system represented by

$$\ddot{y}(t) = au(t) \tag{4.4}$$

where y is the position output, a is the acceleration constant and u is the input to the system. It will be seen later that the dynamics of the actuator of an HDD can be approximated as a double-integrator model. To start with, we rewrite Equation 4.4 as the following state-space model:

$$\dot{x}(t) = Ax(t) + Bu(t) \tag{4.5}$$

with

$$A = \begin{bmatrix} 0 & 1 \\ 0 & 0 \end{bmatrix}, \quad B = \begin{bmatrix} 0 \\ a \end{bmatrix}, \quad x = \begin{pmatrix} x_1 \\ x_2 \end{pmatrix} = \begin{pmatrix} y \\ v \end{pmatrix} = \begin{pmatrix} y \\ \dot{y} \end{pmatrix} \tag{4.6}$$

Note that $v = \dot{y}$ is the velocity of the system. Let the control input be constrained as follows:

$$| u(t) | \leq u_{\max} \tag{4.7}$$

Then, the Hamiltonian (see, *e.g.*, [124]) for such a problem is given by

$$H(x(t), u(t), \lambda(t)) = 1 + \lambda_1(t)x_2(t) + \lambda_2(t)au(t) \tag{4.8}$$

where $\lambda = (\lambda_1 \quad \lambda_2)'$ is a vector of the time-varying Lagrange multipliers. Pontryagin's principle states that the Hamiltonian is minimized by the optimal control, or

$$H(x^*, u^*, \lambda^*) \leq H(x^*, u, \lambda^*) \tag{4.9}$$

where superscript $*$ indicates optimality. Thus, from Equations 4.8 and 4.9, the optimal control is

$$u^*(t) = \left\{ \begin{array}{ll} -u_{\max}, & \text{for } \lambda_2^*(t) > 0 \\ +u_{\max}, & \text{for } \lambda_2^*(t) < 0 \end{array} \right\} := -\operatorname{sgn}(\lambda_2^*(t))u_{\max} \tag{4.10}$$

The calculus of variation (see [124]) yields the following necessary condition for a time-optimal solution:

$$\dot{\lambda}^*(t) = -A'\lambda^*(t) \tag{4.11}$$

which is known as a costate equation in optimal control terminology. The solution to the costate equation is of the form

$$\left. \begin{array}{l} \lambda_1^*(t) = c_1 \\ \lambda_2^*(t) = -c_1 t + c_2 \end{array} \right\} \tag{4.12}$$

where c_1 and c_2 are constants of integration. Equation 4.12 indicates that λ_2^* and, therefore u^* can change sign at most once. Since there can be at most one switching, the optimal control for a specified initial state must be one of the following forms:

$$\left. \begin{array}{l} 1^\circ: \; u^*(t) = +u_{\max}, \quad \forall t \in [t_0, \; t^*] \\ 2^\circ: \; u^*(t) = \left\{ \begin{array}{l} +u_{\max}, \; \forall t \in [t_0, \; t_1) \\ -u_{\max}, \; \forall t \in [t_1, \; t^*] \end{array} \right. \\ 3^\circ: \; u^*(t) = -u_{\max}, \quad \forall t \in [t_0, \; t^*] \\ 4^\circ: \; u^*(t) = \left\{ \begin{array}{l} -u_{\max}, \; \forall t \in [t_0, \; t_1) \\ +u_{\max}, \; \forall t \in [t_1, \; t^*] \end{array} \right. \end{array} \right\} \tag{4.13}$$

Thus, the segment of optimal trajectories can be found by integrating Equation 4.5 with $u = \pm u_{\max}$ to obtain

$$x_2(t) = \pm a\, u_{\max}\, t + c_3 \tag{4.14}$$

$$x_1(t) = \pm \frac{1}{2} a\, u_{\max}\, t^2 + c_3\, t + c_4 \tag{4.15}$$

where c_3 and c_4 are constants of integration. It is to be noted that if the initial state lies on the optimal trajectories defined by Equations 4.14 and 4.15 in the state plane, then the control will be either 1° or 3° in Equation 4.13 depending upon the direction of motion. In HDD servo systems, it will be shown later that the problem is of relative head-positioning control, and hence the initial and final states must be

$$x(0) = \begin{pmatrix} 0 \\ 0 \end{pmatrix}, \quad x(t^*) = \begin{pmatrix} r \\ 0 \end{pmatrix} \tag{4.16}$$

where r is the reference set point. Because of these kinds of initial state in HDD servo systems, the optimal control must be chosen from either 2° or 4° in Equation 4.13. Note that if the control input $+u_{\max}$ produces the acceleration α, then the input $-u_{\max}$ will produce a deceleration of the same magnitude.

Hence, the minimum time performance can be achieved either with maximum acceleration for half of the travel followed by maximum deceleration for an equal amount of time, or by first accelerating and then decelerating the system with maximum effort to follow some predefined optimal velocity trajectory to reach the final destination in minimum time. The former case results in an open-loop form of TOC that uses predetermined time-based acceleration and deceleration inputs, whereas the latter yields a closed-loop form of TOC. We note that if the area under acceleration, which is a function of time, is the same as the area under deceleration, there will be no net change in velocity after the input is removed. The final output velocity and the position will be in a steady state.

In general, the time-optimal performance can be achieved by switching the control between two extreme levels of the input, and we have shown that in the double-integrator system the number of switchings is at most equal to one, *i.e.* one less than the order of dynamics. Thus, if we extend the result to an nth-order system, it will need $n - 1$ switchings between maximum and minimum inputs to achieve a time-optimal performance. Since the control must be switched between two extreme values, the TOC is also known as *bang-bang* control.

In what follows, we discuss the bang-bang control in two versions, *i.e.* in the open-loop and in the closed-loop forms for the double-integrator model characterized by Equation 4.5 with the control constraint represented by Equation 4.7.

4.2.1 Open-loop Bang-bang Control

The open-loop method of bang-bang control uses maximum acceleration and maximum deceleration for a predetermined time period. Thus, the time required for the system to reach the target position in minimum time is predetermined from the above principles and the control input is switched between two extreme levels for this time period. We can precalculate the minimum time t^* for a specified reference set point r. Let the control be

$$u^*(t) = \begin{cases} +u_{\max}, & \text{for } t \in (0, \frac{t^*}{2}] \\ -u_{\max}, & \text{for } t \in (\frac{t^*}{2}, t^*] \end{cases} \tag{4.17}$$

We now solve Equations 4.14 and 4.15 for the accelerating phase with zero initial condition. For the accelerating phase, *i.e.* with $u = +u_{\max}$, we have

$$x_2(t) = a\, u_{\max}\, t, \quad x_1(t) = \frac{1}{2}\, a\, u_{\max}\, t^2 \tag{4.18}$$

At the end of the accelerating phase, *i.e.* at $t = t^*/2$,

$$x_{2a} = a\, u_{\max} \frac{t^*}{2}, \quad x_{1a} = \frac{1}{8} a\, u_{\max} (t^*)^2 \tag{4.19}$$

Similarly, at the end of decelerating phase, we can show that

$$x_{2d} = 0, \quad x_{1d} = \frac{1}{8} a\, u_{\max} (t^*)^2 \tag{4.20}$$

Obviously, the total displacement at the end of bang-bang control must reach the target, *i.e.* the reference set point r. Thus,

$$r = x_{1a} + x_{1d} = \frac{1}{4} a\, u_{\max} (t^*)^2 \tag{4.21}$$

which gives

$$t^* = \sqrt{\frac{4\, r}{a\, u_{\max}}} \tag{4.22}$$

the minimum time required to reach the target set point.

4.2.2 Closed-loop Bang-bang Control

In this method, the velocity of the plant is controlled to follow a predefined trajectory and more specifically the decelerating trajectory. These trajectories can be generated from the phase-plane analysis. This analysis is explained below for the system given by Equation 4.5 and can be extended to higher-order systems (see, *e.g.*, [124]). We will show later that this deceleration trajectory brings the system to the desired set point in finite time. We now move to find the deceleration trajectory.

First, eliminating t from Equations 4.14 and 4.15, we have

$$x_1(t) = \frac{1}{2\, a\, u_{\max}} x_2^2(t) + c_5, \qquad \text{for } u = +u_{\max} \tag{4.23}$$

$$x_1(t) = -\frac{1}{2\, a\, u_{\max}} x_2^2(t) + c_6, \qquad \text{for } u = -u_{\max} \tag{4.24}$$

where c_5 and c_6 are appropriate constants. Note that each of the above equations defines the family of parabolas. Let us define $e(t) := r - x_1(t)$ to be the positioning error with r being the desired final position. Then, if we consider the trajectories between $e(t)$ and $x_2(t)$, our desired final state in $e(t)$ and $x_2(t)$ plane must be

$$\begin{pmatrix} r - x_1 \\ x_2 \end{pmatrix} = \begin{pmatrix} 0 \\ 0 \end{pmatrix} \tag{4.25}$$

In this case, the constants in the above trajectories are equal to zero. Moreover, both of the trajectories given by Equations 4.23 and 4.24 are the decelerating trajectories depending upon the direction of the travel. The mechanism of the TOC can be illustrated in a graphical form as given in Figure 4.1. Clearly, any initial state lying below the curve is to be driven by the positive accelerating force to bring the state to the

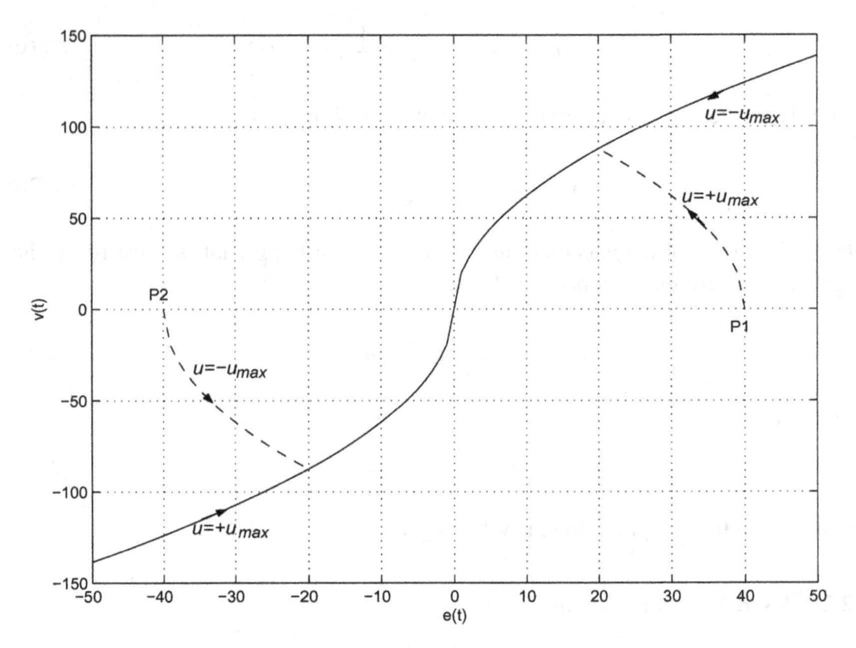

Figure 4.1. Deceleration trajectories for TOC

deceleration trajectory. On the other hand, any initial state lying above the curve is to be accelerated by the negative force to the deceleration trajectory.

Let

$$f_t(e) := \text{sgn}(e)\sqrt{2\,a\,u_{\max}\mid e\mid} \tag{4.26}$$

The control law is then given by

$$u = u_{\max} \cdot \text{sgn}\Big(f_t(e) - v\Big) \tag{4.27}$$

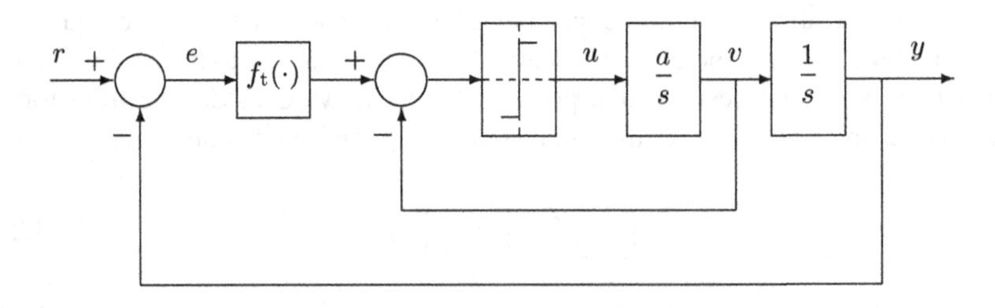

Figure 4.2. Typical scheme of TOC

A block diagram depicting the closed-loop method of bang-bang control is shown in Figure 4.2. Unfortunately, the control law given by Equation 4.27 for the system shown in Figure 4.2, although time-optimal, is not practical. It applies maximum or minimum input to the plant to be controlled even for a small error. Moreover, this algorithm is not suited for disk drive applications for the following reasons:

1. even the smallest system process or measurement noise will cause control "chatter". This will excite the high-frequency modes.
2. any error in the plant model, will cause limit cycles to occur.

As such, the TOC given above has to be modified to suit HDD applications. In the following section, we recall a modified version of the TOC proposed by Workman [30], *i.e.* the PTOS. Such a control scheme is widely used nowadays in designing HDD servo systems.

4.3 Proximate Time-optimal Servomechanism

The infinite gain of the signum function in the TOC causes control chatter, as seen in the previous section. Workman [30], in 1987, proposed a modification of this technique, *i.e.* the so-called PTOS, to overcome such a drawback. The PTOS essentially uses maximum acceleration where it is practical to do so. When the error is small, it switches to a linear control law. To do so, it replaces the signum function in TOC law by a saturation function. In the following sections, we revisit the PTOS method in continuous-time and in discrete-time domains.

4.3.1 Continuous-time Systems

The configuration of the PTOS is shown in Figure 4.3. The function $f_{\mathrm{p}}(\cdot)$ is a finite-slope approximation to the switching function $f_{\mathrm{t}}(\cdot)$ given by Equation 4.26. The PTOS control law for the system in Equation 4.5 is given by

$$u = u_{\max} \cdot \mathrm{sat}\left(\frac{k_2[f_{\mathrm{p}}(e) - v]}{u_{\max}}\right) \tag{4.28}$$

where sat(x) is defined as

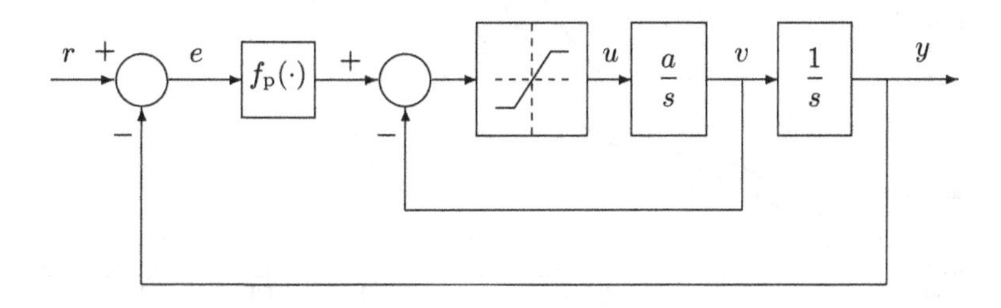

Figure 4.3. Continuous-time PTOS

$$\mathrm{sat}(x) = \begin{cases} +1, & \text{if} \quad x > 1 \\ x, & \text{if} \quad -1 \leq x \leq 1 \\ -1, & \text{if} \quad x < -1 \end{cases} \tag{4.29}$$

and the function $f_{\mathrm{p}}(e)$ is given by

$$f_{\mathrm{p}}(e) = \begin{cases} \dfrac{k_1}{k_2} e & \text{for } |e| \leq y_\ell \\ \mathrm{sgn}(e) \left[\sqrt{2\, a\, u_{\max}\, \alpha \mid e \mid} - \dfrac{u_{\max}}{k_2} \right] & \text{for } |e| > y_\ell \end{cases} \tag{4.30}$$

Here we note that k_1 and k_2 are, respectively, the feedback gains for position and velocity, α is a constant between 0 and 1 and is referred to as the acceleration discount factor, and y_ℓ is the size of the linear region. Since the linear portion of the curve $f_{\mathrm{p}}(\cdot)$ must connect the two disjoint halves of the nonlinear portion, we have constraints on the feedback gains and the linear region to guarantee the continuity of the function $f_{\mathrm{p}}(\cdot)$. It was proved by Workman [30] that

$$y_\ell = \frac{u_{\max}}{k_1}, \quad k_2 = \sqrt{\frac{2k_1}{a\alpha}} \tag{4.31}$$

The control zones in the PTOS are shown in Figure 4.4. The two curves bounding the switching curve (central curve) now redefine the control boundaries and it is termed a linear boundary. Let this region be $\mathbf{U}$. The region below the lower curve is

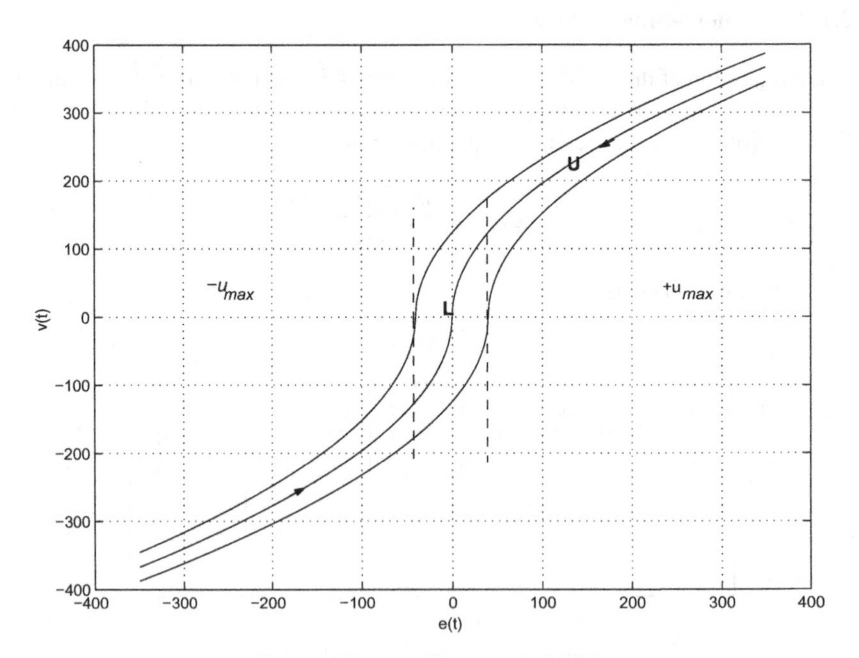

Figure 4.4. Control zones of a PTOS

the region where the control $u = +u_{\max}$, whereas the region above the upper curve is the region where the control $u = -u_{\max}$. It has been proved [30] that once the state trajectory enters the band **U** in Figure 4.4 it remains within **U** and the control signal is below the saturation. The region marked **L** is the region where the linear control is applied.

The presence of the acceleration discount factor α allows us to accommodate uncertainties in the plant accelerating factor at the cost of increase in response time. By approximating the positioning time as the time that it takes the positioning error to be within the linear region, one can show that the percentage increase P in time taken by the PTOS over the time taken by the TOC is given by (see [30]):

$$P = \frac{1}{2}\left(\frac{1}{\sqrt{\alpha}} - 1\right) \times 100\% \tag{4.32}$$

Clearly, larger values of α make the response closer to that of the TOC. As a result of changing the nonlinearity from sgn(·) to sat(·), the control chatter is eliminated.

4.3.2 Discrete-time Systems

The discrete-time PTOS can be derived from its continuous-time counterpart, but with some conditions on sample time to ensure stability. In his seminal work, Workman [30] extended the continuous-time PTOS to discrete-time control of a continuous-time double-integrator plant driven by a zero-order hold as shown in Figure 4.5. As in the continuous-time case, the states are defined as position and velocity. With insignificant calculation delay, the state-space description of the plant given by Equation 4.5 in the discrete-time domain is

$$x(k+1) = \begin{bmatrix} 1 & T_s \\ 0 & 1 \end{bmatrix} x(k) + \begin{bmatrix} aT_s^2/2 \\ aT_s \end{bmatrix} u(k) \tag{4.33}$$

where T_s is the sampling period. The control structure is a discrete-time mapping of the continuous-time PTOS law, but with a constraint on the sampling period to

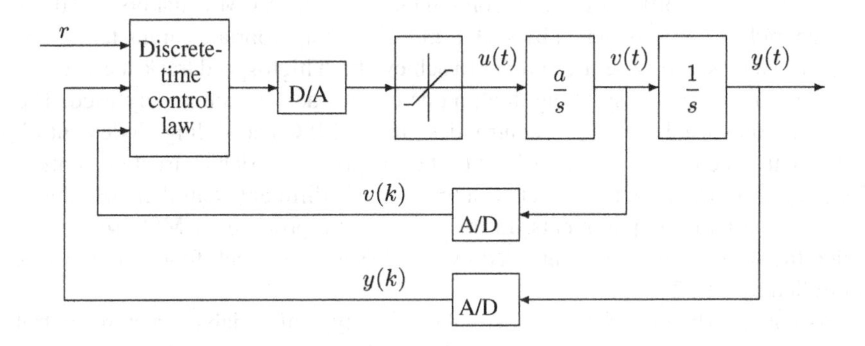

Figure 4.5. Discrete-time PTOS

guarantee that the control does not saturate during the deceleration phase to the target position and also to guarantee its stability. Thus, the mapped control law is

$$u(k) = u_{\max} \cdot \text{sat}\left(\frac{k_2[f_{\text{p}}(e(k)) - v(k)]}{u_{\max}}\right) \tag{4.34}$$

with the following constraint on sampling frequency ω_{s},

$$\frac{\omega_{\text{s}}}{\omega_{\text{n}}} > 6.3 \tag{4.35}$$

where ω_{n} is the desired bandwidth of the closed-loop system.

4.4 Mode-switching Control

In this section, we present a mode-switching control (MSC) design technique for both continuous-time and discrete-time systems, which is a combination of the PTOS of the previous section and the RPT technique given in Chapter 4.

4.4.1 Continuous-time Systems

In this subsection, we follow the development of [125] to introduce the design of an MSC design for a system characterized by a double integrator or in the following state-space equation:

$$\dot{x} = Ax + Bu = \begin{bmatrix} 0 & 1 \\ 0 & 0 \end{bmatrix} x + \begin{bmatrix} 0 \\ a \end{bmatrix} u, \quad x = \begin{pmatrix} y \\ v \end{pmatrix} \tag{4.36}$$

where as usual x is the state, which consists of the displacement y and the velocity v; u is the control input constrained by

$$|u(t)| \leq u_{\max} \tag{4.37}$$

As will be seen shortly in the forthcoming chapters, the VCM actuators of HDDs can generally be approximated by such a model with appropriate parameters a and $u_{\max}$. In HDD servo systems, in order to achieve both high-speed track seeking and highly accurate head positioning, multimode control designs are widely used. The two commonly used multimode control designs are MSC and sliding mode control. Both control techniques in fact belong to the category of variable-structure control. That is, the control is switched between two or more different controllers to achieve the two conflicting requirements. In this section, we propose an MSC scheme in which the seeking mode is controlled by a PTOS and the track-following mode is controlled by a RPT controller.

As noted earlier, the MSC (see, *e.g.*, [15]) is a type of variable structure control systems [126], but the switching is in only one direction. Figure 4.6 shows a basic schematic diagram of MSC. There are track seeking and track following modes.

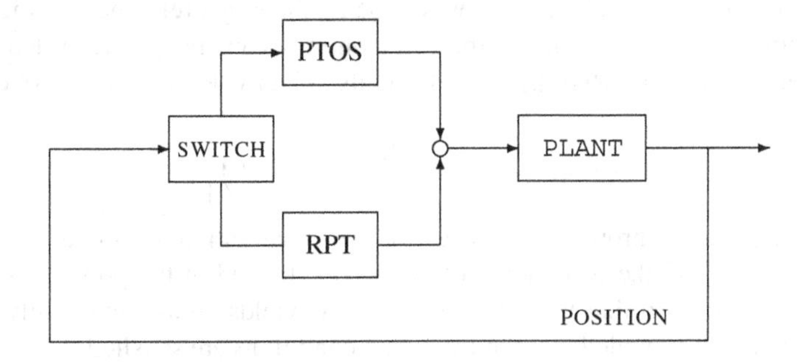

Figure 4.6. Basic schematic diagram of MSC

Each servo mode can be designed independently, and so the main issue in MSC is the design of the switching mechanism.

This design problem has not yet been completely resolved, and many heuristic approaches have been tried so far (see, *e.g.*, [127]). Several methods were proposed for mode switching from one controller to another. In [15], a method called initial value compensation is proposed. Note that, when the switch is transfered from the track-seeking mode to the track-following mode, the final states of the track-seeking controller become the initial states for the track-following controller, and hence affect the settling performance of track-following mode. In order to reduce the impact of these initial values during mode switching, some compensation must be worked out. Here, the initial values are referred to the values of the states during mode switching. However, the RPT controllers developed by Chen and coworkers [74, 106] (see also Chapter 3) have enough robustness against plant variations and are actually independent of initial values. Hence, the use of these controllers in the track-following mode eliminates the need for initial-value compensation during mode switching. Moreover, the RPT controllers in a track-following servo have been proved to be robust against resonance mode changes from disk to disk and work well against runout disturbances.

The MSC law that combines the PTOS and RPT controllers takes the following simple form:

$$u = \begin{cases} u_{\mathrm{P}}, & t < t_1 \\ u_{\mathrm{R}}, & t \geq t_1 \end{cases} \tag{4.38}$$

where u_{P} is a control signal generated by the PTOS control and is given as in Equation 4.28, and u_{R} is a signal generated by the reduced-order RPT control as given in Equation 3.237. Furthermore, t_1, the time that MSC switches from one mode to the other, will be presented in the next subsection together with the stability analysis of the closed loop comprising the given plant and the MSC control law.

In what follows, we show the stability of the MSC and give a set of conditions for mode switching. First, we rewrite the given system in Equation 4.36 as follows:

$$\begin{pmatrix} \dot{e} \\ \dot{v} \end{pmatrix} = \begin{bmatrix} 0 & -1 \\ 0 & 0 \end{bmatrix} \begin{pmatrix} e \\ v \end{pmatrix} + \begin{bmatrix} 0 \\ a \end{bmatrix} u := A_{\mathrm{P}} x_{\mathrm{P}} + B_{\mathrm{P}} u, \quad a > 0 \tag{4.39}$$

where $e = r - y$ is the tracking error with r being the target reference. In the HDD servo systems that we deal with in the forthcoming chapters, y is regarded as the displacement of an HDD R/W head, and v is its velocity. Recall the PTOS control law:

$$u_{\rm P} = u_{\max} \cdot \text{sat} \left[\frac{k_2}{u_{\max}} \Big(f(e) - v \Big) \right] \tag{4.40}$$

where the $u_{\max}$ used throughout this section has a saturation level equal to unity; the function $f(\cdot)$ and the feedback gain k_2 are as defined in the previous section. It has been shown [30] that the PTOS control law yields an asymptotically stable closed-loop system provided that the following conditions are satisfied:

1. $ak_2 > 0$;
2. $f(0) = 0$;
3. $f(e)e > 0$, for any nonzero e;
4. $\lim_{e \to \infty} \int_0^e f(\delta)\, \mathrm{d}\delta = \infty$;
5. $\dot{f}(e)$ exists everywhere; and
6. for any e, $u_{\max} \left[-a + \frac{1}{k_2} \right] \dot{f}(e) < -\dot{f}(e) f(e) < u_{\max} \left[a - \frac{1}{k_2} \dot{f}(e) \right]$.

Generally, as the velocity is not measurable, the PTOS control law has to be modified as follows if it is to be implemented in a real system:

$$\begin{cases} u_{\rm P} = u_{\max}\, \text{sat} \left[\frac{k_2}{u_{\max}} \Big(f(e) - \hat{v} \Big) \right] \\ \dot{z} \;=\; -\kappa z + \kappa^2 e + a\, u_{\rm P} \\ \hat{v} \;=\; z - \kappa e \end{cases} \tag{4.41}$$

where κ is the estimator feedback gain, and z is the estimator state. Next, we let $\tilde{z} = z - v - \kappa e$. Then, the dynamics of the closed-loop system with the above control law can be written as

$$\begin{cases} \dot{e} \;=\; -v \\ \dot{v} \;=\; a\, u_{\max} \text{sat} \left[\frac{k_2}{u_{\max}} \Big(f(e) - \tilde{z} - v \Big) \right] \\ \dot{\tilde{z}} \;=\; -\kappa \tilde{z} \end{cases} \tag{4.42}$$

It can be shown that the closed-loop system comprising the given plant and the modified PTOS control law, in which the velocity is replaced by the above estimation, is asymptotically stable, if conditions 1 to 5 above are satisfied and condition 6 is replaced by

$$a > \frac{1}{k_2} \dot{f}(e) > 0 \tag{4.43}$$

and

$$|\dot{f}(e) f(e)| < u_{\max} \left[a - \frac{1}{k_2} \dot{f}(e) \right] - \left[\kappa + \dot{f}(e) \right] \cdot |\tilde{z}(0)| \tag{4.44}$$

We can show that, under these new conditions, the closed-loop system is stable for the case when the control input is saturated, *i.e.* $|k_2[f(e) - (z - \kappa e)]| > u_{\max}$. For the case when $|k_2[f(e) - (z - \kappa e)]| \leq u_{\max}$, the closed-loop system in Equation 4.42 can be written as

$$\begin{cases} \dot{e} = -v \\ \dot{v} = ak_2[f(e) - \tilde{z} - v] \\ \dot{\tilde{z}} = -\kappa\tilde{z} \end{cases} \tag{4.45}$$

Following the result of [128], we propose the following Lyapunov function for the system in Equation 4.45:

$$V_{\rm P} = \frac{v^2}{2k_2 a} + \int_0^e f(\delta)\, \mathrm{d}\delta + \frac{p_z}{2}\tilde{z}^2 \tag{4.46}$$

where $p_z > 0$ is a scalar constant. The derivative of the above Lyapunov function is given by

$$\dot{V}_{\rm P} = -\left(v + \frac{\tilde{z}}{2}\right)^2 - \tilde{z}^2\left(p_z\kappa - \frac{1}{4}\right) \tag{4.47}$$

The last term is negative for all $p_z > \frac{1}{4\kappa}$. Thus, under this choice of p_z, $\dot{V}_{\rm P} \leq 0$. It follows from LaSalle's Theorem [129] that the closed-loop system comprising the PTOS control law with the estimated velocity and the given plant is asymptotically stable.

It is obvious that the closed-loop system comprising the given plant in Equation 4.39 and the reduced-order RPT control law of Equation 3.237 is asymptotically stable when the control input is not saturated. For completeness, and for the analysis of the overall closed-loop system with the MSC scheme, we proceed to investigate the closed-loop system comprising the plant and the RPT controller, which can be written as

$$\begin{cases} \dot{z} = -\kappa z + \kappa^2 e + a\, u_{\max} \cdot \mathrm{sat}\left(\dfrac{u_{\rm R}}{u_{\max}}\right) \\ u_{\rm R} = (f_1 - \kappa f_2)e + f_2 z \end{cases} \tag{4.48}$$

where κ is again the reduced-order observer gain, which is selected to be exactly the same as that used for the velocity estimation in the PTOS, and $F = [\, f_1 \quad f_2 \,]$ is the feedback gain obtained using the RPT technique of Chapter 3. Again, let $\tilde{z} = z - v - \kappa e$ and rewrite the RPT control law as

$$u_{\rm R} = [\, f_1 \quad f_2 \,]\begin{pmatrix} e \\ v \end{pmatrix} + f_2\tilde{z} = F x_{\rm P} + f_2\tilde{z} \tag{4.49}$$

Let $W_{\rm x} \in \mathbb{R}^{2\times 2}$ be a positive-definite matrix and solve the following Lyapunov equation:

$$(A_{\rm P} + B_{\rm P}F)' P_{\rm x} + P_{\rm x}(A_{\rm P} + B_{\rm P}F) = -W_{\rm x} \tag{4.50}$$

for $P_{\rm x} > 0$. Such a $P_{\rm x}$ always exists as $A_{\rm P} + B_{\rm P}F$ is stable. Next, let

$$w_z > \max\left\{\frac{1}{2},\ f_2 B_{\rm P}' P_{\rm x} W_{\rm x}^{-1} P_{\rm x} B_{\rm P} f_2\right\} > 0 \tag{4.51}$$

and

$$p_z = \frac{w_z}{2\kappa} > \frac{1}{4\kappa} > 0 \tag{4.52}$$

Then, we define a set

$$\boldsymbol{X} := \left\{ \begin{pmatrix} x \\ z \end{pmatrix} : \begin{pmatrix} x \\ z \end{pmatrix}' \begin{bmatrix} P_{\mathrm{x}} & 0 \\ 0 & p_z \end{bmatrix} \begin{pmatrix} x \\ z \end{pmatrix} \leq c \right\} \tag{4.53}$$

where $c > 0$ is the largest positive value such that

$$\begin{pmatrix} x \\ z \end{pmatrix} \in \boldsymbol{X} \Rightarrow \left| [F \quad f_2] \begin{pmatrix} x \\ z \end{pmatrix} \right| \leq u_{\max} \tag{4.54}$$

For all

$$\begin{pmatrix} x_{\mathrm{P}} \\ z \end{pmatrix} \in \boldsymbol{X} \tag{4.55}$$

the resulting closed-loop system can then be written as

$$\begin{pmatrix} \dot{x}_{\mathrm{P}} \\ \dot{\tilde{z}} \end{pmatrix} = \begin{bmatrix} A_{\mathrm{P}} + B_{\mathrm{P}}F & B_{\mathrm{P}}f_2 \\ 0 & -\kappa \end{bmatrix} \begin{pmatrix} x_{\mathrm{P}} \\ \tilde{z} \end{pmatrix} \tag{4.56}$$

Define a Lyapunov function,

$$V_{\mathrm{R}} = \begin{pmatrix} x_{\mathrm{P}} \\ \tilde{z} \end{pmatrix}' \begin{bmatrix} P_{\mathrm{x}} & 0 \\ 0 & p_z \end{bmatrix} \begin{pmatrix} x_{\mathrm{P}} \\ \tilde{z} \end{pmatrix} \tag{4.57}$$

and evaluate its derivative along the trajectories of the closed-loop system in Equation 4.56, *i.e.*

$$\begin{aligned} \dot{V}_{\mathrm{R}} &= \begin{pmatrix} x_{\mathrm{P}} \\ \tilde{z} \end{pmatrix}' \begin{bmatrix} -W_{\mathrm{x}} & P_{\mathrm{x}}B_{\mathrm{P}}f_2 \\ f_2B_{\mathrm{P}}'P_{\mathrm{x}} & -w_z \end{bmatrix} \begin{pmatrix} x_{\mathrm{P}} \\ \tilde{z} \end{pmatrix} \\ &= \begin{pmatrix} \hat{x} \\ \tilde{z} \end{pmatrix}' \begin{bmatrix} -W_{\mathrm{x}} & 0 \\ 0 & -\tilde{w}_z \end{bmatrix} \begin{pmatrix} \hat{x} \\ \tilde{z} \end{pmatrix} \leq 0 \end{aligned} \tag{4.58}$$

where

$$\hat{x} = x_{\mathrm{P}} - W_{\mathrm{x}}^{-1}P_{\mathrm{x}}B_{\mathrm{P}}f_2\tilde{z}, \quad \tilde{w}_z = w_z - f_2B_{\mathrm{P}}'P_{\mathrm{x}}W_{\mathrm{x}}^{-1}P_{\mathrm{x}}B_{\mathrm{P}}f_2 > 0 \tag{4.59}$$

This shows that all trajectories of Equation 4.56 starting from $\boldsymbol{X}$ remain there and converge asymptotically to zero. Hence, the closed-loop system comprising the plant and the reduced-order RPT control law is asymptotically stable provided that the control input is not saturated.

Next, we re-express Equation 4.46 using the Taylor expansion as follows:

$$V_{\mathrm{P}} = \frac{v^2}{2k_2a} + \frac{\dot{f}(\tau)}{2}e^2 + \frac{p_z}{2}\tilde{z}^2 = \begin{pmatrix} e \\ v \\ \tilde{z} \end{pmatrix}' \begin{bmatrix} \frac{\dot{f}(\tau)}{2} & 0 & 0 \\ 0 & \frac{1}{2ak_2} & 0 \\ 0 & 0 & \frac{p_z}{2} \end{bmatrix} \begin{pmatrix} e \\ v \\ \tilde{z} \end{pmatrix}$$

where τ is an appropriate scalar between 0 and e. Let

$$\sigma = \min\left\{\frac{\dot{f}(\tau)}{2}, \frac{1}{2ak_2}, \frac{p_z}{2}\right\} \Big/ \max\left\{\lambda_{\max}(P_{\rm x}), p_z\right\} \tag{4.60}$$

The MSC scheme can be obtained as follows:

$$u(t) = \begin{cases} u_{\rm P}, & t < t_1 \\ u_{\rm R}, & t \geq t_1 \end{cases} \tag{4.61}$$

where t_1 is such that

$$\begin{pmatrix} x_{\rm P}(t_1) \\ z(t_1) \end{pmatrix} \in \boldsymbol{X} \quad \text{and} \quad |e(t_1)| < y_\ell \tag{4.62}$$

and where y_ℓ is the size of the linear region of the PTOS control law. The Lyapunov function for the overall closed-loop system can be chosen as

$$V = V_{\rm P}[1 - 1(t - t_1)] + \sigma V_{\rm R}\, 1(t - t_1) \tag{4.63}$$

where $1(t)$ is the unit step function. It is simple to verify that

$$\dot{V} = \dot{V}_{\rm P}[1 - 1(t - t_1)] + \sigma \dot{V}_{\rm R}\, 1(t - t_1) + (\sigma V_{\rm R} - V_{\rm P})\delta(t - t_1) \tag{4.64}$$

It has already been proved that the derivatives of the functions $V_{\rm P}$ and $V_{\rm R}$ are negative-definite. The last term is always negative in view of the definition of σ in Equation 4.60. Hence, $\dot{V} \leq 0$ and the resulting closed-loop system comprising the given plant and the mode-switching control law is stable. As such, Equation 4.62 can be regarded as the switching condition for the MSC design.

4.4.2 Discrete-time Systems

We now present an MSC design technique for a discrete-time system characterized by a double-integrator or by the following state-space form:

$$x(k+1) = \begin{bmatrix} 1 & T_{\rm s} \\ 0 & 1 \end{bmatrix} x(k) + \begin{bmatrix} aT_{\rm s}^2/2 \\ aT_{\rm s} \end{bmatrix} u(k), \quad x = \begin{pmatrix} y \\ v \end{pmatrix} \tag{4.65}$$

where $T_{\rm s}$ is the sampling period, a is assumed to be a positive scalar for simplicity, y is the displacement and v is the velocity in the context of mechanical servo systems. The PTOS controller for such a discrete-time system is given in Equation 4.34, *i.e.*

$$u_{\rm P}(k) = u_{\max} \cdot \text{sat}\left(\frac{k_2}{u_{\max}}\Big[f(r - y(k)) - v(k)\Big]\right) \tag{4.66}$$

where r is the target reference. To simplify our analysis, we introduce a new variable $\bar{v} = T_{\rm s} v$ and rewrite the system of Equation 4.65 as follows

$$\begin{pmatrix} y(k+1) \\ \bar{v}(k+1) \end{pmatrix} = \begin{bmatrix} 1 & 1 \\ 0 & 1 \end{bmatrix} \begin{pmatrix} y(k) \\ \bar{v}(k) \end{pmatrix} + \begin{bmatrix} \bar{a}/2 \\ \bar{a} \end{bmatrix} u(k) \tag{4.67}$$

and the corresponding PTOS controller as

$$u_{\rm P}(k) = u_{\max} \cdot \operatorname{sat}\left(\frac{\bar{k}_2}{u_{\max}} \left[\bar{f}(r - y(k)) - \bar{v}(k) \right] \right) \tag{4.68}$$

where

$$\bar{a} = T_{\rm s}^2 a, \quad \bar{k}_2 = \frac{k_2}{T_{\rm s}}, \quad \bar{f}(\cdot) = T_{\rm s} f(\cdot) \tag{4.69}$$

The MSC law comprising the PTOS and RPT controllers is given as follows,

$$u(k) = \begin{cases} u_{\rm P}(k), & k < k_{\rm s} \\ u_{\rm R}(k), & k \geq k_{\rm s} \end{cases} \tag{4.70}$$

where $u_{\rm P}(k)$ is a control signal generated by the PTOS controller and $u_{\rm R}(k)$ is a control signal generated by the RPT controller (see Chapter 3 for details). $k_{\rm s}$ is the index that the MSC switches from one mode to the other.

In what follows, we proceed to present the detailed stability analysis and mode switching condition for the above MSC design. Noting that the tracking error is given by $e(k) = r - y(k)$, we can rewrite the system in Equation 4.67 as the following

$$\begin{pmatrix} e(k+1) \\ \bar{v}(k+1) \end{pmatrix} = \begin{bmatrix} 1 & -1 \\ 0 & 1 \end{bmatrix} \begin{pmatrix} e(k) \\ \bar{v}(k) \end{pmatrix} + \begin{bmatrix} -\bar{a}/2 \\ \bar{a} \end{bmatrix} u(k) \tag{4.71}$$

and the corresponding PTOS controller as follows,

$$u_{\rm P}(k) = u_{\max} \cdot \operatorname{sat}\left(\frac{\bar{k}_2}{u_{\max}} \left[\bar{f}(e(k)) - \bar{v}(k) \right] \right) \tag{4.72}$$

It has been shown in [30] that the discrete PTOS control law yields an asymptotically stable closed-loop system provided that the following conditions are satisfied:

1. $\bar{a}\bar{k}_2 \in (0, 2)$;
2. $f(0) = 0$;
3. $f(e)e > 0, \forall e \neq 0$;
4. $\lim_{e \to \infty} \int_0^e f(\delta) \mathrm{d}\delta = \infty$;
5. $\dot{f}(e)$ exists everywhere;
6. For all e, $\left|\dot{f}(e)\right| < \frac{1}{2} T_{\rm s}^{-1}$; and
7. For the unsaturated situation, $|f(e + \Delta e) - (v + \Delta v)| < \dfrac{u_{\max}}{k_2}$, where

$$\Delta e = -\bar{v} - \frac{1}{2} \bar{a} u_{\max} \cdot \operatorname{sat}\left(\frac{k_2}{u_{\max}} [f(e) - v] \right) \tag{4.73}$$

and

$$\Delta \bar{v} = \bar{a} u_{\max} \cdot \operatorname{sat}\left(\frac{k_2}{u_{\max}} [f(e) - v] \right), \quad \Delta v = \frac{\Delta \bar{v}}{T_{\rm s}} \tag{4.74}$$

As the velocity $\bar{v}$ is generally not measurable, the PTOS control law has to be modified as follows,

$$\begin{cases} u_{\mathrm{P}}(k) = u_{\max} \cdot \mathrm{sat}\left(\dfrac{\bar{k}_2}{u_{\max}}\left[\bar{f}(e(k)) - \hat{v}(k)\right]\right) \\ x_{\mathrm{v}}(k+1) = (1-\kappa)x_{\mathrm{v}}(k) + \kappa^2 e(k) + (1-\kappa/2)\bar{a}u_{\mathrm{P}}(k) \\ \hat{v}(k) = x_{\mathrm{v}}(k) - \kappa e(k) \end{cases} \tag{4.75}$$

where $\kappa \in (0,1)$ is the feedback gain of the estimator and $x_{\mathrm{v}}(k)$ is the state variable of the estimator. Next, let

$$\tilde{x}_{\mathrm{v}}(k) = x_{\mathrm{v}}(k) - \bar{v}(k) - \kappa e(k) \tag{4.76}$$

Then, the closed-loop system comprising Equations 4.71 and 4.75 can be rewritten as

$$\begin{cases} e(k+1) = e(k) - \bar{v}(k) - \dfrac{1}{2}\bar{a}u_{\mathrm{P}}(k) \\ \bar{v}(k+1) = \bar{v}(k) + \bar{a}u_{\mathrm{P}}(k) \\ \tilde{x}_{\mathrm{v}}(k+1) = (1-\kappa)\tilde{x}_{\mathrm{v}}(k) \\ u_{\mathrm{P}}(k) = u_{\max}\mathrm{sat}\left(\dfrac{\bar{k}_2}{u_{\max}}[\bar{f}(e(k)) - \bar{v}(k) - \tilde{x}_{\mathrm{v}}(k)]\right) \end{cases} \tag{4.77}$$

For such a case, Condition 7 above has to be replaced by

$$|\bar{f}(e+\Delta e) - (x_{\mathrm{v}} + \Delta x_{\mathrm{v}}) + \kappa(e+\Delta e)| < \frac{u_{\max}}{\bar{k}_2} \tag{4.78}$$

for the unsaturated situation, where

$$\Delta e = -x_{\mathrm{v}} + \kappa e + \tilde{x}_{\mathrm{v}} - \frac{1}{2}\bar{a}u_{\max}\mathrm{sat}\left(\frac{\bar{k}_2}{u_{\max}}[\bar{f}(e) - x_{\mathrm{v}} + \kappa e]\right)$$

and

$$\Delta x_{\mathrm{v}} = -\kappa x_{\mathrm{v}} + \kappa^2 e + \left(1 - \frac{1}{2}\kappa\right)\bar{a}u_{\max} \cdot \mathrm{sat}\left(\frac{\bar{k}_2}{u_{\max}}[\bar{f}(e) - x_{\mathrm{v}} + \kappa e]\right)$$

Following similar lines of reasoning as in [30], we can show that, under the above new conditions, the closed-loop system is stable for the situation when the control input is saturated, *i.e.* $\left|\bar{k}_2[\bar{f}(e(k)) - \bar{v}(k) - \tilde{x}_{\mathrm{v}}(k)]\right| > u_{\max}$. When the control input does not exceed the saturation level, $\left|\bar{k}_2[\bar{f}(e(k)) - \bar{v}(k) - \tilde{x}_{\mathrm{v}}(k)]\right| \le u_{\max}$, the closed-loop system in Equation 4.77 can be written as

$$\begin{cases} e(k+1) = e(k) - \bar{v}(k) - \dfrac{1}{2}\bar{a}u_{\mathrm{P}}(k) \\ \bar{v}(k+1) = \bar{v}(k) + \bar{a}u_{\mathrm{P}}(k) \\ \tilde{x}_{\mathrm{v}}(k+1) = (1-\kappa)\tilde{x}_{\mathrm{v}}(k) \\ u_{\mathrm{P}}(k) = \bar{k}_2\left[\bar{f}(e(k)) - \bar{v}(k) - \tilde{x}_{\mathrm{v}}(k)\right] \end{cases} \tag{4.79}$$

Again, following the result of [30], we propose a Lyapunov function for the system in Equation 4.79:

$$V_{\mathrm{P}}(k) = p_{\mathrm{v}}\bar{v}^2(k) + \int_0^{e(k)} \bar{f}(\delta)\mathrm{d}\delta + p_{\mathrm{s}}\tilde{x}_{\mathrm{v}}^2(k) \tag{4.80}$$

where p_{v} and p_{s} are appropriate positive scalars to be selected later. The increase of the Lyapunov function in Equation 4.80 along the trajectory of the closed-loop system in Equation 4.79 is given by

$$\begin{aligned}
\Delta V_{\mathrm{P}}(k) &= V_{\mathrm{P}}(k+1) - V_{\mathrm{P}}(k) \\
&= p_{\mathrm{v}}\left[\bar{v}^2(k+1) - \bar{v}^2(k)\right] + \int_{e(k)}^{e(k+1)} \bar{f}(\delta)\mathrm{d}\delta + p_{\mathrm{s}}\left[\tilde{x}_{\mathrm{v}}^2(k+1) - \tilde{x}_{\mathrm{v}}^2(k)\right] \\
&= p_{\mathrm{v}}\left[\bar{v}^2(k+1) - \bar{v}^2(k)\right] + p_{\mathrm{s}}(\kappa^2 - 2\kappa)\tilde{x}_{\mathrm{v}}^2(k) \\
&\quad + \bar{f}(e(k))[e(k+1) - e(k)] + \frac{1}{2}\dot{\bar{f}}(\xi)[e(k+1) - e(k)]^2 \\
&\le p_{\mathrm{v}}\left[\bar{v}^2(k+1) - \bar{v}^2(k)\right] + p_{\mathrm{s}}\left(\kappa^2 - 2\kappa\right)\tilde{x}_{\mathrm{v}}^2(k) \\
&\quad + \bar{f}(e(k))[e(k+1) - e(k)] + c_{\xi}[e(k+1) - e(k)]^2
\end{aligned} \tag{4.81}$$

where

$$c_{\xi} := \sup\left|\frac{1}{2}\dot{\bar{f}}(\xi)\right|$$

for ξ between $e(k)$ and $e(k+1)$, $k = 0, 1, \ldots$, and the integration is expressed using the Taylor expansion as follows,

$$\begin{aligned}
\int_{e(k)}^{e(k+1)} \bar{f}(\delta)\mathrm{d}\delta &= \bar{f}(e(k))[e(k+1) - e(k)] + \frac{1}{2}\dot{\bar{f}}(\xi)[e(k+1) - e(k)]^2 \\
&\le \bar{f}(e(k))[e(k+1) - e(k)] + c_{\xi}[e(k+1) - e(k)]^2
\end{aligned} \tag{4.82}$$

By Condition 6, $c_{\xi} \in (0, 1/4)$. Letting $\vartheta = \bar{a}\bar{k}_2$, we have

$$\bar{a}u_{\mathrm{P}}(k) = \vartheta[\bar{f}(e(k)) - \bar{v}(k) - \tilde{x}_{\mathrm{v}}(k)] \tag{4.83}$$

$$\begin{aligned}
\bar{v}^2(k+1) - \bar{v}^2(k) = &(\vartheta^2 - 2\vartheta)\bar{v}^2(k) + \vartheta^2\bar{f}^2(e(k)) + \vartheta^2\tilde{x}_{\mathrm{v}}^2(k) \\
&+ (2\vartheta - 2\vartheta^2)\bar{v}(k)\bar{f}(e(k)) - 2\vartheta^2\tilde{x}_{\mathrm{v}}(k)\bar{f}(e(k)) \\
&+ (2\vartheta^2 - 2\vartheta)\bar{v}(k)\tilde{x}_{\mathrm{v}}(k)
\end{aligned} \tag{4.84}$$

$$\begin{aligned}
\bar{f}(e(k))[e(k+1) - e(k)] = &-\frac{1}{2}\vartheta\bar{f}^2(e(k)) + \left(\frac{1}{2}\vartheta - 1\right)\bar{v}(k)\bar{f}(e(k)) \\
&+ \frac{1}{2}\vartheta\tilde{x}_{\mathrm{v}}(k)\bar{f}(e(k))
\end{aligned} \tag{4.85}$$

and

$$[e(k+1)-e(k)]^2 = \left(1-\frac{1}{2}\vartheta\right)^2 \bar{v}^2(k) + \frac{1}{4}\vartheta^2 \bar{f}^2(e(k)) + \frac{1}{4}\vartheta^2 \tilde{x}_{\mathrm{v}}^2(k) + \left(\vartheta - \frac{1}{2}\vartheta^2\right) \bar{v}(k)\bar{f}(e(k)) - \frac{1}{2}\vartheta^2 \tilde{x}_{\mathrm{v}}(k)\bar{f}(e(k)) + \left(\frac{1}{2}\vartheta^2 - \vartheta\right) \bar{v}(k)\tilde{x}_{\mathrm{v}}(k) \tag{4.86}$$

Thus,

$$\begin{aligned} \Delta V_{\mathrm{P}}(k) \le & \left[p_{\mathrm{v}}(\vartheta^2 - 2\vartheta) + c_\xi \left(1-\frac{1}{2}\vartheta\right)^2\right] \bar{v}^2(k) \\ & + \left(p_{\mathrm{v}}\vartheta^2 - \frac{1}{2}\vartheta + \frac{1}{4}c_\xi\vartheta^2\right) \bar{f}^2(e(k)) \\ & + \left[p_{\mathrm{v}}\left(2\vartheta - 2\vartheta^2\right) + \left(\frac{1}{2}\vartheta - 1\right) + c_\xi\left(\vartheta - \frac{1}{2}\vartheta^2\right)\right] \bar{v}(k)\bar{f}(e(k)) \\ & + \left[p_{\mathrm{v}}\vartheta^2 + \frac{1}{4}c_\xi\vartheta^2 + p_{\mathrm{s}}\left(\kappa^2 - 2\kappa\right)\right] \tilde{x}_{\mathrm{v}}^2(k) \\ & + \left(-2p_{\mathrm{v}}\vartheta^2 + \frac{1}{2}\vartheta - \frac{1}{2}c_\xi\vartheta^2\right) \tilde{x}_{\mathrm{v}}(k)\bar{f}(e(k)) \\ & + \left[p_{\mathrm{v}}\left(2\vartheta^2 - 2\vartheta\right) + c_\xi\left(\frac{1}{2}\vartheta^2 - \vartheta\right)\right] \bar{v}(k)\tilde{x}_{\mathrm{v}}(k) \\ = & \begin{pmatrix} \bar{v}(k) \\ \tilde{x}_{\mathrm{v}}(k) \end{pmatrix}' P \begin{pmatrix} \bar{v}(k) \\ \tilde{x}_{\mathrm{v}}(k) \end{pmatrix} + \begin{pmatrix} \bar{v}(k) \\ \tilde{x}_{\mathrm{v}}(k) \end{pmatrix}' M\bar{f}(e(k)) + N\bar{f}^2(e(k)) \end{aligned} \tag{4.87}$$

where

$$P = \begin{bmatrix} p_{\mathrm{v}}\left(\vartheta^2 - 2\vartheta\right) + c_\xi\left(1-\frac{1}{2}\vartheta\right)^2 & \frac{1}{2}\left\{p_{\mathrm{v}}\left(2\vartheta^2 - 2\vartheta\right) + c_\xi\left(\frac{1}{2}\vartheta^2 - \vartheta\right)\right\} \\ \frac{1}{2}\left\{p_{\mathrm{v}}\left(2\vartheta^2 - 2\vartheta\right) + c_\xi\left(\frac{1}{2}\vartheta^2 - \vartheta\right)\right\} & p_{\mathrm{v}}\vartheta^2 + \frac{1}{4}c_\xi\vartheta^2 + p_{\mathrm{s}}\left(\kappa^2 - 2\kappa\right) \end{bmatrix}$$

$$M = \begin{bmatrix} p_{\mathrm{v}}\left(2\vartheta - 2\vartheta^2\right) + \left(\frac{1}{2}\vartheta - 1\right) + c_\xi\left(\vartheta - \frac{1}{2}\vartheta^2\right) \\ -2p_{\mathrm{v}}\vartheta^2 + \frac{1}{2}\vartheta - \frac{1}{2}c_\xi\vartheta^2 \end{bmatrix} = \begin{bmatrix} m_1 \\ m_2 \end{bmatrix}$$

and

$$N = p_{\mathrm{v}}\vartheta^2 - \frac{1}{2}\vartheta + \frac{1}{4}c_\xi\vartheta^2$$

Obviously, in order to ensure the right-hand side of Equation 4.87 to be negative-definite, we need to select p_{v} and p_{s} such that

$$P < 0, \quad N - \frac{1}{4}M'P^{-1}M \le 0 \tag{4.88}$$

In particular,

$$P = \begin{bmatrix} p_{11} & p_{12} \\ p_{12} & p_{22} \end{bmatrix} < 0$$

is equivalent to

$$p_{11} < 0, \quad p_{11}p_{22} - p_{12}^2 > 0$$

for which the first inequality implies

$$0 < \vartheta < 2, \quad p_{\mathrm{v}} > \frac{2 - \vartheta}{4\vartheta} c_{\xi} \tag{4.89}$$

and the second one is guaranteed if p_{s} is chosen to satisfy the following condition:

$$p_{\mathrm{s}} > \frac{1}{1 - (1 - \kappa)^2} \left(-\frac{p_{12}^2}{p_{11}} + p_{\mathrm{v}}\vartheta^2 + \frac{1}{4}\vartheta^2 c_{\xi} \right) \tag{4.90}$$

To ensure $N - M'P^{-1}M/4 \le 0$, it requires that

$$N \le 0, \quad (4Np_{11} - m_1^2)(4Np_{22} - m_2^2) \ge (4Np_{12} - m_1 m_2)^2 \tag{4.91}$$

Noting that

$$4Np_{11} - m_1^2 = -\left(2p_{\mathrm{v}}\vartheta - 1 + \frac{1}{2}\vartheta \right)^2$$

$$4Np_{22} - m_2^2 = 4Np_{\mathrm{s}}\left(\kappa^2 - 2\kappa\right) - \frac{1}{4}\vartheta^2$$

and

$$(4Np_{12} - m_1 m_2)^2 = \frac{1}{4}\vartheta^2 \left(2p_{\mathrm{v}}\vartheta - 1 + \frac{1}{2}\vartheta \right)^2$$

the inequalities in Equation 4.91 imply

$$Np_{\mathrm{s}}(\kappa^2 - 2\kappa) \left(2p_{\mathrm{v}}\vartheta - 1 + \frac{1}{2}\vartheta \right)^2 \le 0 \tag{4.92}$$

Recall that $\kappa \in (0, 1)$, which implies $\kappa^2 - 2\kappa < 0$, and $p_{\mathrm{s}} > 0$, $N \le 0$. It is clear that Equation 4.92 can only be satisfied when its left-hand side is equal to 0, which implies either $N = 0$ or

$$2p_{\mathrm{v}}\vartheta - 1 + \frac{1}{2}\vartheta = 0 \tag{4.93}$$

i.e.

$$p_{\mathrm{v}} = \frac{2 - \vartheta}{4\vartheta} \tag{4.94}$$

Noting that Condition 1 implies $\vartheta \in (0, 2)$ and Condition 6 implies $c_{\xi} \in (0, 1/4)$, we have

$$p_{\mathrm{v}} = \frac{2 - \vartheta}{4\vartheta} > \frac{2 - \vartheta}{4\vartheta} c_{\xi} \tag{4.95}$$

Hence, by the proper choices of p_v and p_s as in Equations 4.95 and 4.90, respectively, we have $\Delta V_\mathrm{P}(k) \le 0$. As in its continuous-time counterpart, it is straightforward to verify the closed-loop system comprising the modified PTOS control law and the given plant is asymptotically stable.

Next, we employ a reduced-order RPT control law in our MSC framework. The detailed design procedure of the RPT controller is given in Chapter 3. It is given by

$$\begin{cases} x_\mathrm{v}(k+1) = (1-\kappa)x_\mathrm{v}(k) + \kappa^2 e(k) + (1-\kappa/2)\bar{a}u_\mathrm{max} \cdot \mathrm{sat}\left(\dfrac{u_\mathrm{R}(k)}{u_\mathrm{max}}\right) \\ u_\mathrm{R}(k) = (f_1 - \kappa f_2)e(k) + f_2 x_\mathrm{v}(k) \end{cases} \tag{4.96}$$

where κ is the reduced-order observer gain and $F = [\, f_1 \quad f_2 \,]$ is the state feedback gain obtained using the RPT technique. For simplicity, we rewrite the system of Equation 4.71 as

$$x_\mathrm{P}(k+1) = A_\mathrm{P} x_\mathrm{P}(k) + B_\mathrm{P} u(k) \tag{4.97}$$

where

$$x_\mathrm{P}(k) = \begin{pmatrix} e(k) \\ \bar{v}(k) \end{pmatrix}, \quad A_\mathrm{P} = \begin{bmatrix} 1 & -1 \\ 0 & 1 \end{bmatrix}, \quad B_\mathrm{P} = \begin{bmatrix} -\bar{a}/2 \\ \bar{a} \end{bmatrix}$$

Letting $\tilde{x}_\mathrm{v}(k) = x_\mathrm{v}(k) - \bar{v}(k) - \kappa e(k)$, then the RPT controller can be expressed as

$$\tilde{x}_\mathrm{v}(k+1) = (1-\kappa)\tilde{x}_\mathrm{v}(k) \tag{4.98}$$

and

$$u_\mathrm{R}(k) = F x_\mathrm{P}(k) + f_2 \tilde{x}_\mathrm{v}(k) \tag{4.99}$$

Let $W_\mathrm{x} \in \mathbb{R}^{2\times 2}$ be a positive-definite matrix and solve the following Lyapunov equation:

$$P_\mathrm{x} = (A_\mathrm{P} + B_\mathrm{P}F)' P_\mathrm{x} (A_\mathrm{P} + B_\mathrm{P}F) + W_\mathrm{x} \tag{4.100}$$

for $P_\mathrm{x} > 0$. Note that such a P_x always exists as $A_\mathrm{P} + B_\mathrm{P}F$ is asymptotically stable. Next, letting

$$\begin{aligned} w_\mathrm{s} > \max\Big\{ & -\frac{p_{12}^2}{p_{11}} + p_\mathrm{v}\vartheta^2 + \frac{1}{4}\vartheta^2 c_\xi \\ & f_2(B_\mathrm{P}' P_\mathrm{x} B_\mathrm{P} + B_\mathrm{P}' P_\mathrm{x}(A_\mathrm{P} + B_\mathrm{P}F) W_\mathrm{x}^{-1} (A_\mathrm{P} + B_\mathrm{P}F)' P_\mathrm{x} B_\mathrm{P}) f_2 \Big\} \end{aligned} \tag{4.101}$$

and

$$p_\mathrm{s} = \frac{w_\mathrm{s}}{1-(1-\kappa)^2} > \frac{1}{1-(1-\kappa)^2}\left[-\frac{p_{12}^2}{p_{11}} + p_\mathrm{v}\vartheta^2 + \frac{1}{4}\vartheta^2 c_\xi \right] \tag{4.102}$$

we define a set

$$\boldsymbol{X} := \left\{ \begin{pmatrix} x_\mathrm{P}(k) \\ \tilde{x}_\mathrm{v}(k) \end{pmatrix} : \begin{pmatrix} x_\mathrm{P}(k) \\ \tilde{x}_\mathrm{v}(k) \end{pmatrix}' \begin{bmatrix} P_\mathrm{x} & 0 \\ 0 & p_\mathrm{s} \end{bmatrix} \begin{pmatrix} x_\mathrm{P}(k) \\ \tilde{x}_\mathrm{v}(k) \end{pmatrix} \le c \right\} \tag{4.103}$$

where $c > 0$ is the largest positive scalar such that

$$\begin{pmatrix} x_{\rm P}(k) \\ \tilde{x}_{\rm V}(k) \end{pmatrix} \in \boldsymbol{X} \quad \Rightarrow \quad \left| [F \quad f_2] \begin{pmatrix} x_{\rm P}(k) \\ \tilde{x}_{\rm V}(k) \end{pmatrix} \right| \le u_{\rm max} \tag{4.104}$$

For all $\begin{pmatrix} x_{\rm P}(k) \\ \tilde{x}_{\rm V}(k) \end{pmatrix} \in X$, the resulting closed-loop system can then be written as

$$\begin{pmatrix} x_{\rm P}(k+1) \\ \tilde{x}_{\rm V}(k+1) \end{pmatrix} = \begin{bmatrix} A_{\rm P} + B_{\rm P}F & B_{\rm P}f_2 \\ 0 & 1-\kappa \end{bmatrix} \begin{pmatrix} x_{\rm P}(k) \\ \tilde{x}_{\rm V}(k) \end{pmatrix} \tag{4.105}$$

which is clearly an asymptotically stable system. Define a Lyapunov function,

$$V_{\rm R}(k) = \begin{pmatrix} x_{\rm P}(k) \\ \tilde{x}_{\rm V}(k) \end{pmatrix}' \begin{bmatrix} P_{\rm X} & 0 \\ 0 & p_{\rm S} \end{bmatrix} \begin{pmatrix} x_{\rm P}(k) \\ \tilde{x}_{\rm V}(k) \end{pmatrix} \tag{4.106}$$

and evaluate its increase along the trajectories of the closed-loop system in Equation 4.105 as follows,

$$\begin{aligned} \Delta V_{\rm R}(k) &= \begin{pmatrix} x_{\rm P}(k) \\ \tilde{x}_{\rm V}(k) \end{pmatrix}' \begin{bmatrix} -W_{\rm X} & (A_{\rm P}+B_{\rm P}F)'P_{\rm X}B_{\rm P}f_2 \\ f_2 B_{\rm P}' P_{\rm X}(A_{\rm P}+B_{\rm P}F) & -w_{\rm S} + f_2 B_{\rm P}' P_{\rm X} B_{\rm P} f_2 \end{bmatrix} \begin{pmatrix} x_{\rm P}(k) \\ \tilde{x}_{\rm V}(k) \end{pmatrix} \\ &= \begin{pmatrix} \hat{x}(k) \\ \tilde{x}_{\rm V}(k) \end{pmatrix}' \begin{bmatrix} -W_{\rm X} & 0 \\ 0 & -\hat{w}_{\rm S} \end{bmatrix} \begin{pmatrix} \hat{x}(k) \\ \tilde{x}_{\rm V}(k) \end{pmatrix} \\ &\le 0 \end{aligned} \tag{4.107}$$

where

$$\hat{x}(k) = x_{\rm P}(k) - W_{\rm X}^{-1}(A_{\rm P}+B_{\rm P}F)'P_{\rm X}B_{\rm P}f_2\tilde{x}_{\rm V}(k)$$

and

$$\hat{w}_{\rm S} = w_{\rm S} - f_2 \Big[B_{\rm P}'P_{\rm X}B_{\rm P} + B_{\rm P}'P_{\rm X}(A_{\rm P}+B_{\rm P}F)W_{\rm X}^{-1}(A_{\rm P}+B_{\rm P}F)'P_{\rm X}B_{\rm P} \Big] f_2 > 0$$

Hence, all trajectories of the closed-loop system in Equation 4.105 starting from $\boldsymbol{X}$ remain in the set and converge asymptotically to the origin. Hence, the closed-loop system comprising the reduced-order RPT controller is asymptotically stable provided that the control input is not saturated.

Next, we re-express Equation 4.80 using the Taylor expansion as follows,

$$\begin{aligned} V_{\rm P}(k) &= p_{\rm V}\bar{v}^2(k) + \frac{1}{2}\dot{\bar{f}}(\tau)e^2(k) + p_{\rm S}\tilde{x}_{\rm V}^2(k) \\ &= \begin{pmatrix} x_{\rm P}(k) \\ \tilde{x}_{\rm V}(k) \end{pmatrix}' \begin{bmatrix} \dot{\bar{f}}(\tau)/2 & 0 & 0 \\ 0 & p_{\rm V} & 0 \\ 0 & 0 & p_{\rm S} \end{bmatrix} \begin{pmatrix} x_{\rm P}(k) \\ \tilde{x}_{\rm V}(k) \end{pmatrix} \end{aligned} \tag{4.108}$$

where τ is an appropriate scalar between 0 and $e(k)$. Let

$$\sigma = \min\left\{ \frac{1}{2}\dot{\bar{f}}(\tau), p_{\rm V}, p_{\rm S} \right\} \Big/ \max\{\lambda_{\rm max}(P_{\rm X}), p_{\rm S}\} \tag{4.109}$$

The MSC law is given by

$$u(k) = \begin{cases} u_{\mathrm{P}}(k), & k < k_{\mathrm{s}} \\ u_{\mathrm{R}}(k), & k \geq k_{\mathrm{s}} \end{cases} \tag{4.110}$$

where k_{s} is such that

$$\begin{pmatrix} x_{\mathrm{P}}(k_{\mathrm{s}}) \\ \tilde{x}_{\mathrm{V}}(k_{\mathrm{s}}) \end{pmatrix} \in \boldsymbol{X} \quad \text{and} \quad |e(k_{\mathrm{s}})| \leq y_{\ell} \tag{4.111}$$

where y_{ℓ} is the size of the linear region of the PTOS control law. The Lyapunov function for the overall closed-loop system can be chosen as

$$V(k) = V_{\mathrm{P}}(k)[1 - 1(k - k_{\mathrm{s}})] + \sigma V_{\mathrm{R}}(k)1(k - k_{\mathrm{s}}) \tag{4.112}$$

where

$$1(k - k_{\mathrm{s}}) = \begin{cases} 0, & k < k_{\mathrm{s}} \\ 1, & k \geq k_{\mathrm{s}} \end{cases}$$

It is simple to verify that

$$\begin{aligned} \Delta V(k) = {} & \Delta V_{\mathrm{P}}(k)[1 - 1(k + 1 - k_{\mathrm{s}})] + \sigma \Delta V_{\mathrm{R}}(k)1(k + 1 - k_{\mathrm{s}}) \\ & + (\sigma V_{\mathrm{R}}(k) - V_{\mathrm{P}}(k))[1(k + 1 - k_{\mathrm{s}}) - 1(k - k_{\mathrm{s}})] \end{aligned} \tag{4.113}$$

It has already been proved that the increase of the Lyapunov function $V_{\mathrm{P}}(k)$ and $V_{\mathrm{R}}(k)$ are negative-definite when they are effective, respectively. The last item is always negative in view of the definition of σ in Equation 4.109. Hence, $\Delta V(k) \leq 0$ and the resulting closed-loop system comprising the given plant and the MSC control law is asymptotically stable. Furthermore, Equation 4.111 gives the mode-switching condition for the proposed MSC scheme.

5

Composite Nonlinear Feedback Control

5.1 Introduction

The PTOS and MSC schemes discussed in the previous chapter have two controllers, *i.e.* a nonlinear controller and a linear controller, operating in two different time stages. The switching elements in these control schemes generally yield a system response not as smooth as what we would expect and thus limit the overall performance of the corresponding servo systems. Furthermore, both schemes do not pay any special attention to the transient performance of the overall design, which is in fact one of the important issues in tracking control problems. In general, quick response and small overshoot are desirable in most practical situations. However, it is well understood that quick response is often accompanied by a large overshoot. Thus, most of the design schemes have to make a tradeoff between these two transient performance indices. Inspired by a recent study of Lin *et al.* [130], which was introduced to improve the tracking performance under state feedback laws for a class of second-order systems subject to actuator saturation, we have developed in this chapter a nonlinear control technique, the so-called composite nonlinear feedback (CNF) control, for a more general class of systems with measurement feedback. It is applicable to systems with or without external disturbances.

The CNF control consists of a linear feedback law and a nonlinear feedback law without any switching element. Unlike the MSC and PTOS control laws, both the linear and nonlinear controllers in CNF are in operation all the time. The linear feedback part is designed to yield a closed-loop system with a small damping ratio for a quick response, while at the same time not exceeding the actuator limits for the desired command input levels. The nonlinear feedback law is used to increase the damping ratio of the closed-loop system as the system output approaches the target reference to reduce the overshoot caused by the linear part. We would like to note that the design philosophy of the CNF technique is very different from the commonly used antiwindup technique. The CNF control applies more control effort to the system around the final stage when the controlled output is approaching to the target reference. It focuses on reducing overshoot and speeding up its settling time.

On the other hand, the antiwindup design focuses primarily on the initial stage to reduce the effect of saturation.

We show by an example that such a technique could yield a better performance compared to that of the time-optimal control in asymptotic tracking. It is noted that the new control scheme can be utilized to design servo systems that deal with asymptotic target tracking or "point-and-shoot" fast targeting. As will be seen soon in the forthcoming chapters, this new control method has improved the performance of the overall servo system by a great deal.

Since the initiation of CNF in Lin *et al.* [130] for a class of second-order systems, there have been efforts to generalize it to more general systems. For example, Turner *et al.* [131] extended the results of [130] to higher-order and multiple-input systems. This extension was made under a restrictive assumption on the system that excludes many systems including those originally considered in [130]. Also, as in [130], only state feedback is considered in [131]. A fairly complete study dealing with general linear continuous-time systems with measurement feedback has been reported in Chen *et al.* [132], whereas its discrete-time counterpart has been reported in Venkataramanan *et al.* [133]. The technique has recently been extended to solve general multivariable systems with measurement feedback by He *et al.* [134, 135], and to solve a class of nonlinear systems in [136, 137]. The result for tackling systems with external disturbances has recently been reported by Peng *et al.* [138].

Although the CNF technique is fairly mature and complete for solving general linear systems, we would, however, restrict our attention to single-input systems in this chapter for simplicity. The results presented in this chapter are sufficient in designing HDD servo systems in the forthcoming chapters. The results for multi-input systems and nonlinear systems are rather involved. We refer interested readers to the references cited above.

The outline of this chapter is as follows: In Section 5.2, we present the technique of CNF control for general single-input continuous-time systems without and with external disturbances. The discrete-time version is given in Section 5.3. Section 5.4 shows in a numerical example that the CNF control is capable of outperforming the time-optimal control in asymptotic tracking situations. Finally, in Section 5.5, we present a user-friendly software toolkit for the CNF control design for continuous-time systems with and without disturbances. The toolkit is written in MATLAB® together with its simulation package Simulink®. It is available for free downloading at the website `http://hdd.ece.nus.edu.sg/~bmchen`.

5.2 Continuous-time Systems

We first present the result for systems without external disturbances and then extend it to the case when the given system has external disturbances. For the latter, an integrator is introduced in the controller to remove or minimize the steady-state bias due to unknown constant disturbances. The selection of nonlinear design parameters is also presented together with an interpretation of the nonlinear gain introduced in the CNF design under the well-known framework of the classical root-locus theory.

5.2.1 Systems without External Disturbances

Consider a linear continuous-time system Σ with an amplitude-constrained actuator characterized by

$$\begin{cases} \dot{x} = A\ x + B\,\mathrm{sat}(u), & x(0) = x_0 \\ y = C_1\ x \\ h = C_2\ x \end{cases} \tag{5.1}$$

where $x \in \mathbb{R}^n$, $u \in \mathbb{R}$, $y \in \mathbb{R}^p$ and $h \in \mathbb{R}$ are, respectively, the state, control input, measurement output and controlled output of Σ. A, B, C_1 and C_2 are appropriate dimensional constant matrices, and sat: $\mathbb{R} \to \mathbb{R}$ represents the actuator saturation defined as

$$\mathrm{sat}(u) = \mathrm{sgn}(u)\min\{\ u_{\max},\ |\ u\ |\ \} \tag{5.2}$$

with $u_{\max}$ being the saturation level of the input. The following assumptions on the system matrices are required:

1. (A, B) is stabilizable,
2. (A, C_1) is detectable, and
3. (A, B, C_2) is invertible and has no invariant zeros at $s = 0$.

The objective here is to design a CNF control law that causes the output to track a high-amplitude step input rapidly without experiencing large overshoot and without the adverse actuator saturation effects. This is done through the design of a linear feedback law with a small closed-loop damping ratio and a nonlinear feedback law through an appropriate Lyapunov function to cause the closed-loop system to be highly damped as the system output approaches the command input to reduce the overshoot. Similar to the design of linear controllers discussed in Chapter 4, we separate the CNF controller design into three distinct situations 1) the state feedback case, 2) the full-order measurement feedback case, and 3) the reduced-order measurement feedback case.

i. State Feedback Case. We proceed to develop a composite nonlinear feedback control technique for the case when all the states of the plant Σ are measurable, *i.e.* $y = x$. It is done in three steps. In the first step, a linear feedback control law is designed; in the second step, the design of nonlinear feedback control is carried out. Lastly, in the third step, the linear and nonlinear feedback laws are combined to give a composite nonlinear feedback control law. Again, we note that the procedure given for this case follows closely from that reported in Lin *et al.* [130], although our result is applicable to a much larger class of systems.

STEP 5.C.S.1: design a linear feedback law,

$$u_{\mathrm{L}} = Fx + Gr \tag{5.3}$$

where F is chosen such that 1) $A + BF$ is an asymptotically stable matrix, and 2) the closed-loop system $C_2(sI - A - BF)^{-1}B$ has certain desired properties, *e.g.*, having a small damping ratio. We note that such an F can be designed using methods such as the H_2 and H_∞ optimization approaches, as well as the RPT technique given in Chapter 3. Furthermore, G is a scalar and is given by

$$G = -\left[C_2(A + BF)^{-1}B\right]^{-1} \tag{5.4}$$

and r is a step command input. Here we note that G is well defined because $A + BF$ is stable, and the triple (A, B, C_2) is invertible and has no invariant zeros at $s = 0$.

The following lemma determines the magnitude of r that can be tracked by such a control law without exceeding the control limits.

Lemma 5.1. *Given a positive-definite matrix $W \in \mathbb{R}^{n\times n}$, let $P > 0$ be the solution of the following Lyapunov equation:*

$$(A + BF)'P + P(A + BF) = -W \tag{5.5}$$

Such a P exists since $A+BF$ is asymptotically stable. For any $\delta \in (0, 1)$, let $c_\delta > 0$ be the largest positive scalar satisfying the following condition:

$$|\, Fx \,| \le u_{\max}(1 - \delta), \quad \forall x \in \boldsymbol{X}_\delta := \left\{x : x'Px \le c_\delta\right\} \tag{5.6}$$

Also, let

$$H := \left[1 - F(A + BF)^{-1}B\right] G \tag{5.7}$$

and

$$x_{\rm e} := G_{\rm e}\, r := -(A + BF)^{-1}BG\, r \tag{5.8}$$

Then, the control law of Equation 5.3 is capable of driving the controlled output h to track asymptotically a step command input r, provided that the initial state x_0 and r satisfy:

$$\tilde{x}_0 := (x_0 - x_{\rm e}) \in \boldsymbol{X}_\delta \quad \textit{and} \quad |H\, r| \le \delta u_{\max} \tag{5.9}$$

Proof. Let us first define a new state variable

$$\tilde{x} = x - x_{\rm e} \tag{5.10}$$

It is simple to verify that the linear feedback control law of Equation 5.3 can be rewritten as

$$u_{\rm L}(t) = F\tilde{x}(t) + \left[1 - F(A + BF)^{-1}B\right] Gr = F\tilde{x}(t) + H\, r \tag{5.11}$$

and hence for all $\tilde{x} \in \boldsymbol{X}_\delta$ and, provided that $|H\, r| \le \delta u_{\max}$, the closed-loop system is linear and is given by

$$\dot{\tilde{x}} = (A + BF)\tilde{x} + Ax_{\rm e} + BH\, r \tag{5.12}$$

Noting that

$$
\begin{aligned}
Ax_{\rm e} + BHr &= \Big\{B[1-F(A+BF)^{-1}B]G - A(A+BF)^{-1}BG\Big\}r \\
&= \Big\{[I-BF(A+BF)^{-1}]BG - A(A+BF)^{-1}BG\Big\}r \\
&= \Big\{I-BF(A+BF)^{-1} - A(A+BF)^{-1}\Big\}BGr = 0
\end{aligned} \tag{5.13}
$$

the closed-loop system in Equation 5.12 can then be simplified as

$$\dot{\tilde{x}} = (A+BF)\tilde{x} \tag{5.14}$$

Next, we define a Lyapunov function:

$$V(\tilde{x}) = \tilde{x}'P\tilde{x} \tag{5.15}$$

Along the trajectories of the closed-loop system in Equation 5.14, $V(\tilde{x})$ satisfies

$$\dot{V}(\tilde{x}) = -\tilde{x}'W\tilde{x} \tag{5.16}$$

which implies that $V(\tilde{x})$ is a monotonically decreasing function with respect to t along the trajectories of Equation 5.14. Thus, all trajectories of Equation 5.14 starting from $\boldsymbol{X}_\delta$ remain there and converge asymptotically to the origin. For an initial state x_0 and the step command input r that satisfy Equation 5.9, we have

$$\lim_{t\to\infty} x(t) = x_{\rm e} \tag{5.17}$$

and hence

$$\lim_{t\to\infty} h(t) = \lim_{t\to\infty} C_2 x(t) = C_2 x_{\rm e} = -C_2(A+BF)^{-1}BGr = r \tag{5.18}$$

This completes the proof of Lemma 5.1. ◇

Remark 5.2. We would like to note that for the case when $x_0 = 0$, any step command of amplitude r can be asymptotically tracked if

$$|\, r \,| \le \left[c_\delta (G_{\rm e}'PG_{\rm e})^{-1}\right]^{1/2} \quad \text{and} \quad |\, Hr \,| \le \delta \cdot u_{\max} \tag{5.19}$$

Clearly, the trackable amplitudes of reference inputs by the linear feedback control law can be increased by increasing δ and/or decreasing $G_{\rm e}'PG_{\rm e}$ through the choice of W. However, the change in gain F indeed affects the damping ratio of the closed-loop system and hence its rise time. ◇

STEP 5.C.S.2: the nonlinear feedback control law $u_{\rm N}(t)$ is given by

$$u_{\rm N} = \rho(r,y)B'P(x - x_{\rm e}) \tag{5.20}$$

where $\rho(r,y)$ is any nonpositive function locally Lipschitz in y, which is used to change the system closed-loop damping ratio as the output approaches the step command input. The choice of such a $\rho(r,y)$ is to be discussed later.

STEP 5.C.S.3: the linear and nonlinear feedback laws derived in the previous steps are now combined to form a CNF controller:

$$u = u_{\rm L} + u_{\rm N} = Fx + Gr + \rho(r,y)B'P(x - x_{\rm e}) \tag{5.21}$$

We now move to prove that the closed-loop system comprising the given plant in Equation 5.1 and the CNF control law of Equation 5.21 is asymptotically stable.

Theorem 5.3. *Consider the given system in Equation 5.1. Then, for any nonpositive function $\rho(r,y)$, locally Lipschitz in y, the composite nonlinear feedback law in Equation 5.21 internally stabilizes the given plant and drive the system controlled output $h(t)$ to track asymptotically the step command input of amplitude r from an initial state x_0, provided that x_0 and r satisfy Equation 5.9.* ◇

Proof. Again, let $\tilde{x} = x - x_{\rm e}$. Then, the closed-loop system comprising the given plant in Equation 5.1 and the CNF control law of Equation 5.21 can be expressed as

$$\dot{\tilde{x}} = (A + BF)\tilde{x} + Bw \tag{5.22}$$

where

$$w = \text{sat}\,(F\tilde{x} + H\,r + u_{\rm N}) - F\tilde{x} - H\,r \tag{5.23}$$

Clearly, for the given x_0 satisfying Equation 5.9, we have $\tilde{x}_0 = (x_0 - x_{\rm e}) \in \boldsymbol{X}_\delta$. Using the Lyapunov function

$$V = \tilde{x}'P\tilde{x} \tag{5.24}$$

we can evaluate the derivative of V along the trajectories of the closed-loop system in Equation 5.22, *i.e.*

$$\begin{aligned} \dot{V} &= \dot{\tilde{x}}'P\tilde{x} + \tilde{x}'P\dot{\tilde{x}} \\ &= \tilde{x}'(A + BF)'P\tilde{x} + \tilde{x}'P(A + BF)\tilde{x} + 2\tilde{x}'PBw \\ &= -\tilde{x}'W\tilde{x} + 2\tilde{x}'PBw \end{aligned} \tag{5.25}$$

Note that for all

$$\tilde{x} \in \boldsymbol{X}_\delta = \{\tilde{x} : \tilde{x}'P\tilde{x} \le c_\delta\} \ \Rightarrow\ |\, F\tilde{x} \,| \le u_{\max}(1 - \delta) \tag{5.26}$$

We next calculate $\dot{V}$ for three different values of saturation function.

Case 1. If $|F\tilde{x} + Hr + u_{\rm N}| \le u_{\max}$, then $w = u_{\rm N} = \rho B'P\tilde{x}$ and thus

$$\dot{V} = -\tilde{x}'W\tilde{x} + 2\rho\tilde{x}'PBB'P\tilde{x} \le -\tilde{x}'W\tilde{x} \tag{5.27}$$

Case 2. If $F\tilde{x} + Hr + u_{\rm N} > u_{\max}$, and by construction $|\, F\tilde{x} + Hr \,| \le u_{\max}$, we have

$$0 < w = u_{\max} - F\tilde{x} - Hr < u_{\rm N} = \rho B'P\tilde{x} \tag{5.28}$$

which implies that $\tilde{x}'PB < 0$ and hence $\dot{V} = -\tilde{x}'W\tilde{x} + 2\tilde{x}'PBw \le -\tilde{x}'W\tilde{x}$.

Case 3. Finally, if $F\tilde{x} + Hr + u_{\rm N} < -u_{\max}$, we have

$$\rho B' P \tilde{x} = u_{\rm N} < w = -u_{\max} - F\tilde{x} - Hr < 0 \tag{5.29}$$

implying $\tilde{x}' PB > 0$ and hence $\dot{V} \leq -\tilde{x}' W \tilde{x}$.

In conclusion, we have shown that

$$\dot{V} \leq -\tilde{x}' W \tilde{x}, \quad \tilde{x} \in \boldsymbol{X}_{\delta} \tag{5.30}$$

which implies that $\boldsymbol{X}_{\delta}$ is an invariant set of the closed-loop system in Equation 5.22 and all trajectories of Equation 5.22 starting from inside $\boldsymbol{X}_{\delta}$ converge to the origin. This, in turn, indicates that, for all initial states x_0 and the step command input of amplitude r that satisfy Equation 5.9,

$$\lim_{t \to \infty} x(t) = x_{\rm e} \tag{5.31}$$

Therefore,

$$\lim_{t \to \infty} h(t) = \lim_{t \to \infty} C_2 x(t) = C_2 x_{\rm e} = r \tag{5.32}$$

This completes the proof of Theorem 5.3. ◇

Remark 5.4. Theorem 5.3 shows that the additional nonlinear feedback control law $u_{\rm N}$, as given in Equation 5.20, does not affect the ability of the closed-loop system to track the command input. Any command input that can be asymptotically tracked by the linear feedback law of Equation 5.3 can also be asymptotically tracked by the CNF control law in Equation 5.21. However, this additional term $u_{\rm N}$ in the CNF control law can be used to improve the performance of the overall closed-loop system. This is the key property of the CNF control technique. ◇

ii. Full-order Measurement Feedback Case. The assumption that all the states of Σ are measurable is, in general, not practical. For example, in HDD servo systems, the velocity of the actuator is not usually directly measurable. Thus, one has to implement the PTOS controller or the control law obtained in the previous case via a certain velocity estimation. In what follows, we proceed to develop a CNF design using only measurement information.

STEP 5.C.F.1: we first construct a linear full-order measurement feedback control law:

$$\Sigma_{\rm F} : \begin{cases} \dot{x}_{\rm v} = (A + KC_1)x_{\rm v} - Ky + B\,\mathrm{sat}(u_{\rm L}) \\ u_{\rm L} = F(x_{\rm v} - x_{\rm e}) + Hr \end{cases} \tag{5.33}$$

where r is the reference input and $x_{\rm v} \in \mathbb{R}^n$ is the state of the controller. As usual, F and K are gain matrices and are designed such that $A + BF$ and $A + KC_1$ are asymptotically stable and the resulting closed-loop system has the desired properties. As defined in Equations 5.7 and 5.8,

$$H = [1 - F(A + BF)^{-1}B]G, \quad G = -[C_2(A + BF)^{-1}B]^{-1} \tag{5.34}$$

and $x_{\rm e} = -(A + BF)^{-1}BGr$.

We have the following result.

Lemma 5.5. *Given a positive-definite matrix $W_P \in \mathbb{R}^{n\times n}$, let $P > 0$ be the solution to the Lyapunov equation*

$$(A+BF)'P + P(A+BF) = -W_P \tag{5.35}$$

Given another positive-definite matrix $W_Q \in \mathbb{R}^{n\times n}$ with

$$W_Q > F'B'PW_P^{-1}PBF \tag{5.36}$$

let $Q > 0$ be the solution to the Lyapunov equation

$$(A+KC_1)'Q + Q(A+KC_1) = -W_Q \tag{5.37}$$

Note that such P and Q exist as $A+BF$ and $A+KC_1$ are asymptotically stable. For any $\delta \in (0,1)$, let c_δ be the largest positive scalar such that for all

$$\begin{pmatrix} x \\ x_v \end{pmatrix} \in \boldsymbol{X}_{F\delta} := \left\{ \begin{pmatrix} x \\ x_v \end{pmatrix} : \begin{pmatrix} x \\ x_v \end{pmatrix}' \begin{bmatrix} P & 0 \\ 0 & Q \end{bmatrix} \begin{pmatrix} x \\ x_v \end{pmatrix} \le c_\delta \right\} \tag{5.38}$$

we have

$$\left| [F \quad F] \begin{pmatrix} x \\ x_v \end{pmatrix} \right| \le u_{\max}(1-\delta) \tag{5.39}$$

The linear control law in Equation 5.33 drives the system controlled output $h(t)$ to track asymptotically a step command input of amplitude r from an initial state x_0, provided that x_0, $x_{v0} = x_v(0)$ and r satisfy:

$$|Hr| \le \delta \cdot u_{\max} \quad \textit{and} \quad \begin{pmatrix} x_0 - x_e \\ x_{v0} - x_0 \end{pmatrix} \in \boldsymbol{X}_{F\delta} \tag{5.40}$$

Proof. Let us transform the system coordinate by defining

$$\tilde{x} = x - x_e \quad \text{and} \quad \tilde{x}_v = x_v - x \tag{5.41}$$

Then, the linear control law of Equation 5.33 can be written as

$$\dot{\tilde{x}}_v = (A+KC_1)\tilde{x}_v, \quad u_L = [F \quad F] \begin{pmatrix} \tilde{x} \\ \tilde{x}_v \end{pmatrix} + Hr \tag{5.42}$$

Hence, for all states

$$\begin{pmatrix} \tilde{x} \\ \tilde{x}_v \end{pmatrix} \in \boldsymbol{X}_{F\delta} \Rightarrow \left| [F \quad F] \begin{pmatrix} \tilde{x} \\ \tilde{x}_v \end{pmatrix} \right| \le u_{\max}(1-\delta) \tag{5.43}$$

and for any r satisfying

$$|Hr| \le \delta \cdot u_{\max} \tag{5.44}$$

we have

$$|u_L| = \left| [F \quad F] \begin{pmatrix} \tilde{x} \\ \tilde{x}_v \end{pmatrix} + Hr \right| \le \left| [F \quad F] \begin{pmatrix} \tilde{x} \\ \tilde{x}_v \end{pmatrix} \right| + |Hr| = u_{\max} \tag{5.45}$$

Thus, for all $\tilde{x}$ and $\tilde{x}_{\rm v}$ satisfying the condition as given in Equation 5.43, the closed-loop system comprising the given plant and the linear control law of Equation 5.33 can be rewritten as

$$\begin{pmatrix} \dot{\tilde{x}} \\ \dot{\tilde{x}}_{\rm v} \end{pmatrix} = \begin{bmatrix} A+BF & BF \\ 0 & A+KC_1 \end{bmatrix} \begin{pmatrix} \tilde{x} \\ \tilde{x}_{\rm v} \end{pmatrix} \tag{5.46}$$

Next, we define a Lyapunov function for the closed-loop system in Equation 5.46:

$$V = \begin{pmatrix} \tilde{x} \\ \tilde{x}_{\rm v} \end{pmatrix}' \begin{bmatrix} P & 0 \\ 0 & Q \end{bmatrix} \begin{pmatrix} \tilde{x} \\ \tilde{x}_{\rm v} \end{pmatrix} \tag{5.47}$$

Along the trajectories of the closed-loop system in Equation 5.46 the derivative of the Lyapunov function is given by

$$\begin{aligned} \dot{V} &= \begin{pmatrix} \tilde{x} \\ \tilde{x}_{\rm v} \end{pmatrix}' \begin{bmatrix} -W_{\rm P} & PBF \\ F'B'P & -W_{\rm Q} \end{bmatrix} \begin{pmatrix} \tilde{x} \\ \tilde{x}_{\rm v} \end{pmatrix} \\ &= \begin{pmatrix} \hat{x} \\ \tilde{x}_{\rm v} \end{pmatrix}' \begin{bmatrix} -W_{\rm P} & 0 \\ 0 & -\tilde{W}_{\rm Q} \end{bmatrix} \begin{pmatrix} \hat{x} \\ \tilde{x}_{\rm v} \end{pmatrix} \end{aligned} \tag{5.48}$$

where

$$\hat{x} = \tilde{x} - W_{\rm P}^{-1} PBF\tilde{x}_{\rm v}, \quad \tilde{W}_{\rm Q} = W_{\rm Q} - F'B'PW_{\rm P}^{-1}PBF \tag{5.49}$$

With the choice of $W_{\rm Q}$ satisfying Equation 5.36, it is obvious that $\dot{V} \le 0$. This shows that $\boldsymbol{X}_{\rm F\delta}$ is an invariant set of the closed-loop system in Equation 5.46 and all trajectories starting from the set converge asymptotically to the origin. Thus, for the initial states of x_0 and $x_{\rm v0}$ and step command inputs that satisfy Equation 5.40,

$$\lim_{t\to\infty} \tilde{x}_{\rm v}(t) = 0 \quad \text{and} \quad \lim_{t\to\infty} x(t) = x_{\rm e} \tag{5.50}$$

which imply

$$\lim_{t\to\infty} h(t) = \lim_{t\to\infty} C_2 x(t) = C_2 x_{\rm e} = r \tag{5.51}$$

This completes the proof of Lemma 5.5. ◇

STEP 5.C.F.2: as in the state feedback case, the linear control law of Equation 5.33 obtained in the above step is to be combined with a nonlinear control law to form the following CNF controller:

$$\begin{cases} \dot{x}_{\rm v} = (A+KC_1)x_{\rm v} - Ky + B\,\mathrm{sat}(u) \\ u = F(x_{\rm v} - x_{\rm e}) + Hr + \rho(r,y)B'P(x_{\rm v} - x_{\rm e}) \end{cases} \tag{5.52}$$

where $\rho(r,y)$ is a nonpositive scalar function, locally Lipschitz in y, and is to be chosen to improve the performance of the closed-loop system.

It turns out that, for the measurement feedback case, the choice of $\rho(r,y)$, the nonpositive scalar function, is not totally free. It is subject to certain constraints. We have the following theorem.

Theorem 5.6. *Consider the given system in Equation 5.1. Then, there exists a scalar $\rho^* > 0$ such that for any nonpositive function $\rho(r,y)$, locally Lipschitz in y and $|\rho(r,y)| \le \rho^*$, the CNF control law of Equation 5.52 internally stabilizes the given plant and drive the system controlled output $h(t)$ to track asymptotically the step command input of amplitude r from an initial state x_0, provided that x_0, x_{v0} and r satisfy Equation 5.40.* ◇

Proof. Again, let $\tilde{x} = x - x_e$ and $\tilde{x}_v = x_v - x$. For simplicity, we drop r and h in $\rho(r,y)$ throughout the rest of the proof of this theorem. Then, the closed-loop system with the CNF control law of Equation 5.52 can be expressed as

$$\begin{pmatrix} \dot{\tilde{x}} \\ \dot{\tilde{x}}_v \end{pmatrix} = \begin{bmatrix} A+BF & BF \\ 0 & A+KC_1 \end{bmatrix} \begin{pmatrix} \tilde{x} \\ \tilde{x}_v \end{pmatrix} + \begin{bmatrix} B \\ 0 \end{bmatrix} w \tag{5.53}$$

where

$$w = \text{sat}\left[[F \quad F] \begin{pmatrix} \tilde{x} \\ \tilde{x}_v \end{pmatrix} + Hr + \rho [B'P \quad B'P] \begin{pmatrix} \tilde{x} \\ \tilde{x}_v \end{pmatrix} \right] - [F \quad F] \begin{pmatrix} \tilde{x} \\ \tilde{x}_v \end{pmatrix} - Hr \tag{5.54}$$

Clearly, for the given x_0 and x_{v0} satisfying Equation 5.40, we have

$$\begin{pmatrix} \tilde{x}(0) \\ \tilde{x}_v(0) \end{pmatrix} \in \boldsymbol{X}_{F\delta} \tag{5.55}$$

Using the following Lyapunov function:

$$V = \begin{pmatrix} \tilde{x} \\ \tilde{x}_v \end{pmatrix}' \begin{bmatrix} P & 0 \\ 0 & Q \end{bmatrix} \begin{pmatrix} \tilde{x} \\ \tilde{x}_v \end{pmatrix} \tag{5.56}$$

we evaluate the derivative of V along the trajectories of the closed-loop system in Equation 5.53, *i.e.*

$$\dot{V} = \begin{pmatrix} \tilde{x} \\ \tilde{x}_v \end{pmatrix}' \begin{bmatrix} -W_P & PBF \\ F'B'P & -W_Q \end{bmatrix} \begin{pmatrix} \tilde{x} \\ \tilde{x}_v \end{pmatrix} + 2\tilde{x}'PBw \tag{5.57}$$

Note that for all

$$\begin{pmatrix} \tilde{x} \\ \tilde{x}_v \end{pmatrix} \in \boldsymbol{X}_{F\delta} \Rightarrow \left| [F \quad F] \begin{pmatrix} \tilde{x} \\ \tilde{x}_v \end{pmatrix} \right| \le u_{\max}(1-\delta) \tag{5.58}$$

Again, as is done in the full state feedback case, let us find the above derivative of V for three different cases.

Case 1. If

$$\left| [F \quad F] \begin{pmatrix} \tilde{x} \\ \tilde{x}_v \end{pmatrix} + Hr + \rho [B'P \quad B'P] \begin{pmatrix} \tilde{x} \\ \tilde{x}_v \end{pmatrix} \right| \le u_{\max} \tag{5.59}$$

then

$$w = \rho\,[\,B'P \quad B'P\,]\begin{pmatrix} \tilde{x} \\ \tilde{x}_{\mathrm{V}} \end{pmatrix} \tag{5.60}$$

which implies

$$\begin{aligned}\dot{V} &= \begin{pmatrix} \tilde{x} \\ \tilde{x}_{\mathrm{V}} \end{pmatrix}' \begin{bmatrix} -W_{\mathrm{P}} & PB(F+\rho B'P) \\ (F+\rho B'P)'B'P & -W_{\mathrm{Q}} \end{bmatrix} \begin{pmatrix} \tilde{x} \\ \tilde{x}_{\mathrm{V}} \end{pmatrix} + 2\rho\tilde{x}'PBB'P\tilde{x} \\ &\leq \begin{pmatrix} \hat{x} \\ \tilde{x}_{\mathrm{V}} \end{pmatrix}' \begin{bmatrix} -W_{\mathrm{P}} & 0 \\ 0 & -\tilde{W}_{\mathrm{Q}} \end{bmatrix} \begin{pmatrix} \hat{x} \\ \tilde{x}_{\mathrm{V}} \end{pmatrix}\end{aligned} \tag{5.61}$$

where

$$\hat{x} = \tilde{x} - W_{\mathrm{P}}^{-1}PB(F+\rho B'P)\tilde{x}_{\mathrm{V}}, \tag{5.62}$$

$$\tilde{W}_{\mathrm{Q}} = W_{\mathrm{Q}} - (F+\rho B'P)'B'PW_{\mathrm{P}}^{-1}PB(F+\rho B'P) \tag{5.63}$$

Noting Equation 5.36, *i.e.* $W_{\mathrm{Q}} > F'B'PW_{\mathrm{P}}^{-1}PBF$, and $\rho(r,y)$ is locally Lipschitz, it is clear that there exists a $\rho_1^* > 0$ such that for any scalar function with $|\rho(r,y)| \leq \rho_1^*$ we have $\tilde{W}_{\mathrm{Q}} > 0$ and hence $\dot{V} \leq 0$.

Case 2. If

$$[\,F \quad F\,]\begin{pmatrix} \tilde{x} \\ \tilde{x}_{\mathrm{V}} \end{pmatrix} + Hr + \rho\,[\,B'P \quad B'P\,]\begin{pmatrix} \tilde{x} \\ \tilde{x}_{\mathrm{V}} \end{pmatrix} > u_{\max} \tag{5.64}$$

then for the trajectories inside $\boldsymbol{X}_{\mathrm{F}\delta}$,

$$\left|[\,F \quad F\,]\begin{pmatrix} \tilde{x} \\ \tilde{x}_{\mathrm{V}} \end{pmatrix} + Hr\right| \leq u_{\max} \tag{5.65}$$

which implies that

$$0 < w < \rho\,[\,B'P \quad B'P\,]\begin{pmatrix} \tilde{x} \\ \tilde{x}_{\mathrm{V}} \end{pmatrix} \tag{5.66}$$

Next, let us express

$$w = q\rho\,[\,B'P \quad B'P\,]\begin{pmatrix} \tilde{x} \\ \tilde{x}_{\mathrm{V}} \end{pmatrix} \tag{5.67}$$

for an appropriate positive piecewise continuous function $q(t)$, bounded by 1 for all t. In this case, the derivative of V becomes

$$\begin{aligned}\dot{V} &= \begin{pmatrix} \tilde{x} \\ \tilde{x}_{\mathrm{V}} \end{pmatrix}' \begin{bmatrix} -W_{\mathrm{P}} & PB(F+q\rho B'P) \\ (F+q\rho B'P)'B'P & -W_{\mathrm{Q}} \end{bmatrix} \begin{pmatrix} \tilde{x} \\ \tilde{x}_{\mathrm{V}} \end{pmatrix} + 2q\rho\tilde{x}'PBB'P\tilde{x} \\ &\leq \begin{pmatrix} \hat{x}_+ \\ \tilde{x}_{\mathrm{V}} \end{pmatrix}' \begin{bmatrix} -W_{\mathrm{P}} & 0 \\ 0 & -\tilde{W}_{\mathrm{Q}+} \end{bmatrix} \begin{pmatrix} \hat{x}_+ \\ \tilde{x}_{\mathrm{V}} \end{pmatrix}\end{aligned} \tag{5.68}$$

where

$$\hat{x}_+ = \tilde{x} - W_{\mathrm{P}}^{-1} PB(F + q\rho B'P)\tilde{x}_{\mathrm{v}} \tag{5.69}$$

$$\tilde{W}_{\mathrm{Q}+} = W_{\mathrm{Q}} - (F + q\rho B'P)'B'PW_{\mathrm{P}}^{-1}PB(F + q\rho B'P) \tag{5.70}$$

Again, noting Equation 5.36, it can be shown that there exists a ρ_2^* such that for any $\rho(r, y)$ satisfying $|\rho(r, y)| \le \rho_2^*$ we have $\tilde{W}_{\mathrm{Q}+} > 0$ and hence $\dot{V} \le 0$.

Case 3. Similarly, for the case when

$$[F \quad F]\begin{pmatrix} \tilde{x} \\ \tilde{x}_{\mathrm{v}} \end{pmatrix} + Hr + \rho[B'P \quad B'P]\begin{pmatrix} \tilde{x} \\ \tilde{x}_{\mathrm{v}} \end{pmatrix} < -u_{\max} \tag{5.71}$$

we can show that there exists a $\rho_3^* > 0$ such that for any $\rho(r, y)$ satisfying $|\rho(r, y)| \le \rho_3^*$, we have $\dot{V} \le 0$ for all the trajectories in $\boldsymbol{X}_{\mathrm{F}\delta}$.

Finally, let $\rho^* = \min\{\rho_1^*, \rho_2^*, \rho_3^*\}$. Then, we have for any nonpositive scalar function $\rho(r, y)$ satisfying $|\rho(r, y)| \le \rho^*$,

$$\dot{V} \le 0, \quad \forall \begin{pmatrix} \tilde{x} \\ \tilde{x}_{\mathrm{v}} \end{pmatrix} \in \boldsymbol{X}_{\mathrm{F}\delta} \tag{5.72}$$

Thus, $\boldsymbol{X}_{\mathrm{F}\delta}$ is an invariant set of the closed-loop system in Equation 5.53, and all trajectories starting from $\boldsymbol{X}_{\mathrm{F}\delta}$ remain inside and asymptotically converge to the origin. This, in turn, indicates that, for the initial state of the given system x_0, the initial state of the controller $x_{\mathrm{v}0}$, and step command input r that satisfy Equation 5.40,

$$\lim_{t\to\infty} \tilde{x}_{\mathrm{v}}(t) = 0 \quad \text{and} \quad \lim_{t\to\infty} x(t) = x_{\mathrm{e}} \tag{5.73}$$

and hence

$$\lim_{t\to\infty} h(t) = \lim_{t\to\infty} C_2 x(t) = C_2 x_{\mathrm{e}} = r \tag{5.74}$$

This completes the proof of Theorem 5.6. ◊

iii. Reduced-order Measurement Feedback Case. For the given system in Equation 5.1, it is clear that there are p states of the system measurable if C_1 is of maximal rank. Thus, in general, it is not necessary to estimate these measurable states in measurement feedback laws. As such, we design a dynamic controller that has a dynamical order less than that of the given plant. We now proceed to construct such a control law under the CNF control framework.

For simplicity of presentation, we assume that C_1 is already in the form

$$C_1 = [I_p \quad 0] \tag{5.75}$$

Then, the system in Equation 5.1 can be rewritten as

$$\begin{cases} \begin{pmatrix} \dot{x}_1 \\ \dot{x}_2 \end{pmatrix} = \begin{bmatrix} A_{11} & A_{12} \\ A_{21} & A_{22} \end{bmatrix} \begin{pmatrix} x_1 \\ x_2 \end{pmatrix} + \begin{bmatrix} B_1 \\ B_2 \end{bmatrix} \mathrm{sat}(u), \ x_0 = \begin{pmatrix} x_{10} \\ x_{20} \end{pmatrix} \\ y = [\ I_p \quad 0\ ] \begin{pmatrix} x_1 \\ x_2 \end{pmatrix} \\ h = C_2 \begin{pmatrix} x_1 \\ x_2 \end{pmatrix} \end{cases} \tag{5.76}$$

where the original state x is partitioned into two parts, x_1 and x_2 with $y \equiv x_1$. Thus, we only need to estimate x_2 in the reduced-order measurement feedback design. Next, we let F be chosen such that 1) $A + BF$ is asymptotically stable, and 2) $C_2(sI - A - BF)^{-1}B$ has the desired properties, and let K_R be chosen such that $A_{22} + K_R A_{12}$ is asymptotically stable. Here, we note that it was shown in Chen [110] that (A_{22}, A_{12}) is detectable if and only if (A, C_1) is detectable. Thus, there exists a stabilizing K_R. Again, such F and K_R can be designed using any of the linear control techniques presented in Chapter 3. We then partition F in conformity with x_1 and x_2:

$$F = [\, F_1 \quad F_2 \,] \tag{5.77}$$

As defined in Equations 5.7 and 5.8,

$$H = [1 - F(A + BF)^{-1}B]G, \quad G = -[C_2(A + BF)^{-1}B]^{-1} \tag{5.78}$$

and $x_e = -(A + BF)^{-1}BGr$.

The reduced-order CNF controller is given by

$$\begin{aligned}\dot{x}_v = (A_{22} + K_R A_{12})x_v + \Big[A_{21} + K_R A_{11} - (A_{22} + K_R A_{12})K_R\Big]y \\ + (B_2 + K_R B_1)\,\mathrm{sat}(u)\end{aligned} \tag{5.79}$$

and

$$u = F\left[\begin{pmatrix} y \\ x_v - K_R y \end{pmatrix} - x_e\right] + Hr + \rho(r, y)B'P\left[\begin{pmatrix} y \\ x_v - K_R y \end{pmatrix} - x_e\right] \tag{5.80}$$

where $\rho(r, y)$ is a nonpositive scalar function locally Lipschitz in y subject to certain constraints to be discussed later.

Next, given a positive-definite matrix $W_P \in \mathbb{R}^{n \times n}$, let $P > 0$ be the solution to the Lyapunov equation

$$(A + BF)'P + P(A + BF) = -W_P \tag{5.81}$$

Given another positive-definite matrix $W_R \in \mathbb{R}^{(n-p)\times(n-p)}$ with

$$W_R > F_2' B' P W_P^{-1} P B F_2 \tag{5.82}$$

let $Q_R > 0$ be the solution to the Lyapunov equation

$$(A_{22} + K_R A_{12})'Q_R + Q_R(A_{22} + K_R A_{12}) = -W_R \tag{5.83}$$

Note that such P and Q_R exist as $A + BF$ and $A_{22} + K_R A_{12}$ are asymptotically stable. For any $\delta \in (0, 1)$, let c_δ be the largest positive scalar such that for all

$$\begin{pmatrix} x \\ x_v \end{pmatrix} \in \boldsymbol{X}_{R\delta} := \left\{ \begin{pmatrix} x \\ x_v \end{pmatrix} : \begin{pmatrix} x \\ x_v \end{pmatrix}' \begin{bmatrix} P & 0 \\ 0 & Q_R \end{bmatrix} \begin{pmatrix} x \\ x_v \end{pmatrix} \le c_\delta \right\} \tag{5.84}$$

we have

$$\left| [\, F \quad F_2 \,] \begin{pmatrix} x \\ x_v \end{pmatrix} \right| \le u_{\max}(1 - \delta) \tag{5.85}$$

We have the following theorem.

Theorem 5.7. *Consider the given system in Equation 5.1. Then, there exists a scalar $\rho^* > 0$ such that for any nonpositive function $\rho(r, y)$, locally Lipschitz in y satisfying $|\rho(r, y)| \leq \rho^*$, the reduced-order CNF control law given by Equations 5.79 and 5.80 internally stabilizes the given plant and drives the system controlled output $h(t)$ to track asymptotically the step command input of amplitude r from an initial state x_0, provided that x_0, x_{v0} and r satisfy*

$$\begin{pmatrix} x_0 - x_e \\ x_{v0} - x_{20} - K_R x_{10} \end{pmatrix} \in \boldsymbol{X}_{R\delta} \quad \text{and} \quad |Hr| \leq \delta \cdot u_{\max} \tag{5.86}$$

Proof. Let $\tilde{x} = x - x_e$ and $\tilde{x}_v = x_v - x_2 - K_R x_1$. Then, the closed-loop system comprising the given plant in Equation 5.1 and the reduced-order CNF control law of Equations 5.79 and 5.80 can be expressed as

$$\begin{pmatrix} \dot{\tilde{x}} \\ \dot{\tilde{x}}_v \end{pmatrix} = \begin{bmatrix} A + BF & BF_2 \\ 0 & A_{22} + K_R A_{12} \end{bmatrix} \begin{pmatrix} \tilde{x} \\ \tilde{x}_v \end{pmatrix} + \begin{bmatrix} B \\ 0 \end{bmatrix} w \tag{5.87}$$

where

$$w = \operatorname{sat}\left\{ [F \quad F_2] \begin{pmatrix} \tilde{x} \\ \tilde{x}_v \end{pmatrix} + Hr + \rho(r, y) B' P \left[\tilde{x} + \begin{pmatrix} 0 \\ \tilde{x}_v \end{pmatrix} \right] \right\} - [F \quad F_2] \begin{pmatrix} \tilde{x} \\ \tilde{x}_v \end{pmatrix} - Hr \tag{5.88}$$

The rest of the proof follows along similar lines to the reasoning given in the full-order measurement feedback case. ◇

5.2.2 Systems with External Disturbances

We introduce in this subsection an enhanced version of the CNF control design, which is capable of removing constant bias in servo systems. When the given system has disturbances, the resulting system output generally does not asymptotically match the target reference without knowing *a priori* the level of bias. A common approach for removing bias resulting from constant disturbances is to add an integrator to the controller. In what follows, we propose an enhanced CNF design scheme by introducing an additional integration action in the design. The new approach retains the fast rise time property of the original CNF control and at the same time has an additional capacity of eliminating steady-state bias due to disturbances.

Specifically, we consider a linear system with an amplitude constrained actuator, characterized by

$$\begin{cases} \dot{x} = A\,x + B\,\text{sat}(u) + Ew, \quad x(0) = x_0 \\ y = C_1\,x \\ h = C_2\,x \end{cases} \tag{5.89}$$

where $x \in \mathbb{R}^n$, $u \in \mathbb{R}$, $y \in \mathbb{R}^p$, $h \in \mathbb{R}$ and $w \in \mathbb{R}$ are, respectively, the state, control input, measurement output, controlled output and disturbance input of the system. A,

B, C_1, C_2 and E are appropriate dimensional constant matrices. The function, sat: $\mathbb{R} \to \mathbb{R}$, represents the actuator saturation defined as

$$\text{sat}(u) = \text{sgn}(u) \min\{ u_{\max}, |u| \} \tag{5.90}$$

with $u_{\max}$ being the saturation level of the input. The assumptions on the given system are made:

1. (A, B) is stabilizable,
2. (A, C_1) is detectable,
3. (A, B, C_2) is invertible and has no invariant zero at $s = 0$,
4. w is bounded unknown constant disturbance, and
5. h is part of y, *i.e.* h is also measurable.

Note that all these assumptions are fairly standard for tracking control. We aim to design an enhanced CNF control law for the system with disturbances such that the resulting controlled output would track a target reference (set point), say r, as fast and as smooth as possible without having steady-state error. We first follow the usual practice to augment an integrator into the given system. Such an integrator eventually becomes part of the final control law. To be more specific, we define an auxiliary state variable,

$$\dot{x}_{\text{i}} := \kappa_{\text{i}} e := \kappa_{\text{i}}(h - r) = \kappa_{\text{i}} C_2 x - \kappa_{\text{i}} r \tag{5.91}$$

which is implementable as h is assumed to be measurable, and κ_{i} is a positive scalar to be selected to yield an appropriate integration action. The augmented system is then given as follows,

$$\begin{cases} \dot{\bar{x}} = \bar{A}\, \bar{x} + \bar{B} \text{ sat}(u) + \bar{B}_{\text{r}}\, r + \bar{E}\, w \\ \bar{y} = \bar{C}_1\, \bar{x} \\ h = \bar{C}_2\, \bar{x} \end{cases} \tag{5.92}$$

where

$$\bar{x} = \begin{pmatrix} x_{\text{i}} \\ x \end{pmatrix}, \quad \bar{x}_0 = \begin{pmatrix} 0 \\ x_0 \end{pmatrix}, \quad \bar{y} = \begin{pmatrix} x_{\text{i}} \\ y \end{pmatrix} \tag{5.93}$$

$$\bar{A} = \begin{bmatrix} 0 & \kappa_{\text{i}} C_2 \\ 0 & A \end{bmatrix}, \quad \bar{B} = \begin{bmatrix} 0 \\ B \end{bmatrix}, \quad \bar{B}_{\text{r}} = \begin{bmatrix} -\kappa_{\text{i}} \\ 0 \end{bmatrix} \tag{5.94}$$

and

$$\bar{E} = \begin{bmatrix} 0 \\ E \end{bmatrix}, \quad \bar{C}_1 = \begin{bmatrix} 1 & 0 \\ 0 & C_1 \end{bmatrix}, \quad \bar{C}_2 = [\, 0 \quad C_2 \,] \tag{5.95}$$

It is straightforward to show that Assumptions 1 and 3 imply that the pair $(\bar{A}, \bar{B})$ is stabilizable.

Next, we proceed to carry out the design of enhanced CNF control laws for two different cases, *i.e.* the state feedback case and the reduced-order measurement feedback case. The full-order measurement feedback case is straightforward once the result for the reduced-order case is established. We note that the procedure for designing the enhanced CNF control laws is a natural extension of that given in the previous subsection.

i. State Feedback Case. We first investigate the case when all the state variables of the plant in Equation 5.92 are measurable, *i.e.* $\bar{y} = \bar{x}$. The procedure that generates an enhanced CNF state feedback law is done in three steps. That is, in the first step, a linear feedback control law is designed, in the second step, the design of nonlinear feedback control is carried out, and lastly, in the final step, the linear and nonlinear feedback laws are combined to form an enhanced CNF control law.

STEP 5.C.W.S.1: design a linear feedback control law,

$$u_{\mathrm{L}} = F\bar{x} + G\,r \tag{5.96}$$

where F is chosen such that 1) $\bar{A} + \bar{B}F$ is an asymptotically stable matrix, and 2) the closed-loop system $\bar{C}_2(sI - \bar{A} - \bar{B}F)^{-1}\bar{B}$ has certain desired properties. Let us partition $F = [\,F_{\mathrm{i}} \quad F_{\mathrm{x}}\,]$ in conformity with x_{i} and x. The general guideline in the design of such an F is to place the closed-loop pole of $\bar{A} + \bar{B}F$ corresponding to the integration mode, x_{i}, to be sufficiently closer to the imaginary axis compared to the rest eigenvalues, which implies that F_{i} is a relatively small scalar. The remaining closed-loop poles of $\bar{A} + \bar{B}F$ are placed to have a dominating pair with a small damping ratio, which in turn would yield a fast rise time in the closed-loop system response. Finally, G is chosen as

$$G = -[C_2(A + BF_{\mathrm{x}})^{-1}B]^{-1} \tag{5.97}$$

which is well defined as (A, B, C_2) is assumed to have no invariant zeros at $s = 0$ and $A+BF_{\mathrm{x}}$ is nonsingular whenever $\bar{A}+\bar{B}F$ is stable and F_{i} is relatively small.

STEP 5.C.W.S.2: given a positive-definite matrix $W \in \mathbb{R}^{(n+1)\times(n+1)}$, we solve the following Lyapunov equation:

$$(\bar{A} + \bar{B}F)'P + P(\bar{A} + \bar{B}F) = -W \tag{5.98}$$

for $P > 0$. Such a solution is always existent as $\bar{A}+\bar{B}F$ is asymptotically stable. The nonlinear feedback portion of the enhanced CNF control law, u_{N}, is given by

$$u_{\mathrm{N}} = \rho(e)\bar{B}'P(\bar{x} - \bar{x}_{\mathrm{e}}) \tag{5.99}$$

where $\rho(e)$, with $e = h - r$ being the tracking error, is a smooth and nonpositive function of $|e|$, and tends to a constant as $t \to \infty$. It is used to gradually change the system closed-loop damping ratio to yield a better tracking performance. The choices of the design parameters, $\rho(e)$ and W, will be discussed later. Next, we define

$$G_{\mathrm{e}} := \begin{bmatrix} 0 \\ -(A + BF_{\mathrm{x}})^{-1}BG \end{bmatrix} \quad \text{and} \quad \bar{x}_{\mathrm{e}} := G_{\mathrm{e}}\,r \tag{5.100}$$

STEP 5.C.W.S.3: the linear feedback control law and nonlinear feedback portion derived in the previous steps are now combined to form an enhanced CNF control law,

$$u = u_{\mathrm{L}} + u_{\mathrm{N}} = F\bar{x} + G\,r + \rho(e)\bar{B}'P(\bar{x} - \bar{x}_{\mathrm{e}}) \tag{5.101}$$

We have the following result.

Theorem 5.8. *Consider the given system in Equation 5.89 with $y = x$ and the disturbance w being bounded by a non-negative scalar τ_w, i.e. $|w| \le \tau_w$. Let*

$$\gamma := 2\tau_w \lambda_{\max}(PW^{-1}) \left(\bar{E}'P\bar{E}\right)^{1/2} \tag{5.102}$$

Then, for any $\rho(e)$, which is a smooth and nonpositive function of $|e|$ and tends to a constant as $t \to \infty$, the enhanced CNF control law in Equation 5.101 internally stabilizes the given plant and drives the system controlled output h to track the step reference of amplitude r from an initial state $\bar{x}_0$ asymptotically without steady-state error, provided that the following conditions are satisfied:

1. *There exist positive scalars $\delta \in (0,1)$ and $c_\delta > \gamma^2$ such that*

$$\forall \bar{x} \in \boldsymbol{X}(F, c_\delta) := \{\, \bar{x} \,:\, \bar{x}'P\bar{x} \le c_\delta \,\} \quad \Rightarrow \quad |F\bar{x}| \le (1-\delta)u_{\max} \tag{5.103}$$

2. *The initial condition, $\bar{x}_0$, satisfies*

$$\bar{x}_0 - \bar{x}_{\mathrm{e}} \in \boldsymbol{X}(F, c_\delta) \tag{5.104}$$

3. *The level of the target reference, r, satisfies*

$$|H\, r| \le \delta u_{\max} \tag{5.105}$$

where $H := FG_{\mathrm{e}} + G$.

Proof. For simplicity, we drop the variable e in $\rho(e)$ throughout this proof. Noting that

$$\begin{aligned}
\bar{A}\bar{x}_{\mathrm{e}} + \bar{B}Hr + \bar{B}_{\mathrm{r}}r &= \bar{A}\bar{x}_{\mathrm{e}} + \bar{B}(FG_{\mathrm{e}} + G)r + \bar{B}_{\mathrm{r}}r \\
&= (\bar{A} + \bar{B}F)\bar{x}_{\mathrm{e}} + \bar{B}_{\mathrm{r}}r + \bar{B}Gr \\
&= \begin{bmatrix} 0 & \kappa_{\mathrm{i}}C_2 \\ BF_{\mathrm{i}} & A + BF_{\mathrm{x}} \end{bmatrix} \begin{bmatrix} 0 \\ -(A+BF_{\mathrm{x}})^{-1}BG \end{bmatrix} r + \begin{bmatrix} -\kappa_{\mathrm{i}} \\ 0 \end{bmatrix} r + \begin{bmatrix} 0 \\ BG \end{bmatrix} r \\
&= \begin{bmatrix} -\kappa_{\mathrm{i}}C_2(A+BF_{\mathrm{x}})^{-1}BG \\ -BG \end{bmatrix} r + \begin{bmatrix} -\kappa_{\mathrm{i}} \\ BG \end{bmatrix} r = 0
\end{aligned} \tag{5.106}$$

and letting $\tilde{x} = \bar{x} - \bar{x}_{\mathrm{e}}$, then the dynamics equation of the augmented plant in Equation 5.92 can be expressed as,

$$\begin{aligned}
\dot{\tilde{x}} = \dot{\bar{x}} &= \bar{A}(\tilde{x} + \bar{x}_{\mathrm{e}}) + \bar{B}\,\mathrm{sat}(u) + \bar{B}_{\mathrm{r}}\,r + \bar{E}w \\
&= (\bar{A} + \bar{B}F)\tilde{x} + \bar{A}\bar{x}_{\mathrm{e}} - \bar{B}F\tilde{x} + \bar{B}\,\mathrm{sat}(u) + \bar{B}_{\mathrm{r}}\,r + \bar{E}w \\
&= (\bar{A} + \bar{B}F)\tilde{x} + \bar{B}\left[\mathrm{sat}(u) - F\tilde{x} - Hr\right] + \left(\bar{A}\bar{x}_{\mathrm{e}} + \bar{B}Hr + \bar{B}_{\mathrm{r}}\,r\right) + \bar{E}w \\
&= (\bar{A} + \bar{B}F)\tilde{x} + \bar{B}\,v + \bar{E}w
\end{aligned} \tag{5.107}$$

where $v := \mathrm{sat}(u) - F\tilde{x} - H\,r$. Following the similar lines of reasoning as those in the previous subsection, we can show that for $\tilde{x} \in \boldsymbol{X}(F, c_\delta)$ and $|H\,r| \le \delta u_{\max}$,

v can be expressed as $v = q\rho\bar{B}'P\tilde{x}$, for some non-negative variable $q \in [0,1]$. Thus, for the case when $\tilde{x} \in \boldsymbol{X}(F, c_\delta)$ and $|Hr| \leq \delta u_{\max}$, the closed-loop system comprising the augmented plant in Equation 5.92 and the enhanced CNF control law in Equation 5.101 can be expressed as follows:

$$\dot{\tilde{x}} = (\bar{A} + \bar{B}F + q\rho\bar{B}\bar{B}'P)\tilde{x} + \bar{E}w \tag{5.108}$$

In what follows, we show that the system in Equation 5.108 is stable provided that the initial condition, $\bar{x}_0$, the target reference, r, and the disturbance, w, satisfy those conditions listed in the theorem. Let us define a Lyapunov function

$$V = \tilde{x}'P\tilde{x} \tag{5.109}$$

For easy derivation, we introduce a matrix S such that $P = S'S$. We then obtain the derivative of V calculated along the trajectory of the system in Equation 5.108,

$$\begin{aligned}
\dot{V} &= \dot{\tilde{x}}'P\tilde{x} + \tilde{x}'P\dot{\tilde{x}} \\
&= \tilde{x}'[(\bar{A} + \bar{B}F)'P + P(\bar{A} + \bar{B}F)]\tilde{x} + 2q\rho\tilde{x}'P\bar{B}\bar{B}'P\tilde{x} + 2\tilde{x}'P\bar{E}w \\
&\leq -\tilde{x}'W\tilde{x} + 2\tilde{x}'P\bar{E}w \\
&= -\tilde{x}'S'SP^{-1}WP^{-1}S'S\tilde{x} + 2\tilde{x}'S'S\bar{E}w \\
&\leq -\lambda_{\min}(SP^{-1}WP^{-1}S')\tilde{x}'S'S\tilde{x} + 2\|S\tilde{x}\| \cdot \|S\bar{E}\|\tau_w \\
&= -\lambda_{\min}(P^{-1}W)\tilde{x}'P\tilde{x} + 2\tau_w(\tilde{x}'P\tilde{x})^{1/2}(\bar{E}'P\bar{E})^{1/2} \\
&= -\lambda_{\min}(P^{-1}W)(\tilde{x}'P\tilde{x})^{1/2}\left[(\tilde{x}'P\tilde{x})^{1/2} - 2\tau_w\lambda_{\max}(PW^{-1})(\bar{E}'P\bar{E})^{1/2}\right] \\
&= -\lambda_{\min}(P^{-1}W)(\tilde{x}'P\tilde{x})^{1/2}\left[(\tilde{x}'P\tilde{x})^{1/2} - \gamma\right]
\end{aligned} \tag{5.110}$$

We note that we have used the following matrix properties: i) $z'Xz \geq \lambda_{\min}(X)z'z$, where X is a symmetric matrix; ii) $\lambda(XY) = \lambda(YX)$, if both X and Y are square matrices; and iii) $\lambda(XY) > 0$ if $X > 0$ and $Y > 0$. Clearly, the closed-loop system in the absence of the disturbance, w, has $\dot{V} < 0$ and thus is asymptotically stable.

With the presence of the unknown constant disturbance, w, and with the initial condition $\tilde{x}(0) = \bar{x}_0 - \bar{x}_{\mathrm{e}} \in \boldsymbol{X}(F, c_\delta)$, where $c_\delta > \gamma^2$, the corresponding trajectory of Equation 5.108 remains in $\boldsymbol{X}(F, c_\delta)$ and converges to a point on a ball $\{\tilde{x} : \tilde{x}'P\tilde{x} \leq \tilde{\gamma}^2\}$ with $\tilde{\gamma} \leq \gamma$. Since $x_{\mathrm{i}}(t) = \int_0^t \kappa_{\mathrm{i}} e(\tau)d\tau$ converges to a constant, it is clear that the tracking error $e(t) \to 0$ as $t \to \infty$. This completes the proof of Theorem 5.8. ◇

ii. Measurement Feedback Case. Next, we proceed to design an enhanced CNF control law using only information measurable from the plant. In principle, we can design either a full-order measurement feedback control law, for which its dynamical order is identical to that of the given plant, or a reduced-order measurement feedback control law, in which we make a full use of the measurement output and estimate only the unknown part of the state variable. As such, the dynamical order of the controller is reduced. It is more feasible to implement controllers with smaller dynamical order. The procedure below on the enhanced CNF control using reduced-order measurement feedback follows closely from that given in the previous subsection.

For simplicity of presentation, we assume that C_1 in the measurement output of the given plant in Equation 5.89 is already in the form,

$$C_1 = [\, I_p \quad 0\,] \tag{5.111}$$

The augmented plant in Equation 5.92 can then be partitioned as follows:

$$\left\{\begin{array}{l} \begin{pmatrix} \dot{x}_{\rm i} \\ \dot{x}_1 \\ \dot{x}_2 \end{pmatrix} = \begin{bmatrix} 0 & \kappa_{\rm i} C_{21} & \kappa_{\rm i} C_{22} \\ 0 & A_{11} & A_{12} \\ 0 & A_{21} & A_{22} \end{bmatrix} \begin{pmatrix} x_{\rm i} \\ x_1 \\ x_2 \end{pmatrix} + \begin{bmatrix} 0 \\ B_1 \\ B_2 \end{bmatrix} \mathrm{sat}(u) + \begin{bmatrix} -\kappa_{\rm i} \\ 0 \\ 0 \end{bmatrix} r + \begin{bmatrix} 0 \\ E_1 \\ E_2 \end{bmatrix} w \\ \bar{y} = \begin{bmatrix} 1 & 0 & 0 \\ 0 & I_p & 0 \end{bmatrix} \begin{pmatrix} x_{\rm i} \\ x_1 \\ x_2 \end{pmatrix} \\ h = [\,0 \quad C_{21} \quad C_{22}\,] \begin{pmatrix} x_{\rm i} \\ x_1 \\ x_2 \end{pmatrix} \end{array}\right. \tag{5.112}$$

where

$$\begin{pmatrix} x_{\rm i} \\ x_1 \\ x_2 \end{pmatrix} = \bar{x}, \quad \begin{pmatrix} x_{\rm i}(0) \\ x_1(0) \\ x_2(0) \end{pmatrix} = \begin{pmatrix} 0 \\ x_{10} \\ x_{20} \end{pmatrix} = \bar{x}_0 \tag{5.113}$$

and

$$\bar{y} = \begin{pmatrix} x_{\rm i} \\ y \end{pmatrix} = \begin{pmatrix} x_{\rm i} \\ x_1 \end{pmatrix} \tag{5.114}$$

Clearly, $x_{\rm i}$ and x_1 are readily available and need not be estimated. We only need to estimate x_2. There are two main step in designing a reduced-order measurement feedback control laws: i) the construction of a full state feedback gain matrix F; and ii) the construction of a reduced-order observer gain matrix $K_{\rm R}$. The construction of the gain matrix F is totally identical to that given in the previous subsection, which can be partitioned in conformity with $x_{\rm i}$, x_1 and x_2, as follows:

$$F = [\, F_{\rm i} \quad F_1 \quad F_2\,] \tag{5.115}$$

The reduced-order observer gain matrix $K_{\rm R}$ is chosen such that the closed-loop poles of $A_{22} + K_{\rm R} A_{12}$ are placed in appropriate locations in the open-left half plane.

The reduced-order enhanced CNF control law is then given by,

$$\begin{aligned} \dot{x}_{\rm v} = (A_{22} + K_{\rm R} A_{12}) x_{\rm v} &+ (B_2 + K_{\rm R} B_1)\, \mathrm{sat}(u) \\ &+ [A_{21} + K_{\rm R} A_{11} - (A_{22} + K_{\rm R} A_{12}) K_{\rm R}]\, y \end{aligned} \tag{5.116}$$

and

$$u = F \begin{pmatrix} x_{\rm i} \\ x_1 \\ x_{\rm v} - K_{\rm R} y \end{pmatrix} + G r + \rho(e) \bar{B}' P \left[\begin{pmatrix} x_{\rm i} \\ x_1 \\ x_{\rm v} - K_{\rm R} y \end{pmatrix} - \bar{x}_{\rm e} \right] \tag{5.117}$$

where G is as defined in Equation 5.97 and $\rho(e)$ is the smooth, nonpositive and nondecreasing function of $|e|$, to be chosen to yield a desired performance.

Next, given a positive-definite matrix $W \in \mathbb{R}^{(n+1)\times(n+1)}$, let $P > 0$ be the solution to the Lyapunov equation

$$(\bar{A} + \bar{B}F)'P + P(\bar{A} + \bar{B}F) = -W \tag{5.118}$$

Given another positive-definite matrix $W_{\mathrm{R}} \in \mathbb{R}^{(n-p)\times(n-p)}$ with

$$W_{\mathrm{R}} > F_2'\bar{B}'PW^{-1}P\bar{B}F_2 \tag{5.119}$$

let $Q_{\mathrm{R}} > 0$ be the solution to the Lyapunov equation

$$(A_{22} + K_{\mathrm{R}}A_{12})'Q_{\mathrm{R}} + Q_{\mathrm{R}}(A_{22} + K_{\mathrm{R}}A_{12}) = -W_{\mathrm{R}} \tag{5.120}$$

Note that such P and Q_{R} exist as $\bar{A} + \bar{B}F$ and $A_{22} + K_{\mathrm{R}}A_{12}$ are both asymptotically stable. We have the following result.

Theorem 5.9. *Consider the given system in Equation 5.89 with w being bounded by a scalar $\tau_w \geq 0$, i.e. $|w| \leq \tau_w$. Let*

$$\gamma_{\mathrm{R}} := 2\tau_w \lambda_{\max}\left(\begin{bmatrix} P & 0 \\ 0 & Q_{\mathrm{R}} \end{bmatrix}\begin{bmatrix} W & -P\bar{B}F_2 \\ -F_2'\bar{B}'P & W_{\mathrm{R}} \end{bmatrix}^{-1}\right) \times \left[\bar{E}'P\bar{E} + (E_2 + K_{\mathrm{R}}E_1)'Q_{\mathrm{R}}(E_2 + K_{\mathrm{R}}E_1)\right]^{1/2} \tag{5.121}$$

Then, there exists a scalar $\rho^* > 0$ such that for any $\rho(e)$, a smooth and nonpositive function of $|e|$ with $|\rho(e)| \leq \rho^*$ and tending to a constant as $t \to \infty$, the enhanced reduced-order CNF control law of Equations 5.116 and 5.117 internally stabilizes the given plant and drives the system controlled output h to track the step reference of amplitude r asymptotically without steady-state error, provided that the following conditions are satisfied:

1. *There exist positive scalars $\delta \in (0,1)$ and $c_{\mathrm{R}\delta} > \gamma_{\mathrm{R}}^2$ such that*

$$\forall \bar{x} \in \boldsymbol{X}(F, c_{\mathrm{R}\delta}) := \left\{\bar{x} \,:\, \bar{x}'\begin{bmatrix} P & 0 \\ 0 & Q_{\mathrm{R}} \end{bmatrix}\bar{x} \leq c_{\mathrm{R}\delta}\right\} \Rightarrow |[F \quad F_2]\,\bar{x}| \leq (1-\delta)u_{\max} \tag{5.122}$$

2. *The initial conditions, $\bar{x}_0$ and $x_{\mathrm{v}0} = x_{\mathrm{v}}(0)$, satisfy*

$$\begin{pmatrix} \bar{x}_0 - \bar{x}_{\mathrm{e}} \\ x_{\mathrm{v}0} - x_{20} - K_{\mathrm{R}}\,x_{10} \end{pmatrix} \in \boldsymbol{X}(F, c_{\mathrm{R}\delta}) \tag{5.123}$$

3. *The level of the target reference, r, satisfies*

$$|H\,r| \leq \delta u_{\max} \tag{5.124}$$

where H is the same as that defined in Theorem 5.8.

Proof. The result follows from similar lines of reasoning as those in Theorem 5.8 and those for the measurement feedback case in the previous subsection. ◇

5.2.3 Selection of Nonlinear Feedback Parameters

Basically, the freedom to choose the function ρ in either the usual CNF design or the enhanced CNF design is used to tune the control laws so as to improve the performance of the closed-loop system as the controlled output, h, approaches the set point, r. Since the main purpose of adding the nonlinear part to the CNF or the enhanced CNF controllers is to shorten the settling time, or equivalently to contribute a significant value to the control input when the tracking error, e, is small. The nonlinear part, in general, is set in action when the control signal is far away from its saturation level, and thus it does not cause the control input to hit its limits. For simplicity, we now focus our attention on the case when the given system has external disturbances. The following analysis is equally applicable to the case when the given system does not have disturbances. Under such circumstances, the closed-loop system comprising the augmented plant in Equation 5.92 and the enhanced CNF control law can be expressed as:

$$\dot{\tilde{x}} = (\bar{A} + \bar{B}F + \rho\bar{B}\bar{B}'P)\tilde{x} + \bar{E}w \tag{5.125}$$

We note that the additional term ρ does not affect the stability of the estimators. It is now clear that eigenvalues of the closed-loop system in Equation 5.125 can be changed by the function ρ. Such a mechanism can be interpreted using the classical feedback control concept as shown in Figure 5.1, where the auxiliary system $G_{\text{aux}}(s)$ is defined as:

$$G_{\text{aux}}(s) := C_{\text{aux}}(sI - A_{\text{aux}})^{-1}B_{\text{aux}} := \bar{B}'P(sI - \bar{A} - \bar{B}F)^{-1}\bar{B} \tag{5.126}$$

$G_{\text{aux}}(s)$ has the following interesting properties.

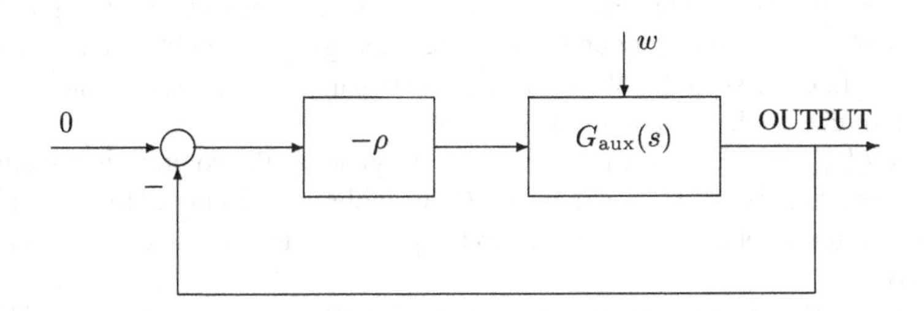

Figure 5.1. Interpretation of the nonlinear function $\rho(e)$

Theorem 5.10. *The auxiliary system $G_{\text{aux}}(s)$ defined in Equation 5.126 is stable and invertible with a relative degree equal to* 1, *and is of minimum phase with n stable invariant zeros.* $\diamondsuit$

Proof. First, it is obvious to see that $G_{\text{aux}}(s)$ is stable since $\bar{A} + \bar{B}F$ is a stable matrix. Next, since $P > 0$ and $\bar{B} \neq 0$, we have

$$C_{\text{aux}} B_{\text{aux}} = \bar{B}' P \bar{B} > 0 \tag{5.127}$$

which implies that $G_{\text{aux}}(s)$ is invertible and has a relative degree equal to 1 (or an infinite zero of order 1). Furthermore, $G_{\text{aux}}(s)$ has n invariant zeros, as it is a SISO system.

The last property of $G_{\text{aux}}(s)$, *i.e.* the invariant zeros of $G_{\text{aux}}(s)$ are stable and hence it is of minimum phase, can be shown by using the well-known classical root-locus theory. Observing the block diagram in Figure 5.1, it follows from the classical feedback control theory (see, *e.g.*, [1]) that the poles of the closed-loop system of Equation 5.125, which are the functions of the tuning parameter ρ, start from the open-loop poles, *i.e.* the eigenvalues of $\bar{A} + \bar{B}F$, when $\rho = 0$, and end up at the open-loop zeros (including the zero at the infinity) as $|\rho| \to \infty$. It then follows from the proof of Theorem 5.3 that the closed-loop system remains asymptotically stable for any nonpositive ρ, which implies that all the invariant zeros of the open-loop system, *i.e.* $G_{\text{aux}}(s)$, must be stable. ♢

It is clear from Theorem 5.10 and its proof that the invariant zeros of $G_{\text{aux}}(s)$ play an important role in selecting the poles of the closed-loop system of Equation 5.125. The poles of the closed-loop system approach the locations of the invariant zeros of $G_{\text{aux}}(s)$ as $|\rho|$ becomes larger and larger. We would like to note that there is freedom in preselecting the locations of these invariant zeros. This can actually be done by selecting an appropriate $W > 0$ in Equation 5.98. In general, we should select the invariant zeros of $G_{\text{aux}}(s)$, which are corresponding to the closed-loop poles for larger $|\rho|$, such that the dominated ones have a large damping ratio, which in turn yields a smaller overshoot. The following procedure can be used as a guideline for the selection of such a W:

1. Given the pair $(A_{\text{aux}}, B_{\text{aux}})$ and the desired locations of the invariant zeros of $G_{\text{aux}}(s)$, we follow the result of Chen and Zheng [139] (see also Chapter 9 of Chen *et al.* [71]) on finite and infinite zero assignment to obtain an appropriate matrix C_{aux} such that the resulting matrix triple $(A_{\text{aux}}, B_{\text{aux}}, C_{\text{aux}})$ has the desired relative degree and invariant zeros.
2. Solve $C_{\text{aux}} = \bar{B}' P$ for a $P = P' > 0$. In general, the solution is nonunique as there are $n(n+1)/2$ elements in P available for selection. However, if the solution does not exist, we go back to the previous step to reselect the invariant zeros.
3. Calculate W using Equation 5.98 and check if W is positive-definite. If W is not positive-definite, we go back to the previous step to choose another solution of P or go to the first step to reselect the invariant zeros.

Generally, the above procedure would yield a desired result. The selection of the nonlinear function ρ is relatively simple once the desired invariant zeros of $G_{\text{aux}}(s)$ are obtained. Assuming the tracking error e is available, the following choice of $\rho(e)$ is a smooth and nonpositive function of $|e|$:

$$\rho(e) = -\beta \left| \mathrm{e}^{-\alpha|e|} - \mathrm{e}^{-\alpha|h(0)-r|} \right| \tag{5.128}$$

where α and β are appropriate positive scalars that can be chosen to yield a desired performance, *i.e.* fast settling time and small overshoot. This function $\rho(e)$ changes from 0 to $\rho_0 = -\beta \left|1 - e^{-\alpha|h(0)-r|}\right|$ as the tracking error approaches zero. At the initial stage, when the controlled output, h, is far away from the final set point, $\rho(e)$ is small and the effect of the nonlinear part on the overall system is very limited. When the controlled output, h, approaches the set point, $\rho(e) \approx \rho_0$, and the nonlinear control law becomes effective. In general, the parameter ρ_0 is chosen such that the poles of $\bar{A} + \bar{B}F + \rho_0 \bar{B}\bar{B}'P$ are in the desired locations, *e.g.*, the dominated poles have a large damping ratio, which would reduce the overshoot of the output response. Note that the choice of $\rho(e)$ is nonunique. Any function would work so long as it has similar properties of that given in Equation 5.128.

5.2.4 An Illustrative Example

We illustrate the enhanced CNF control technique for continuous-time systems in the following example. We consider a continuous-time system of Equation 5.89 with

$$A = \begin{bmatrix} 0 & 1 & 0 \\ 0 & 0 & 1 \\ -1 & -1 & -1 \end{bmatrix}, \quad B = \begin{bmatrix} 0 \\ 0 \\ 1 \end{bmatrix}, \quad E = \begin{bmatrix} 1 \\ 0 \\ 0 \end{bmatrix}, \quad x_0 = \begin{pmatrix} 0 \\ 0.5 \\ 0 \end{pmatrix} \tag{5.129}$$

$$C_1 = I_3, \quad C_2 = [\,1 \quad 0 \quad 0\,] \tag{5.130}$$

and $u_{\max} = 2$. The disturbance w is unknown. For simulation purpose, we assume $w = -0.1$. Our goal is to design an enhanced CNF state feedback control law that would yield a good transient performance in tracking a target reference $r = 1$.

Following the procedure given in the previous subsection, we select an integration gain $\kappa_{\rm i} = 1$ and obtain an appropriate augmented system. After a few tries, we found that the following state feedback gain to the augmented system would yield a good performance for our problem:

$$F = [\,-0.6 \quad -2.8 \quad -3.8 \quad -3.2\,] \tag{5.131}$$

which places the poles of $\bar{A} + \bar{B}F$ at -0.2, -3, $-0.5 \pm j0.866$. We note that the first one corresponds to the integrator. Both the linear state feedback control and enhanced CNF control share the same integration dynamics:

$$\dot{x}_{\rm i} = h - r \tag{5.132}$$

The linear state feedback control law is given by

$$u = [\,-2.8 \quad -3.8 \quad -3.2\,]\, x - 0.6x_{\rm i} + 3.8r \tag{5.133}$$

Letting $W = \text{diag}\{0.04,\ 0.1,\ 5,\ 0.1\}$, we obtain a positive-definite solution P for Equation 5.98, which is given by

$$P = \begin{bmatrix} 0.2403 & 0.6694 & 0.2684 & 0.0333 \\ 0.6694 & 3.8665 & 1.5752 & 0.1893 \\ 0.2684 & 1.5752 & 4.4039 & 0.8490 \\ 0.0333 & 0.1893 & 0.8490 & 0.2140 \end{bmatrix} \tag{5.134}$$

and an enhanced CNF state feedback law:

$$u = [-2.8 \quad -3.8 \quad -3.2]\,x - 0.6x_{\rm i} + 3.8r + \rho(e)$$

$$\times [0.0333 \quad 0.1893 \quad 0.8490 \quad 0.2140] \left[\begin{pmatrix} x_{\rm i} \\ x \end{pmatrix} - \begin{pmatrix} 0 \\ 2 \\ 0 \\ 0 \end{pmatrix} \right] \tag{5.135}$$

where $\rho(e)$ is as given in (5.128) with $\alpha = 1$ and $\beta = 8$. The simulation results given in Figures 5.2 and 5.3 clearly show that the CNF control has outperformed the linear control.

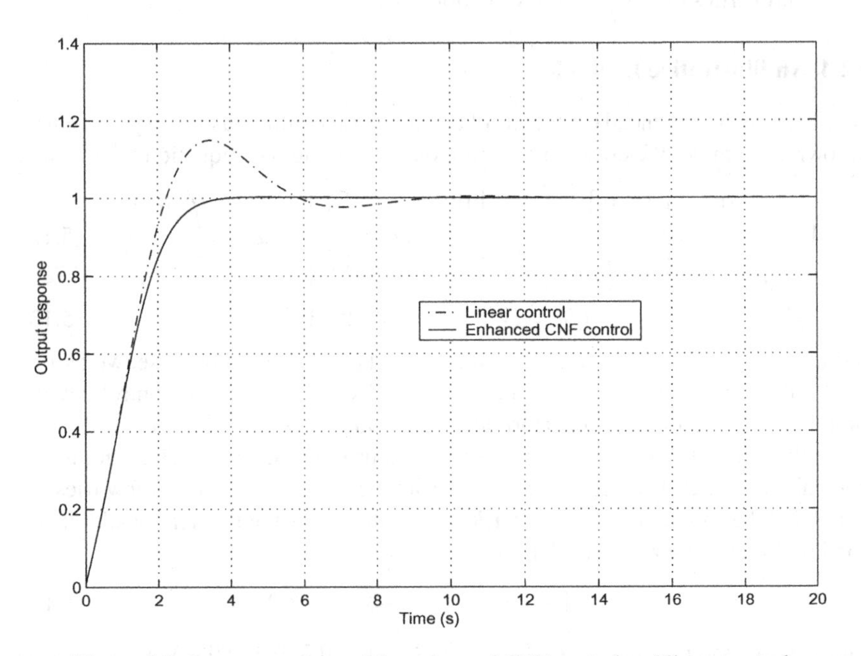

Figure 5.2. Output responses of the enhanced CNF control and linear control

5.3 Discrete-time Systems

As in the continuous-time case, we present in this section the CNF design technique for systems without and with external disturbances. Selection and interpretation of nonlinear gain design parameters are also discussed.

5.3.1 Systems without External Disturbances

Let us now consider a linear discrete-time system Σ with an amplitude-constrained actuator characterized by

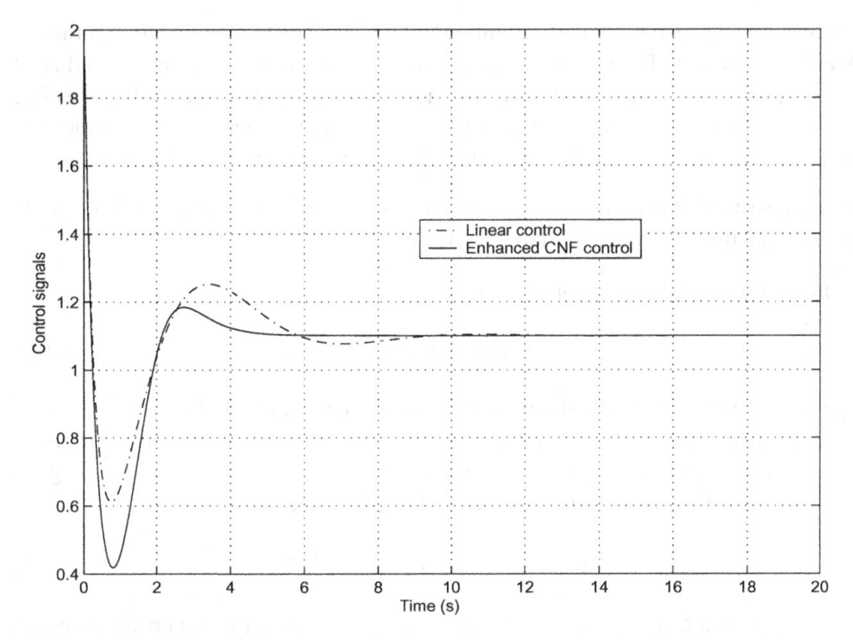

Figure 5.3. Control signals of the enhanced CNF control and linear control

$$\begin{cases} x(k+1) = A\ x(k) + B\,\text{sat}(u(k)), & x(0) = x_0 \\ y(k) = C_1\ x(k) \\ h(k) = C_2\ x(k) \end{cases} \tag{5.136}$$

where $x \in \mathbb{R}^n$, $u \in \mathbb{R}$, $y \in \mathbb{R}^p$ and $h \in \mathbb{R}$ are, respectively, the state, control input, measurement output and controlled output of Σ. A, B, C_1 and C_2 are appropriate dimensional constant matrices, and sat: $\mathbb{R} \to \mathbb{R}$ represents the actuator saturation defined as

$$\text{sat}(u) = \text{sgn}(u) \min\{\ u_{\text{max}},\ |\ u\ |\ \} \tag{5.137}$$

with u_{max} being the saturation level of the input. The following assumptions on the system matrices are required:

1. (A, B) is stabilizable,
2. (A, C_1) is detectable, and
3. (A, B, C_2) is invertible and has no invariant zeros at $z = 1$.

We now extend the results of the continuous-time composite nonlinear control method to the discrete-time system in Equation 5.136. Thus, the objective here is to design a discrete-time CNF control law that causes the output to track a high-amplitude step input rapidly without experiencing large overshoot and without the adverse actuator saturation effects. This can be done through the design of a discrete-time linear feedback law with a small closed-loop damping ratio and a nonlinear feedback law through an appropriate Lyapunov function to cause the closed-loop system to be highly damped as system output approaches the command input to reduce the overshoot. The result of this discrete-time version is analogous to that of its continuous-time counterpart. Here, we again separate the design of discrete-time CNF control into three distinct situations, *i.e.* 1) the state feedback case, 2) the full-order measurement case, and 3) the reduced-order measurement feedback case.

i. State Feedback Case. We consider the case when $y = x$, *i.e.* all the state variables of Σ of Equation 5.136 are available for feedback.

STEP 5.D.S.1: design a linear feedback law,

$$u_{\rm L}(k) = Fx(k) + G\,r \tag{5.138}$$

where r is the input command, and F is chosen such that $A + BF$ has all its eigenvalues in $\mathbb{C}^{\odot}$ and the closed-loop system $C_2(zI - A - BF)^{-1}B$ meets certain design specifications. We note again that such an F can be designed using any of the techniques reported in Chapter 3. Furthermore,

$$G = \left[C_2(I - A - BF)^{-1}B\right]^{-1} \tag{5.139}$$

We note that G is well defined because $A + BF$ has all its eigenvalues in $\mathbb{C}^{\odot}$, and (A, B, C_2) is invertible and has no invariant zeros at $z = 1$.

The following lemma determines the magnitude of r that can be tracked by such a control law without exceeding the control limits.

Lemma 5.11. *Given a positive-definite matrix $W \in \mathbb{R}^{n\times n}$, let $P > 0$ be the solution of the following Lyapunov equation:*

$$P = (A + BF)'P(A + BF) + W. \tag{5.140}$$

Such a P exists as $A + BF$ is asymptotically stable. For any $\delta \in (0, 1)$, let $c_\delta > 0$ be the largest positive scalar such that

$$|\ Fx(k)\ | \le u_{\max}(1 - \delta), \quad \forall\, x(k) \in \boldsymbol{X}_\delta := \left\{x : x'Px \le c_\delta\right\} \tag{5.141}$$

Also, let

$$H := \left[1 + F(I - A - BF)^{-1}B\right] G \tag{5.142}$$

and

$$x_{\rm e} := G_{\rm e}\, r := (I - A - BF)^{-1}BG\, r \tag{5.143}$$

Then, the control law in Equation 5.138 is capable of driving the system controlled output $h(k)$ to track asymptotically a step command input of amplitude r, provided that the initial state x_0 and r satisfy:

$$(x_0 - x_{\rm e}) \in \boldsymbol{X}_\delta \quad \textit{and} \quad |Hr| \le \delta \cdot u_{\max} \tag{5.144}$$

Proof. Let $\tilde{x} = x - x_e$. Then, the linear feedback control law u_L can be rewritten as

$$u_L(k) = F\tilde{x}(k) + \left[1 + F(I - A - BF)^{-1}B\right]Gr = F\tilde{x}(k) + Hr \tag{5.145}$$

Hence, for all

$$\tilde{x}(k) \in \boldsymbol{X}_\delta \Rightarrow \mid F\tilde{x}(k) \mid \leq u_{\max}(1 - \delta) \tag{5.146}$$

and for any r satisfying

$$|Hr| \leq \delta \cdot u_{\max} \tag{5.147}$$

the linear control law can be written as

$$\mid u_L(k) \mid = |F\tilde{x}(k) + Hr| \leq \mid F\tilde{x}(k) \mid + |Hr| \leq u_{\max} \tag{5.148}$$

which indicates that the control signal $u_L(k)$ never exceeds the saturation. Next, let us move to verify the asymptotic stability of the closed-loop system comprising the given plant in Equation 5.136 and the linear feedback law in Equation 5.138, which can be expressed as follows:

$$\tilde{x}(k+1) = (A + BF)\tilde{x}(k) \tag{5.149}$$

Let us define a Lyapunov function for the closed-loop system in Equation 5.149 as

$$V(k) = \tilde{x}'(k)P\tilde{x}(k) \tag{5.150}$$

Along the trajectories of the closed-loop system in Equation 5.149 the increment of the Lyapunov function in Equation 5.150 is given by

$$\begin{aligned} \Delta V(k) &= \tilde{x}'(k+1)P\tilde{x}(k+1) - \tilde{x}'(k)P\tilde{x}(k) \\ &= \tilde{x}'(k)(A + BF)'P(A + BF)\tilde{x}(k) - \tilde{x}'(k)P\tilde{x}(k) \\ &= -\tilde{x}'(k)W\tilde{x}(k) \\ &\leq 0 \end{aligned} \tag{5.151}$$

This shows that $\boldsymbol{X}_\delta$ is an invariant set of the the closed-loop system in Equation 5.149 and all trajectories of Equation 5.149 starting from $\boldsymbol{X}_\delta$ converge to the origin. Thus, for any initial state x_0 and the step command input r that satisfy Equation 5.144, we have

$$\lim_{k \to \infty} x(k) = x_e \tag{5.152}$$

and hence

$$\lim_{k \to \infty} h(k) = \lim_{k \to \infty} C_2 x(k) = C_2 x_e = r \tag{5.153}$$

This completes the proof of Lemma 5.11. ◇

Remark 5.12. We would like to note that, for the case when $x_0 = 0$, any step command of amplitude r can be tracked asymptotically provided that

$$\mid r \mid \leq \left[c_\delta (G_e' P G_e)^{-1}\right]^{1/2} \quad \text{and} \quad \mid Hr \mid \leq \delta \cdot u_{\max} \tag{5.154}$$

This input command amplitude can be increased by increasing δ and/or decreasing $G_e' P G_e$ through the choice of W. However, the change in F affects the damping ratio of the closed-loop system and hence its rise time. ◇

STEP 5.D.S.2: the nonlinear feedback control law $u_{\mathrm{N}}(k)$ is given by

$$u_{\mathrm{N}}(k) = \rho(r, y)B'P(A + BF)\Big[x(k) - x_{\mathrm{e}}\Big] \tag{5.155}$$

where $\rho(r, y)$ is a nonpositive scalar function, locally Lipschitz in y, and is to be used to change the system closed-loop damping ratio as the output approaches the step command input. The choice of $\rho(r, y)$ will be discussed later in detail.

STEP 5.D.S.3: the linear and nonlinear components derived above are now combined to form a discrete-time CNF control law:

$$u(k) = u_{\mathrm{L}}(k) + u_{\mathrm{N}}(k) \tag{5.156}$$

We have the following result.

Theorem 5.13. *Consider the discrete-time system in Equation 5.136. Then, for any nonpositive $\rho(r, y)$, locally Lipschitz in y and $|\rho(r, y)| \le \rho^* := 2(B'PB)^{-1}$, the CNF control law in Equation 5.156 is capable of stabilizing the given plant and driving the system controlled output $h(k)$ to track the step command input of amplitude r from an initial state x_0, provided that x_0 and r satisfy the properties in Equation 5.144.* ◊

Proof. Let $\tilde{x} = x - x_{\mathrm{e}}$. Then, the closed-loop system can be written as

$$\tilde{x}(k + 1) = (A + BF)\tilde{x}(k) + Bw(k) \tag{5.157}$$

where

$$w(k) = \mathrm{sat}\Big[F\tilde{x}(k) + Hr + u_{\mathrm{N}}(k)\Big] - F\tilde{x}(k) - Hr \tag{5.158}$$

Equation 5.144 implies that $\tilde{x}_0 \in \boldsymbol{X}_\delta$. Define a Lyapunov function

$$V(k) = \tilde{x}'(k)P\tilde{x}(k) \tag{5.159}$$

Noting that

$$\tilde{x}(k) \in \boldsymbol{X}_\delta \quad \Rightarrow \quad | F\tilde{x}(k) | \le u_{\max}(1 - \delta) \tag{5.160}$$

we can evaluate the increment of $V(k)$ along the trajectories of the closed-loop system in Equation 5.157 as follows:

$$\begin{aligned}\Delta V(k) &= \tilde{x}'(k + 1)P\tilde{x}(k + 1) - \tilde{x}'(k)P\tilde{x}(k)\\ &= \tilde{x}'(k)(A + BF)'P(A + BF)\tilde{x}(k) - \tilde{x}'(k)P\tilde{x}(k) + w'(k)B'PBw(k)\\ &\qquad +2\,\tilde{x}'(k)(A + BF)'PBw(k)\\ &= -\tilde{x}'(k)W\tilde{x}(k)+2\tilde{x}'(k)(A+BF)'PBw(k)+w'(k)B'PBw(k)\end{aligned} \tag{5.161}$$

Next, we proceed to find the increment of $V(k)$ for three different cases, as is done in continuous-time systems.

If $|F\tilde{x}(k) + Hr + u_{\mathrm{N}}(k)| \le u_{\max}$, then

$$w(k) = u_{\rm N}(k) = \rho B'P(A + BF)\tilde{x}(k) \tag{5.162}$$

Thus,

$$\begin{aligned}\Delta V(k) &= -\tilde{x}'(k)W\tilde{x}(k) + 2\tilde{x}'(k)(A + BF)'PBw(k) + w'(k)B'PBw(k) \\ &= -\tilde{x}'(k)W\tilde{x}(k) \\ &\quad + \rho\tilde{x}'(k)(A + BF)'PB(2 + \rho B'PB)B'P(A + BF)\tilde{x}(k)\end{aligned} \tag{5.163}$$

For any nonpositive $\rho(r, y)$ with $|\rho(r, y)| \le \rho^*$, it is clear that the increment $\Delta V(k) \le -\tilde{x}'(k)W\tilde{x}(k) \le 0$

If $F\tilde{x}(k) + Hr + u_{\rm N}(k) > u_{\max}$, then $| F\tilde{x}(k) + Hr |\le u_{\max}$ implies that $0 < w(k) < u_{\rm N}(k)$ and $\rho(r, y) < 0$ Hence,

$$\begin{aligned}\Delta V(k) &= -\tilde{x}'(k)W\tilde{x}(k) + 2\tilde{x}'(k)(A + BF)'PBw(k) + w'(k)B'PBw(k) \\ &= -\tilde{x}'(k)W\tilde{x}(k) + w'(k)[2\rho^{-1}u_{\rm N}(k) + B'PBw(k)] \\ &< -\tilde{x}'(k)W\tilde{x}(k) + w'(k)[2\rho^{-1}u_{\rm N}(k) + B'PBu_{\rm N}(k)] \\ &= -\tilde{x}'(k)W\tilde{x}(k) + w'(k)(2\rho^{-1} + B'PB)u_{\rm N}(k)\end{aligned} \tag{5.164}$$

Thus, for all $-\rho^* \le \rho(r, y) < 0$, we have $2\rho^{-1} + B'PB \le 0$, and hence

$$\Delta V(k) \le -\tilde{x}'(k)W\tilde{x}(k) \le 0 \tag{5.165}$$

Similarly, for the case when $F\tilde{x}(k) + Hr + u_{\rm N}(k) < -u_{\max}$, it can be shown that $\Delta V(k) \le -\tilde{x}'(k)W\tilde{x}(k) \le 0$.

Thus, $\boldsymbol{X}_\delta$ is an invariant set of the closed-loop system in Equation 5.157 and all trajectories of Equation 5.157 starting from $\boldsymbol{X}_\delta$ remain there and converge to the origin. This, in turn, indicates that, for all initial states x_0 and the step command input of amplitude r that satisfy Equation 5.144,

$$\lim_{k\to\infty} x(k) = x_{\rm e} \tag{5.166}$$

and

$$\lim_{k\to\infty} y(k) = \lim_{k\to\infty} C_2 x(k) = C_2 x_{\rm e} = r \tag{5.167}$$

This completes the proof of Theorem 5.13. ◇

Remark 5.14. Theorem 5.13 shows that the addition of the nonlinear feedback control law $u_{\rm N}$ as given in Equation 5.155 does not affect the ability to track the class of command inputs. Any command input that can be tracked by the linear feedback law in Equation 5.138 can also be tracked by the CNF control law in Equation 5.156. The composite feedback law in Equation 5.156 does not reduce the level of the trackable command input for any choice of the function $\rho(r, y)$. This freedom can be used to improve the performance of the overall system. The choice of $\rho(r, y)$ will be discussed in the forthcoming subsection. ◇

ii. Full-order Measurement Feedback Case. We proceed to construct a discrete-time full-order CNF control law in the following.

STEP 5.D.F.1: we first construct a linear full-order measurement feedback control law

$$\begin{cases} x_{\rm v}(k+1) = (A+KC_1)x_{\rm v}(k) - Ky(k) + B\,{\rm sat}(u_{\rm L}(k)) \\ u_{\rm L}(k) = F[x_{\rm v}(k) - x_{\rm e}] + H\,r \end{cases} \tag{5.168}$$

where r is the command input, $x_{\rm v} \in \mathbb{R}^n$ is the state of the controller, F and K are chosen such that $A+BF$ and $A+KC_1$ have all their eigenvalues in $\mathbb{C}^{\odot}$, *i.e.* both are stable matrices, and, furthermore, the resulting closed-loop system has met certain design specifications. As usual, we let

$$x_{\rm e} = (I - A - BF)^{-1}B \cdot [C_2(I - A - BF)^{-1}B]^{-1}r \tag{5.169}$$

and

$$H = [1 + F(I - A - BF)^{-1}B] \cdot [C_2(I - A - BF)^{-1}B]^{-1} \tag{5.170}$$

We note that both $x_{\rm e}$ and H are well defined.

Lemma 5.15. *Given a positive-definite matrix $W_{\rm P} \in \mathbb{R}^{n\times n}$, let $P > 0$ be the solution to the Lyapunov equation*

$$P = (A+BF)'P(A+BF) + W_{\rm P} \tag{5.171}$$

Given another positive-definite matrix $W_{\rm Q} \in \mathbb{R}^{n\times n}$ with

$$W_{\rm Q} > F'B'P(A+BF)W_{\rm P}^{-1}(A+BF)'PBF \tag{5.172}$$

let $Q > 0$ be the solution to the Lyapunov equation

$$Q = (A+KC_1)'Q(A+KC_1) + W_{\rm Q} \tag{5.173}$$

Note that such P and Q exist as $A+BF$ and $A+KC_1$ are asymptotically stable. For any $\delta \in (0,1)$, let c_δ be the largest positive scalar such that for all

$$\begin{pmatrix} x \\ x_{\rm v} \end{pmatrix} \in \boldsymbol{X}_{{\rm F}\delta} := \left\{ \begin{pmatrix} x \\ x_{\rm v} \end{pmatrix} : \begin{pmatrix} x \\ x_{\rm v} \end{pmatrix}' \begin{bmatrix} P & 0 \\ 0 & Q \end{bmatrix} \begin{pmatrix} x \\ x_{\rm v} \end{pmatrix} \le c_\delta \right\} \tag{5.174}$$

we have

$$\left| [F \;\; F] \begin{pmatrix} x \\ x_{\rm v} \end{pmatrix} \right| \le u_{\max}(1-\delta) \tag{5.175}$$

The linear control law in Equation 5.168 drives the system controlled output $h(k)$ to track asymptotically a step command input of amplitude r from an initial state x_0, provided that x_0, $x_{\rm v0} = x_{\rm v}(0)$ and r satisfy:

$$|Hr| \le \delta \cdot u_{\max} \quad \text{and} \quad \begin{pmatrix} x_0 - x_{\rm e} \\ x_{\rm v0} - x_0 \end{pmatrix} \in \boldsymbol{X}_{{\rm F}\delta} \tag{5.176}$$

Proof. This follows along similar lines to the reasoning given in the proofs of Lemmas 5.5 and 5.11. ♢

STEP 5.D.F.2: the discrete-time full-order measurement composite nonlinear feedback control law is given by

$$x_{\rm v}(k+1) = (A + KC_1)x_{\rm v}(k) - Ky(k) + B{\rm sat}(u(k)) \tag{5.177}$$

and

$$u(k) = F[x_{\rm v}(k) - x_{\rm e}] + Hr + \rho(r, y)B'P(A + BF)[x_{\rm v}(k) - x_{\rm e}] \tag{5.178}$$

where $\rho(r, y)$ is a nonpositive scalar function, locally Lipschitz in y, and is to be chosen to improve the performance of the closed-loop system.

We have the following result.

Theorem 5.16. *Consider the given discrete-time system in Equation 5.136. Then, there exists a scalar* $0 < \rho^* \le 2(B'PB)^{-1}$ *such that for any nonpositive function* $\rho(r, y)$*, locally Lipschitz in* y *and* $|\rho(r, y)| \le \rho^*$*, the discrete-time CNF control law in Equations 5.177 and 5.178 internally stabilizes the given plant and drive the system controlled output* $h(k)$ *to track asymptotically the step command input of amplitude* r *from an initial state* x_0*, provided that* x_0*,* $x_{\rm v0}$ *and* r *satisfy the conditions in Equation 5.176.* ♢

Proof. The proof of this theorem follows along similar lines to the reasoning given in Theorems 5.6 and 5.13. ♢

iii. Reduced-order Measurement Feedback Case. As in its continuous-time counterpart, we now proceed to design a reduced-order measurement feedback controller. For the given system in Equation 5.136, it is clear that p states of the system are measurable if C_1 is of maximal rank. As such, we could design a dynamic controller that has a dynamical order less than that of the given plant. We now proceed to construct such a control law under the CNF control framework.

For simplicity of presentation, we assume that C_1 is already in the form

$$C_1 = [\, I_p \quad 0 \,] \tag{5.179}$$

Then, the system in Equation 5.136 can be rewritten as

$$\begin{cases} \begin{pmatrix} x_1(k+1) \\ x_2(k+1) \end{pmatrix} = \begin{bmatrix} A_{11} & A_{12} \\ A_{21} & A_{22} \end{bmatrix} \begin{pmatrix} x_1(k) \\ x_2(k) \end{pmatrix} + \begin{bmatrix} B_1 \\ B_2 \end{bmatrix} {\rm sat}(u(k)) \\ y(k) = [\ I_p \quad 0\ ] \begin{pmatrix} x_1(k) \\ x_2(k) \end{pmatrix} \\ h(k) = C_2 \begin{pmatrix} x_1(k) \\ x_2(k) \end{pmatrix} \end{cases} \tag{5.180}$$

and

$$x_0 = \begin{pmatrix} x_{10} \\ x_{20} \end{pmatrix} \tag{5.181}$$

where the original state x is partitioned into two parts, x_1 and x_2 with $y \equiv x_1$. Thus, we only need to estimate x_2 in the reduced-order measurement feedback design. Next, we let F be chosen such that 1) $A + BF$ is asymptotically stable, and 2) $C_2(sI - A - BF)^{-1}B$ has the desired properties, and let K_R be chosen such that $A_{22} + K_R A_{12}$ is asymptotically stable. Again, it follows from Chen [110] that (A_{22}, A_{12}) is detectable if and only if (A, C_1) is detectable. Thus, there exists a stabilizing K_R. Again, such F and K_R can be designed using any of the linear control techniques presented in Chapter 3. We then partition F in conformity with x_1 and x_2:

$$F = [\, F_1 \quad F_2 \,] \tag{5.182}$$

As defined in Equations 5.169 and 5.169, we let

$$x_e = (I - A - BF)^{-1}B \cdot [C_2(I - A - BF)^{-1}B]^{-1}r \tag{5.183}$$

and

$$H = [1 + F(I - A - BF)^{-1}B] \cdot [C_2(I - A - BF)^{-1}B]^{-1} \tag{5.184}$$

The reduced-order CNF controller is given by

$$\begin{aligned} x_v(k+1) &= (A_{22} + K_R A_{12})x_v(k) + (B_2 + K_R B_1)\,\mathrm{sat}(u(k)) \\ &\quad + \big[A_{21} + K_R A_{11} - (A_{22} + K_R A_{12})K_R\big]y(k) \end{aligned} \tag{5.185}$$

and

$$\begin{aligned} u(k) &= F\left[\begin{pmatrix} y \\ x_v(k) - K_R y(k) \end{pmatrix} - x_e\right] + Hr \\ &\quad + \rho(r,y)B'P(A+BF)\left[\begin{pmatrix} y(k) \\ x_v - K_R y(k) \end{pmatrix} - x_e\right] \end{aligned} \tag{5.186}$$

where $\rho(r, y)$ is a nonpositive scalar function locally Lipschitz in y subject to certain constraints to be discussed later.

Next, given a positive-definite matrix $W_P \in \mathbb{R}^{n \times n}$, let $P > 0$ be the solution to the Lyapunov equation

$$P = (A + BF)'P(A + BF) + W_P \tag{5.187}$$

Given another positive-definite matrix $W_R \in \mathbb{R}^{(n-p)\times(n-p)}$ with

$$W_R > F_2'B'P(A + BF)W_P^{-1}(A + BF)'PBF_2 \tag{5.188}$$

let $Q_R > 0$ be the solution to the Lyapunov equation

$$Q_R = (A_{22} + K_R A_{12})'Q_R(A_{22} + K_R A_{12}) + W_R \tag{5.189}$$

Note that such P and $Q_{\rm R}$ exist as $A + BF$ and $A_{22} + K_{\rm R}A_{12}$ are asymptotically stable. For any $\delta \in (0,1)$, let c_δ be the largest positive scalar such that for all

$$\begin{pmatrix} x(k) \\ x_{\rm v}(k) \end{pmatrix} \in \boldsymbol{X}_{{\rm R}\delta} := \left\{ \begin{pmatrix} x \\ x_{\rm v} \end{pmatrix} : \begin{pmatrix} x \\ x_{\rm v} \end{pmatrix}' \begin{bmatrix} P & 0 \\ 0 & Q_{\rm R} \end{bmatrix} \begin{pmatrix} x \\ x_{\rm v} \end{pmatrix} \le c_\delta \right\} \tag{5.190}$$

we have

$$\left| [F \quad F_2] \begin{pmatrix} x(k) \\ x_{\rm v}(k) \end{pmatrix} \right| \le u_{\max}(1-\delta) \tag{5.191}$$

We have the following theorem.

Theorem 5.17. *Consider the system given in Equation 5.1. Then, there exists a scalar $0 < \rho^* \le 2(B'PB)^{-1}$ such that for any nonpositive function $\rho(r, y)$, locally Lipschitz in y and $|\rho(r,y)| \le \rho^*$, the reduced-order CNF control law given by Equations 5.185 and 5.186 internally stabilizes the given plant and drives the system controlled output $h(k)$ to track asymptotically the step command input of amplitude r from an initial state x_0, provided that x_0, $x_{\rm v0}$ and r satisfy*

$$\begin{pmatrix} x_0 - x_{\rm e} \\ x_{\rm v0} - x_{20} - K_{\rm R}x_{10} \end{pmatrix} \in \boldsymbol{X}_{{\rm R}\delta} \quad \text{and} \quad |Hr| \le \delta \cdot u_{\max} \tag{5.192}$$

Proof. Again, the proof of this theorem is similar to those given earlier. ◇

5.3.2 Systems with External Disturbances

We consider a linear discrete-time system with actuator saturation and disturbances characterized by

$$\begin{cases} x(k+1) = A\,x(k) + B\,\text{sat}(u(k)) + E\,w, & x(0) = x_0 \\ y(k) = C_1\,x(k) \\ h(k) = C_2\,x(k) \end{cases} \tag{5.193}$$

where $x \in \mathbb{R}^n$, $u \in \mathbb{R}$, $y \in \mathbb{R}^p$, $h \in \mathbb{R}$ and $w \in \mathbb{R}$ are, respectively, the state, control input, measurement output, controlled output and disturbance input of the system. A, B, C_1, C_2 and E are appropriate dimensional constant matrices. The function, sat: $\mathbb{R} \to \mathbb{R}$, represents the actuator saturation defined as

$$\text{sat}(u(k)) = \text{sgn}(u(k)) \min\{\, u_{\max},\ |u(k)|\,\} \tag{5.194}$$

with $u_{\max}$ being the input saturation level. The following assumptions on the given system are made:

1. (A, B) is stabilizable,
2. (A, C_1) is detectable,
3. (A, B, C_2) is invertible with no invariant zero at $z = 1$,
4. w is bounded unknown constant disturbance, and

5. $h(k)$ is part of $y(k)$, *i.e.* $h(k)$ is also measurable.

We aim to design a discrete enhanced CNF control law for the system with input saturation and disturbances to track a step reference, say r, neither violating the input saturation nor having steady-state bias. An equivalent discrete integration, which eventually becomes part of the final control law, is defined as follows,

$$x_{\rm i}(k+1) = x_{\rm i}(k) + \kappa_{\rm i} e(k) = x_{\rm i}(k) + \kappa_{\rm i} C_2 x(k) - \kappa_{\rm i} r \tag{5.195}$$

where the tracking error $e(k) := h(k) - r$ is available for feedback as $h(k)$ is assumed to be measurable and $\kappa_{\rm i}$ is a positive scalar to be selected to yield an appropriate integration speed. By integrating Equation 5.195 into the given system, we obtain the following augmented system

$$\begin{cases} \bar{x}(k+1) = \bar{A}\,\bar{x}(k) + \bar{B}\,\mathrm{sat}(u(k)) + \bar{B}_{\rm r}\,r + \bar{E}\,w \\ \bar{y}(k) = \bar{C}_1\,\bar{x}(k) \\ h(k) = \bar{C}_2\,\bar{x}(k) \end{cases} \tag{5.196}$$

where

$$\bar{x}(k) = \begin{pmatrix} x_{\rm i}(k) \\ x(k) \end{pmatrix}, \quad \bar{x}_0 = \begin{pmatrix} 0 \\ x_0 \end{pmatrix}, \quad \bar{y}(k) = \begin{pmatrix} x_{\rm i}(k) \\ y(k) \end{pmatrix} \tag{5.197}$$

$$\bar{A} = \begin{bmatrix} 1 & \kappa_{\rm i} C_2 \\ 0 & A \end{bmatrix}, \quad \bar{B} = \begin{bmatrix} 0 \\ B \end{bmatrix}, \quad \bar{B}_{\rm r} = \begin{bmatrix} -\kappa_{\rm i} \\ 0 \end{bmatrix} \tag{5.198}$$

and

$$\bar{E} = \begin{bmatrix} 0 \\ E \end{bmatrix}, \quad \bar{C}_1 = \begin{bmatrix} 1 & 0 \\ 0 & C_1 \end{bmatrix}, \quad \bar{C}_2 = [\,0 \quad C_2\,] \tag{5.199}$$

We note that under Assumptions 1 and 3, it is straightforward to verify that the pair $(\bar{A}, \bar{B})$ is stabilizable.

In what follows, we proceed to design an enhanced CNF control laws for the given system for two different cases, *i.e.* the state feedback case and the reduced-order measurement feedback case. The full-order measurement feedback case can be solved in a straightforward manner once the result for the reduced-order case is established.

i. State Feedback Case. We consider in the following the situation when all the state variables of the given system in Equation 5.193 are measurable, *i.e.* $y = x$. The procedure that generates an enhanced CNF state feedback law is done in three steps. That is, in the first step, a linear feedback control law with appropriate properties is designed, then in the second step, the design of nonlinear feedback portion is carried out, and lastly, in the final step, the linear and nonlinear feedback laws are combined to form an enhanced CNF control law.

STEP 5.D.W.S.1: Design a linear feedback control law,

$$u_{\rm L}(k) = F\bar{x}(k) + G\,r \tag{5.200}$$

where F is chosen such that i) $\bar{A} + \bar{B}F$ is asymptotically stable, and ii) the closed-loop system $\bar{C}_2(zI - \bar{A} - \bar{B}F)^{-1}\bar{B}$ has certain desired properties. Let us partition $F = [\, F_{\rm i} \quad F_{\rm x} \,]$ in conformity with $x_{\rm i}(k)$ and $x(k)$. The general guideline in designing such a state feedback gain F is to place the closed-loop pole of $\bar{A} + \bar{B}F$ corresponding to $x_{\rm i}(k)$ to be sufficiently closer to $z = 1$ compared to the other eigenvalues, which implies that $F_{\rm i}$ is a relatively small scalar. The remaining closed-loop poles of $\bar{A} + \bar{B}F$ are placed to have a dominating pair with a small damping ratio, which in turn would yield a fast rise time in the closed-loop system response. Finally, G is chosen as

$$G = [C_2(I - A - BF_{\rm x})^{-1}B]^{-1} \tag{5.201}$$

which is well defined as (A, B, C_2) is assumed to have no invariant zeros at $z = 1$ and $A + BF_{\rm x}$ is nonsingular whenever $\bar{A} + \bar{B}F$ is stable and $F_{\rm i}$ is relatively small.

STEP 5.D.W.S.2: Given an appropriate positive-definite matrix $W \in \mathbb{R}^{(n+1)\times(n+1)}$, we solve the following Lyapunov equation:

$$P = (\bar{A} + \bar{B}F)'P(\bar{A} + \bar{B}F) + W \tag{5.202}$$

for $P > 0$. Such a solution is always existent as $\bar{A}+\bar{B}F$ is asymptotically stable. The nonlinear feedback portion of the enhanced CNF control law, $u_{\rm N}(k)$, is then given by

$$u_{\rm N}(k) = \rho(e(k))\bar{B}'P(\bar{A} + \bar{B}F)[\bar{x}(k) - \bar{x}_{\rm e}] \tag{5.203}$$

where $\rho(e(k))$, with $e = h - r$, is a nonpositive function of $|e|$ and tends to a finite scalar as $k \to \infty$. It is to be used to gradually change the system closed-loop damping ratio to yield a better tracking performance. The choices of the design parameters, $\rho(e(k))$ and W, will be discussed later. Next, we define

$$G_{\rm e} := \begin{bmatrix} 0 \\ (I - A - BF_{\rm x})^{-1}BG \end{bmatrix}, \quad \bar{x}_{\rm e} := G_{\rm e}\, r \tag{5.204}$$

STEP 5.D.W.S.3: the linear and nonlinear feedback control laws derived in the previous steps are now combined to form an enhanced CNF control law,

$$u(k) = F\bar{x}(k) + Gr + \rho(e(k))\bar{B}'P(\bar{A}+\bar{B}F)[\bar{x}(k) - \bar{x}_{\rm e}] \tag{5.205}$$

We have the following result.

Theorem 5.18. *Consider the given system in Equation 5.193 with $y = x$ and the disturbance w being bounded by a non-negative scalar τ_w, i.e. $|w| \le \tau_w$. Let*

$$\gamma := \left[1 - \left(1 - \lambda_{\min}(WP^{-1})\right)^{1/2}\right]^{-1} \left(\bar{E}'P\bar{E}\right)^{1/2} \tau_w \tag{5.206}$$

Then, for any $\rho(e(k)) \in [-2(\bar{B}'P\bar{B})^{-1}, 0]$, which is a nonpositive function of $|e|$ and tends to a constant as $k \to \infty$, the enhanced CNF control law in Equation 5.205 internally stabilizes the given plant and drives the system controlled output $h(k)$ to track the step reference of amplitude r from an initial state $\bar{x}_0$ asymptotically without steady-state error, provided that the following conditions are satisfied:

1. There exist scalars $\delta \in (0,1)$ *and* $c_\delta > \gamma^2$ *such that*

$$\forall \bar{x}(k) \in \boldsymbol{X}(F, c_\delta) := \{\bar{x}(k) : \bar{x}'(k) P \bar{x}(k) \le c_\delta\} \\ \Rightarrow |F\bar{x}(k)| \le (1-\delta) u_{\max} \tag{5.207}$$

2. The initial condition, $\bar{x}_0$*, satisfies*

$$\bar{x}_0 - \bar{x}_{\rm e} \in \boldsymbol{X}(F, c_\delta) \tag{5.208}$$

3. The level of the target reference, r*, satisfies*

$$|H\,r| \le \delta u_{\max} \tag{5.209}$$

where $H := FG_{\rm e} + G$*. Note that* $\lambda_{\min}(WP^{-1}) \in (0,1)$*.*

Proof. For simplicity, we drop $e(k)$ in the nonlinear function ρ throughout the following proof. First, it is straightforward to verify that

$$\bar{A}\bar{x}_{\rm e} + \bar{B}Hr + \bar{B}_{\rm r} r = \bar{x}_{\rm e}. \tag{5.210}$$

Letting $\tilde{x}(k) = \bar{x}(k) - \bar{x}_{\rm e}$, the augmented system in Equation 5.196 can be expressed as

$$\tilde{x}(k+1) = (\bar{A} + \bar{B}F)\tilde{x}(k) + \bar{B}\,v(k) + \bar{E}\,w \tag{5.211}$$

where

$$v(k) := \text{sat}(u(k)) - F\tilde{x}(k) - H\,r \tag{5.212}$$

and the control law in Equation 5.205 can be rewritten as

$$u(k) = F\tilde{x}(k) + H\,r + \rho\bar{B}'P(\bar{A} + \bar{B}F)\tilde{x}(k) \tag{5.213}$$

Next, for $\tilde{x}(k) \in \boldsymbol{X}(F, c_\delta)$ and $|H\,r| \le \delta u_{\max}$, we have

$$|F\tilde{x}(k) + H\,r| \le |F\tilde{x}(k)| + |H\,r| \le u_{\max} \tag{5.214}$$

which implies

$$v(k) = \rho\bar{B}'P(\bar{A}+\bar{B}F)\tilde{x}(k) \tag{5.215}$$

if $|u(k)| \le u_{\max}$, or

$$0 < v(k) < \rho\bar{B}'P(\bar{A}+\bar{B}F)\tilde{x}(k) \tag{5.216}$$

if $u(k) > u_{\max}$, or

$$\rho\bar{B}'P(\bar{A} + \bar{B}F)\tilde{x}(k) < v(k) < 0 \tag{5.217}$$

if $u(k) < -u_{\max}$. Obviously, for all possible situations, $v(k)$ can be written as

$$v(k) = q\rho\bar{B}'P(\bar{A} + \bar{B}F)\tilde{x}(k) \tag{5.218}$$

with some appropriate $q \in [0,1]$. Thus, for $\tilde{x}(k) \in \boldsymbol{X}(F, c_\delta)$ and $|Hr| \le \delta u_{\max}$, the closed-loop system comprising the augmented system in Equation 5.196 and the CNF control law in Equation 5.205 can be expressed as follows

$$\tilde{x}(k+1) = [\bar{A} + \bar{B}F + q\rho\bar{B}\bar{B}'P(\bar{A} + \bar{B}F)]\tilde{x}(k) + \bar{E}\,w \tag{5.219}$$

Defining a discrete-time Lyapunov function, $V(k) = \tilde{x}'(k)P\tilde{x}(k)$, and factoring $P > 0$ as $P = S'S$, the increment of $V(k)$ along the trajectory of the system in Equation 5.219 can be calculated as

$$\begin{aligned}\Delta V(k) &= -\tilde{x}'(k)P\tilde{x}(k) + \tilde{x}'(k)(\bar{A} + \bar{B}F)'P(\bar{A} + \bar{B}F)\tilde{x}(k)\\ &\quad+\tilde{x}'(k)(\bar{A} + \bar{B}F)'P\bar{B}(2q\rho + q^2\rho^2\bar{B}'P\bar{B})\bar{B}'P(\bar{A} + \bar{B}F)\tilde{x}(k)\\ &\quad+w'\bar{E}'P\bar{E}w + 2\tilde{x}'(k)(\bar{A} + \bar{B}F)'(I+q\rho\bar{B}\bar{B}'P)'P\bar{E}w\\ &\le -\tilde{x}'(k)P\tilde{x}(k) + \tilde{x}'(k)(P-W)\tilde{x}(k) + \bar{E}'P\bar{E}\tau_w^2\\ &\quad+2\|\tilde{x}'(k)(\bar{A} + \bar{B}F)'(I + q\rho\bar{B}\bar{B}'P)'S'\| \cdot \|S\bar{E}\|\tau_w\end{aligned} \tag{5.220}$$

Noting that

$$\begin{aligned}&\|\tilde{x}'(k)(\bar{A} + \bar{B}F)'(I + q\rho\bar{B}\bar{B}'P)'S'\|\\ &\qquad\le \Big[\tilde{x}'(k)(\bar{A} + \bar{B}F)'P(\bar{A} + \bar{B}F)\tilde{x}(k)\Big]^{1/2}\end{aligned} \tag{5.221}$$

for $\rho \in [-2(\bar{B}'P\bar{B})^{-1}, 0]$, we have

$$\begin{aligned}\Delta V(k) &\le -\tilde{x}'(k)P\tilde{x}(k) + \tilde{x}'(k)(P - W)\tilde{x}(k) + \bar{E}'P\bar{E}\tau_w^2\\ &\quad+2\Big[\tilde{x}'(k)(\bar{A} + \bar{B}F)'P(\bar{A} + \bar{B}F)\tilde{x}(k)\Big]^{1/2}(\bar{E}'P\bar{E})^{1/2}\tau_w\\ &= -\Big\{(\tilde{x}'(k)P\tilde{x}(k))^{1/2} + [\tilde{x}'(k)(P - W)\tilde{x}(k)]^{1/2} + (\bar{E}'P\bar{E})^{1/2}\tau_w\Big\}\\ &\quad\times\Big\{(\tilde{x}'(k)P\tilde{x}(k))^{1/2} - [\tilde{x}'(k)(P - W)\tilde{x}(k)]^{1/2} - (\bar{E}'P\bar{E})^{1/2}\tau_w\Big\}\\ &= -\Big\{(\tilde{x}'(k)P\tilde{x}(k))^{1/2} + [\tilde{x}'(k)(P - W)\tilde{x}(k)]^{1/2} + (\bar{E}'P\bar{E})^{1/2}\tau_w\Big\}\\ &\quad\times\left\{-(\bar{E}'P\bar{E})^{1/2}\tau_w+(\tilde{x}'(k)P\tilde{x}(k))^{1/2}\left[1-\left(1-\frac{\tilde{x}'(k)W\tilde{x}(k)}{\tilde{x}'(k)P\tilde{x}(k)}\right)^{1/2}\right]\right\}\\ &\le -\Big\{(\tilde{x}'(k)P\tilde{x}(k))^{1/2} + [\tilde{x}'(k)(P - W)\tilde{x}(k)]^{1/2} + (\bar{E}'P\bar{E})^{1/2}\tau_w\Big\}\\ &\quad\times\Big\{-(\bar{E}'P\bar{E})^{1/2}\tau_w+(\tilde{x}'(k)P\tilde{x}(k))^{1/2}\left[1-\left(1-\lambda_{\min}(WP^{-1})\right)^{1/2}\right]\Big\}\\ &= -\Big\{(\tilde{x}'(k)P\tilde{x}(k))^{1/2} + [\tilde{x}'(k)(P - W)\tilde{x}(k)]^{1/2} + (\bar{E}'P\bar{E})^{1/2}\tau_w\Big\}\\ &\quad\times\left[1 - \left(1 - \lambda_{\min}(WP^{-1})\right)^{1/2}\right]\left[(\tilde{x}'(k)P\tilde{x}(k))^{1/2} - \gamma\right]\end{aligned} \tag{5.222}$$

Note that we have used the following property:

$$\lambda_{\min}(WP^{-1}) = \min_{x\neq 0}\frac{x'Wx}{x'Px} \tag{5.223}$$

as both W and P are positive-definite matrices. Clearly, the closed-loop system in the absence of the disturbance, w, has $\Delta V(k) < 0$ and thus is asymptotically stable.

With the presence of the disturbance, w, and with $\tilde{x}(0) = \bar{x}_0 - \bar{x}_e \in \boldsymbol{X}(F, c_\delta)$, where $c_\delta > \gamma^2$, the corresponding trajectory of Equation 5.219 remains in $\boldsymbol{X}(F, c_\delta)$ and converges to a ball characterized by $\{\tilde{x} : \tilde{x}' P \tilde{x} \le \tilde{\gamma}^2\}$ with $\tilde{\gamma} \le \gamma$.

Note that ρ is chosen such that it tends to a constant as $k \to \infty$. Also, for large k, the control signal $u(k)$ is under its saturation level. Thus, the closed-loop system in Equation 5.219 becomes an almost linear time-invariant system. We can conclude that the corresponding trajectory of Equation 5.219 converges asymptotically to a point, *i.e.* $\tilde{x}(k)$ tends to a constant. Thus,

$$\lim_{k \to \infty} e(k) = \lim_{k \to \infty} \frac{1}{\kappa_i} \Big[x_i(k+1) - x_i(k) \Big] = 0 \tag{5.224}$$

This completes the proof of Theorem 5.18. ◇

ii. Measurement Feedback Case. Next, we consider the general measurement feedback situation, in which there is only part of state variables available for feedback. As usual, for such a situation, one could either design a full-order or a reduced-order measurement feedback control law. In what follows, we focus on designing a reduced-order controller. Without loss of generality, we assume that C_1 in the measurement output of the given plant in Equation 5.193 is in the form:

$$C_1 = [\, I_p \quad 0 \,] \tag{5.225}$$

The augmented plant in Equation 5.196 can then be partitioned as

$$\begin{pmatrix} x_i(k+1) \\ x_1(k+1) \\ x_2(k+1) \end{pmatrix} = \begin{bmatrix} 1 & \kappa_i C_{21} & \kappa_i C_{22} \\ 0 & A_{11} & A_{12} \\ 0 & A_{21} & A_{22} \end{bmatrix} \begin{pmatrix} x_i(k) \\ x_1(k) \\ x_2(k) \end{pmatrix} + \begin{bmatrix} 0 \\ B_1 \\ B_2 \end{bmatrix} \mathrm{sat}(u(k)) + \begin{bmatrix} -\kappa_i \\ 0 \\ 0 \end{bmatrix} r + \begin{bmatrix} 0 \\ E_1 \\ E_2 \end{bmatrix} w \tag{5.226}$$

$$\bar{y}(k) = \begin{bmatrix} 1 & 0 & 0 \\ 0 & I_p & 0 \end{bmatrix} \begin{pmatrix} x_i \\ x_1 \\ x_2 \end{pmatrix} (k) \tag{5.227}$$

and

$$h(k) = [\, 0 \quad C_{21} \quad C_{22} \,] \begin{pmatrix} x_i \\ x_1 \\ x_2 \end{pmatrix} (k) \tag{5.228}$$

with

$$\begin{pmatrix} x_i(0) \\ x_1(0) \\ x_2(0) \end{pmatrix} = \begin{pmatrix} 0 \\ x_{10} \\ x_{20} \end{pmatrix} = \bar{x}_0 \tag{5.229}$$

We only need to estimate $x_2(k)$. There are two main steps in designing a reduced-order measurement feedback control law: i) the construction of a full state feedback gain matrix F, which is totally identical to that given in the previous subsection; and

ii) the reduced-order observer gain matrix $K_{\rm R}$, which is chosen such that the poles of $A_{22} + K_{\rm R}A_{12}$ are placed at appropriate locations inside the unit circle. To derive the reduced-order measurement feedback controller, we partitioned $F = [\, F_{\rm i} \quad F_1 \quad F_2 \,]$ in conformity with $x_{\rm i}(k)$, $x_1(k)$ and $x_2(k)$. The reduced-order enhanced CNF control law is then given by,

$$x_{\rm v}(k+1) = (A_{22} + K_{\rm R}A_{12})x_{\rm v}(k) + (B_2 + K_{\rm R}B_1)\text{sat}(u(k)) \\ + [A_{21} + K_{\rm R}A_{11} - (A_{22} + K_{\rm R}A_{12})K_{\rm R}]y(k) \quad (5.230)$$

and

$$u(k) = Fx_{\rm r}(k) + Gr + \rho(e(k))\bar{B}'P(\bar{A} + \bar{B}F)[x_{\rm r}(k) - \bar{x}_{\rm e}] \quad (5.231)$$

where

$$x_{\rm r}(k) = \begin{pmatrix} x_{\rm i}(k) \\ x_1(k) \\ x_{\rm v}(k) - K_{\rm R}y(k) \end{pmatrix}$$

G is as defined in Equation 5.201, $\bar{x}_{\rm e}$ is as defined in Equation 5.204 and $\rho(e(k))$ is the nonpositive function of $|e(k)|$, which tends to a constant as $k \to \infty$.

Next, given a positive-definite matrix $W \in \mathbb{R}^{(n+1)\times(n+1)}$, let $P > 0$ be the solution to the Lyapunov equation

$$P = (\bar{A} + \bar{B}F)'P(\bar{A} + \bar{B}F) + W \quad (5.232)$$

Let $W_{\rm R} \in \mathbb{R}^{(n-p)\times(n-p)}$ be a positive-definite matrix such that

$$W_{\rm R} > F_2'[\bar{B}'P\bar{B} + \bar{B}'P(\bar{A}+\bar{B}F)W^{-1}(\bar{A}+\bar{B}F)'P\bar{B}]F_2 \quad (5.233)$$

and let $Q_{\rm R} > 0$ be the solution to the Lyapunov equation

$$Q_{\rm R} = (A_{22} + K_{\rm R}A_{12})'Q_{\rm R}(A_{22} + K_{\rm R}A_{12}) + W_{\rm R} \quad (5.234)$$

Note that such P and $Q_{\rm R}$ exist as $\bar{A} + \bar{B}F$ and $A_{22} + K_{\rm R}A_{12}$ are both asymptotically stable. Next, let

$$\lambda_{\rm R} := \lambda_{\min}\left\{ \begin{bmatrix} P & 0 \\ 0 & Q_{\rm R} \end{bmatrix}^{-1} \begin{bmatrix} W & -(\bar{A}+\bar{B}F)'P\bar{B}F_2 \\ -F_2'\bar{B}'P(\bar{A}+\bar{B}F) & W_{\rm R} - F_2'\bar{B}'P\bar{B}F_2 \end{bmatrix} \right\} \quad (5.235)$$

It is noted that $\lambda_{\rm R} \in (0, 1)$. We have the following result.

Theorem 5.19. *Consider the given system in Equation 5.193 with the disturbance w being bounded by a scalar $\tau_w > 0$, i.e. $|w| \le \tau_w$. Let*

$$\gamma_{\rm R} := \left[1 - (1-\lambda_{\rm R})^{1/2}\right]^{-1} \left[\bar{E}'P\bar{E} + (E_2 + K_{\rm R}E_1)'Q_{\rm R}(E_2 + K_{\rm R}E_1)\right]^{1/2} \tau_w \quad (5.236)$$

Then, there exists a $\rho^\star \in (0, 2(\bar{B}'P\bar{B})^{-1}]$ such that for any $\rho(e(k)) \in [-\rho^\star, 0]$, which is a nonpositive function of $|e(k)|$ and tends to a constant as $k \to \infty$, the *reduced-order enhanced CNF control law of Equations 5.230 and 5.231 internally stabilizes the given plant and drives the system controlled output $h(k)$ to track the step reference of amplitude r asymptotically without steady-state error, provided that the following conditions are satisfied:*

1. *There exist positive scalars $\delta \in (0,1)$ and $c_{R\delta} > \gamma_R^2$ such that*

$$\forall \bar{x}(k) \in \boldsymbol{X}(F, c_{R\delta}) := \left\{ \bar{x}(k) : \bar{x}'(k) \begin{bmatrix} P & 0 \\ 0 & Q_R \end{bmatrix} \bar{x}(k) \le c_{R\delta} \right\}$$
$$\Rightarrow \quad |[\, F \quad F_2 \,]\, \bar{x}(k)| \le (1-\delta) u_{\max} \tag{5.237}$$

2. *The initial conditions, $\bar{x}_0$ and $x_{v0} = x_v(0)$, satisfy*

$$\begin{pmatrix} \bar{x}_0 - \bar{x}_e \\ x_{v0} - x_{20} - K_R\, x_{10} \end{pmatrix} \in \boldsymbol{X}(F, c_{R\delta}) \tag{5.238}$$

3. *The level of the target reference, r, satisfies*

$$|H\, r| \le \delta u_{\max} \tag{5.239}$$

where H is the same as that defined in Theorem 5.18.

Proof. It follows from the similar lines of reasoning as those in Theorem 5.18 and the similar arguments for the measurement feedback case reported in its continuous-time counterpart in the previous section. ◇

5.3.3 Selection of Nonlinear Feedback Parameters

As in its continuous-time counterpart, we focus our attention on the situation for the case when the given system has external disturbances. Since the main purpose of adding the nonlinear part to the CNF controller is to shorten the settling time and to reduce the overshoot, or equivalently to contribute a significant value to the control input when the tracking error, $h - r$, is small, it is appropriate for us to select a nonlinear gain matrix such that the nonlinear part is in action when the control signal is far from its saturation level, and thus it does not cause the control input to hit its limits. Under such a circumstance, it is straightforward to verify that the closed-loop system comprising the augmented plant in Equation 5.196 and the CNF control law in Equation 5.205 can be expressed as

$$\tilde{x}(k+1) = (\bar{A} + \bar{B}F)\tilde{x}(k) + \rho \bar{B}\bar{B}' P(\bar{A} + \bar{B}F)\tilde{x}(k) + \bar{E}w \tag{5.240}$$

As in its continuous-time counterpart, the eigenvalues of the closed-loop system in Equation 5.240 can be changed by the nonlinear function ρ. Assuming that h is available, we propose the following nonlinear gain function

$$\rho(e(k)) = -\beta(\bar{B}' P \bar{B})^{-1} \frac{2}{\pi} \arctan\left(\alpha \Big| |h(k) - r| - |h(0) - r| \Big| \right) \tag{5.241}$$

with $0 \le \beta \le 2$. We note that ρ starts from 0 and gradually decreases to a constant $-\beta(\bar{B}'P\bar{B})^{-1}2\arctan(\alpha|h(0)-r|)/\pi > -\beta(\bar{B}'P\bar{B})^{-1}$ as h approaches the target reference r. The parameter α is used to determine the speed of change in ρ.

To examine the behavior of the closed-loop system in Equation 5.240 more explicitly, we define an auxiliary system $G_{\text{aux}}(z)$ characterized by

$$G_{\text{aux}}(z) := C_{\text{aux}}(zI - A_{\text{aux}})^{-1}B_{\text{aux}} := \bar{B}'P(zI - \bar{A} - \bar{B}F)^{-1}\bar{B} \tag{5.242}$$

Clearly, $G_{\text{aux}}(z)$ is stable. Note that $C_{\text{aux}}B_{\text{aux}} = \bar{B}'P\bar{B} > 0$, which implies $G_{\text{aux}}(z)$ is a square, invertible and uniform rank system with a relative degree of 1 and with n invariant zeros. We show that this auxiliary system is in fact of minimum phase, *i.e.* all its invariant zeros are stable. We note that for such a system, it follows from the result reported in Chapter 5 of Chen *et al.* [71] that there exist nonsingular transformations $\Gamma_{\text{s}} \in \mathbb{R}^{(n+1)\times(n+1)}$, $\Gamma_{\text{i}} \in \mathbb{R}$ and $\Gamma_{\text{o}} \in \mathbb{R}$ such that the transformed system has the following special form,

$$\left(\Gamma_{\text{s}}^{-1}A_{\text{aux}}\Gamma_{\text{s}},\ \Gamma_{\text{s}}^{-1}B_{\text{aux}}\Gamma_{\text{i}},\ \Gamma_{\text{o}}^{-1}C_{\text{aux}}\Gamma_{\text{s}}\right) = \left(\begin{bmatrix} A_{\text{aa}} & L_{\text{ad}} \\ E_{\text{da}} & A_{\text{dd}} \end{bmatrix}, \begin{bmatrix} 0 \\ 1 \end{bmatrix}, [0 \quad 1]\right)$$

where the eigenvalues of A_{aa} are the invariant zeros of the auxiliary system $G_{\text{aux}}(z)$, L_{ad}, E_{da} and A_{dd} are some constant matrices. Next, we proceed to show that all the eigenvalues of A_{aa} are inside the unit circle and thus $G_{\text{aux}}(z)$ is of minimum phase. We note that at the steady state when $h = r$, the nonlinear function matrix ρ of Equation 5.241 with an appropriately chosen β can be set to $\rho = -(\bar{B}'P\bar{B})^{-1}$ and the closed-loop system of Equation 5.240 can be expressed as

$$\begin{aligned}
\tilde{x}(k+1) &= (\bar{A} + \bar{B}F)\tilde{x}(k) - \bar{B}(\bar{B}'P\bar{B})^{-1}\bar{B}'P(\bar{A} + \bar{B}F)\tilde{x}(k) \\
&= [I - \bar{B}(\bar{B}'P\bar{B})^{-1}\bar{B}'P](\bar{A} + \bar{B}F)\tilde{x}(k) \\
&= [I - B_{\text{aux}}(C_{\text{aux}}B_{\text{aux}})^{-1}C_{\text{aux}}]A_{\text{aux}}\tilde{x}(k) \\
&= \left[I - \Gamma_{\text{s}}\begin{bmatrix} 0 \\ 1 \end{bmatrix}\Gamma_{\text{i}}^{-1}\left(\Gamma_{\text{o}}[0 \quad 1]\Gamma_{\text{s}}^{-1}\Gamma_{\text{s}}\begin{bmatrix} 0 \\ 1 \end{bmatrix}\Gamma_{\text{i}}^{-1}\right)^{-1}\Gamma_{\text{o}}[0 \quad 1]\Gamma_{\text{s}}^{-1}\right] \\
&\quad \times \Gamma_{\text{s}}\begin{bmatrix} A_{\text{aa}} & L_{\text{ad}} \\ E_{\text{da}} & A_{\text{dd}} \end{bmatrix}\Gamma_{\text{s}}^{-1}\tilde{x}(k) \\
&= \left(\Gamma_{\text{s}}\begin{bmatrix} A_{\text{aa}} & L_{\text{ad}} \\ 0 & 0 \end{bmatrix}\Gamma_{\text{s}}^{-1}\right)\tilde{x}(k)
\end{aligned} \tag{5.243}$$

Clearly, the closed-loop system has n eigenvalues at $\lambda(A_{\text{aa}})$ and one at 0. Thus, the stability of the closed-loop system with $\rho = -(\bar{B}'P\bar{B})^{-1}$ implies the eigenvalues of A_{aa} are all inside the unit circle. This shows that $G_{\text{aux}}(z)$ is indeed of minimum phase.

It is noted that there is freedom in preselecting the locations of these invariant zeros by choosing an appropriate W in Equation 5.202. In general, we should select the invariant zeros of $G_{\text{aux}}(z)$, which correspond to the closed-loop poles of Equation 5.240 for the steady-state nonlinear gain matrix, with dominating ones having a large damping ratio, which in turn generally yield a smaller overshoot. The following procedure might be used for such a purpose.

1. Given a set of n self-conjugated complex scalars, which include all the uncontrollable modes, if any, of $(\bar{A}, \bar{B})$, we are to determine a $W > 0$ such that the resulting auxiliary system $G_{\mathrm{aux}}(z)$ has its invariant zeros placed exactly at the locations given in the set.
First, use the singular value decomposition technique to find a unitary matrix $U \in \mathbb{R}^{n\times n}$ and a nonsingular matrix $T_{\mathrm{i}} \in \mathbb{R}^{m\times m}$ such that

$$\tilde{B}_{\mathrm{aux}} = U' B_{\mathrm{aux}} T_{\mathrm{i}} = U' \bar{B} T_{\mathrm{i}} = \begin{bmatrix} 0 \\ 1 \end{bmatrix}$$

and partition accordingly

$$\tilde{A}_{\mathrm{aux}} = U' A_{\mathrm{aux}} U = U'(\bar{A} + \bar{B}F)U = \begin{bmatrix} A_{11} & A_{12} \\ A_{21} & A_{22} \end{bmatrix}$$

It is straightforward to verify that the stabilizability of $(\bar{A}, \bar{B})$ implies the stabilizability of (A_{11}, A_{12}). In fact, their uncontrollable modes, if any, are identical. Next, for determining an appropriate matrix $P = P' > 0$, we partition it accordingly as follows

$$\tilde{P} = U' P U = \begin{bmatrix} P_{11} & P_{21}' \\ P_{21} & P_{22} \end{bmatrix} \tag{5.244}$$

Then, C_{aux} can be expressed as

$$\begin{aligned} C_{\mathrm{aux}} &= \bar{B}' P = (T_{\mathrm{i}}^{-1})' [0 \quad 1] U' U \begin{bmatrix} P_{11} & P_{21}' \\ P_{21} & P_{22} \end{bmatrix} U' \\ &= (T_{\mathrm{i}}^{-1})' [P_{21} \quad P_{22}] U' \\ &= [(T_{\mathrm{i}}^{-1})' P_{22}] [P_{22}^{-1} P_{21} \quad 1] U' \\ &:= T_{\mathrm{o}} [P_{22}^{-1} P_{21} \quad 1] U' \end{aligned} \tag{5.245}$$

Using the results of Chen *et al.* [71] (see *e.g.*, Chapters 8 and 9), we can show that the invariant zeros of the auxiliary system $G_{\mathrm{aux}}(z)$ are given by the eigenvalues of $A_{11} - A_{12} P_{22}^{-1} P_{21}$. Since (A_{11}, A_{12}) is stabilizable and the given set of complex scalars include all its uncontrollable modes, there exists a constant matrix, say F_*, such that $A_{11} - A_{12} F_*$ has its eigenvalues placed exactly at the locations given in the set. Obviously, we can select P_{22} and P_{21} such that $P_{22}^{-1} P_{21} = F_*$
2. Select appropriately dimensional matrices $P_{22} = P_{22}' > 0$, $P_{21} = P_{22} F_*$, and $P_{11} = P_{11}' > P_{21}' P_{22}^{-1} P_{21}$ to ensure that

$$P = U \begin{bmatrix} P_{11} & P_{21}' \\ P_{21} & P_{22} \end{bmatrix} U' > 0 \tag{5.246}$$

3. Compute

$$W = P - (\bar{A} + \bar{B}F)' P (\bar{A} + \bar{B}F) \tag{5.247}$$

If W is not positive-definite, we go back to Step 2 to choose another solution of P or go to the first step to reselect another set of desired invariant zeros.

Another method for selecting W is based on a trial and error approach by limiting the choice of W to be a diagonal matrix and adjusting its diagonal weights through simulation. Generally, such an approach would yield a satisfactory result as well.

5.3.4 An Illustrative Example

We illustrate the enhanced CNF control technique for discrete-time systems in the following example. We consider a discrete-time system of Equation 5.193 with

$$A = \begin{bmatrix} 0 & 1 & 0 \\ 0 & 0 & 1 \\ 1 & -3 & 3 \end{bmatrix}, \quad B = \begin{bmatrix} 0 \\ 0 \\ 1 \end{bmatrix}, \quad E = \begin{bmatrix} 1 \\ 0 \\ 1 \end{bmatrix}, \quad x_0 = \begin{pmatrix} 0 \\ 0 \\ 0.5 \end{pmatrix} \tag{5.248}$$

$$C_1 = I_3, \quad C_2 = [\,1 \quad 0 \quad 0\,] \tag{5.249}$$

and $u_{\max} = 1$. The disturbance w is unknown. For simulation purpose, we assume $w = -0.1$. Our goal is to design a CNF state feedback control law that would yield a good transient performance in tracking a target reference $r = 2$.

Following the procedure given in the previous subsection, we select an integration gain $\kappa_{\mathrm{i}} = 0.3$ and obtain an appropriate augmented system. After a few tries, we found that the following state feedback gain to the augmented system would yield a good performance for our problem:

$$F = [\,-0.1 \quad -0.85 \quad 2.14 \quad -1.7\,] \tag{5.250}$$

which places the poles of $\bar{A} + \bar{B}F$ at 0.9, 0.4, $0.5 \pm j0.5$. We note that the first one corresponds to the integrator. Both the linear state feedback control and enhanced CNF control share the same integration dynamics:

$$x_{\mathrm{i}}(k+1) = x_{\mathrm{i}}(k) + 0.3[h(k) - r] \tag{5.251}$$

The linear state feedback control law is given by

$$u(k) = [\,-0.85 \quad 2.14 \quad -1.7\,]\,x(k) - 0.1x_{\mathrm{i}}(k) + 0.41r \tag{5.252}$$

Letting $W = \mathrm{diag}\{0.1, 1, 1, 1\}$, we obtain a positive-definite solution P for Equation 5.202, which is given by

$$P = \begin{bmatrix} 2.6201 & 0.7258 & 0.6543 & 1.1678 \\ 0.7258 & 1.6414 & -1.4825 & 2.1586 \\ 0.6543 & -1.4825 & 8.8061 & -7.3158 \\ 1.1678 & 2.1586 & -7.3158 & 13.3551 \end{bmatrix} \tag{5.253}$$

and a CNF state feedback law:

$$\begin{aligned} u(k) = {} & [\,-0.85 \quad 2.14 \quad -1.7\,]\,x(k) - 0.1x_{\mathrm{i}}(k) + 0.41r + \rho(e(k)) \\ & \times [\,-0.1678 \quad 2.3536 \quad -9.3268 \quad 10.0458\,] \left[\begin{pmatrix} x_{\mathrm{i}}(k) \\ x(k) \end{pmatrix} - \begin{pmatrix} 0 \\ 2 \\ 2 \\ 2 \end{pmatrix} \right] \end{aligned} \tag{5.254}$$

where ρ is as given in (5.241) with $\alpha = 20$ and $\beta = 1.5$. The simulation results given in Figures 5.4 and 5.5 clearly show that the CNF control has outperformed the linear control.

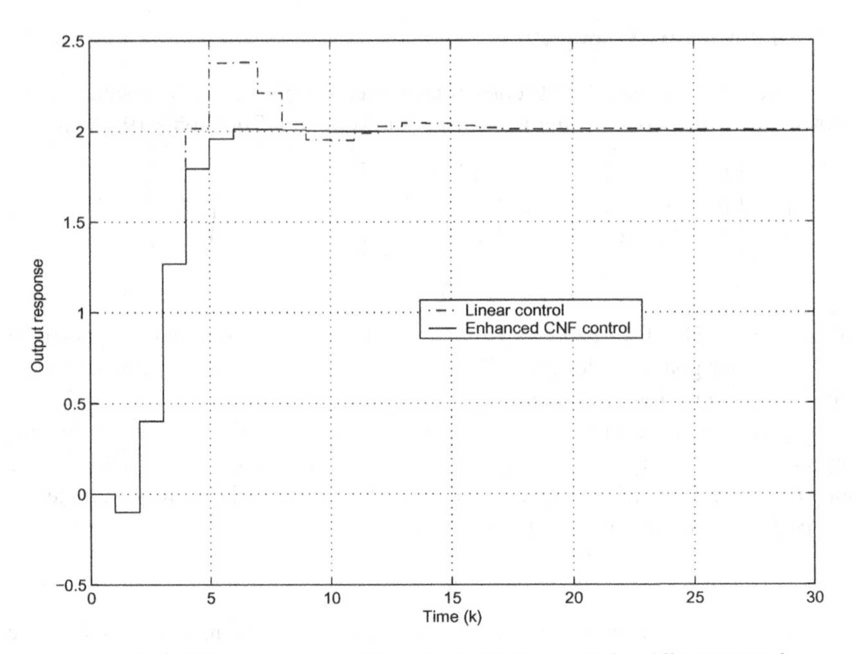

Figure 5.4. Output responses of the enhanced CNF control and linear control

5.4 Can We Beat Time-optimal Control?

So far, we have presented quite a number of control techniques in Chapters 4 and 5 that can be used to design control laws to track certain target references for systems with actuator saturations. The TOC technique is believed to be nonrobust to system uncertainties and noise, and thus cannot be used in tackling real problems. Unfortunately, it has also been regarded as a method that would, at least theoretically, yield the best performance in terms of settling time.

Can we design a control system that would beat the performance of the TOC? Obviously, the answer to this question is no if it is required to have a precise point-to-point tracking, *i.e.* to track a target reference precisely from a given initial point. However, surprisingly, the answer would be yes if we consider an asymptotic tracking situation, *i.e.* if we consider the settling time to be the total time that the controlled system output takes to get from its initial position to reach a predetermined neighborhood of the target reference. The reason that we are interested in this issue is that asymptotic tracking is widely used in almost all practical situations.

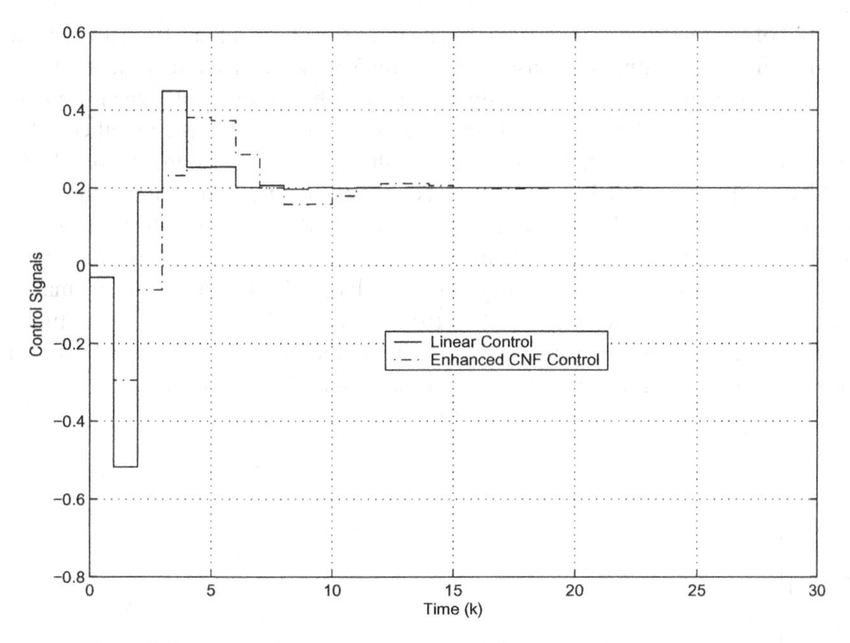

Figure 5.5. Control signals of the enhanced CNF control and linear control

In what follows, we show the above observation in an example. Let us consider a system characterized by a double integrator, *i.e.*

$$\dot{x} = \begin{bmatrix} 0 & 1 \\ 0 & 0 \end{bmatrix} x + \begin{bmatrix} 0 \\ 1 \end{bmatrix} \mathrm{sat}(u), \quad y = x, \quad h = [\,1 \quad 0\,]\, x \tag{5.255}$$

where as usual x is the state, u is the input, and y and h are, respectively, the measurement and controlled outputs. Moreover, we assume that

$$\mathrm{sat}(u) = \mathrm{sgn}(u) \cdot \min\{\, 1,\ |u|\, \} \tag{5.256}$$

Let the initial state $x(0) = 0$ and the target reference $r = 1$. Then, it follows from Equation 4.22 that the minimum time required for the controlled output to reach precisely the target reference under TOC control is exactly 2 s. Let us now consider an asymptotic tracking situation instead. As is commonly accepted in the literature (see, *e.g.*, [86]), we define the settling time to be the total time that it takes for the control output h to enter the $\pm 1\%$ region of the target reference. The following control law, obtained from a variation form of the CNF control technique, would give a faster settling time than that of the TOC,

$$\begin{aligned} u = {} & [\,-6.5 \quad -1\,]\, x + 6.5\, r \\ & - \left(\mathrm{e}^{-|1-h|} - 0.36788 \right) [\,1.4481 \quad 10.8609\,] \left(x - \begin{bmatrix} 1 \\ 0 \end{bmatrix} \right) \end{aligned} \tag{5.257}$$

Figures 5.6 and 5.7, respectively, show the resulting controlled output responses and the control signals of the TOC and the modified CNF control. The resulting output response of the modified CNF control has an overshoot of less than 1%. However, if we zoom in on the output responses (see Figure 5.8), we will see that the modified CNF control clearly has a faster settling time than that of the TOC when it enters the target region, *i.e.* $0.99 \leq h \leq 1.01$. It can be computed that the modified CNF control has a settling time of 1.8453 s whereas the TOC has a settling time of 1.8586 s. Although the difference is not much, since we have not tried to optimize the solution of the modified CNF control, it is, however, significant enough to address one interesting issue: *there are control laws that can achieve a faster settling time than that of the TOC in asymptotic tracking situations*. It can also be shown that, no matter how small the target region is, say $1 \pm \varepsilon$ for any small $\varepsilon > 0$, we can always find a suitable control law that beats the TOC in settling time. Nonetheless, we believe that it would be interesting to carry out some further studies in this subject.

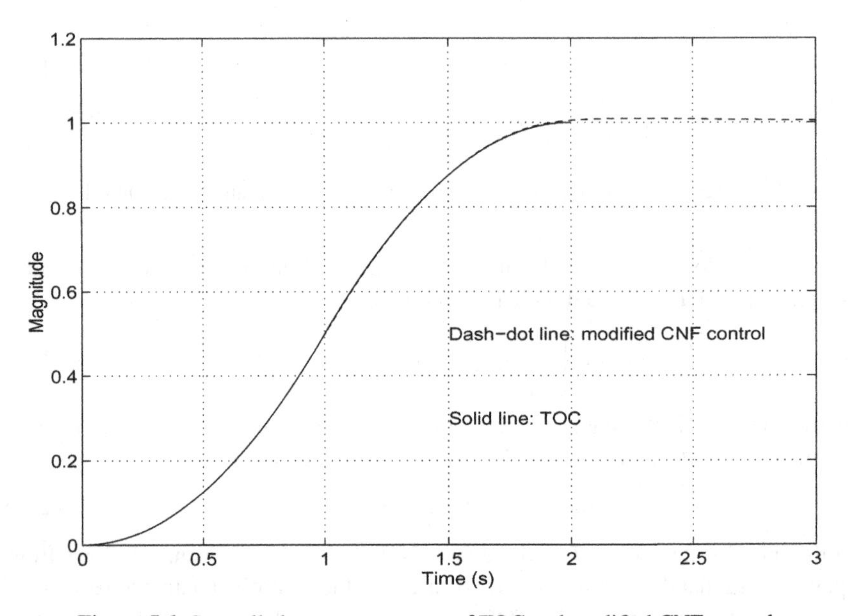

Figure 5.6. Controlled output responses of TOC and modified CNF control

5.5 CNF Control Software Toolkit

The CNF design involves selecting quite a number of design parameters, especially those parameters for forming the nonlinear feedback law. Some tuning and retuning are needed in order to obtain the best possible solutions. In what follows, we present a MATLAB® toolkit with a user-friendly graphical interface for the CNF control system design. With the help of the rich collection of m-functions in MATLAB®

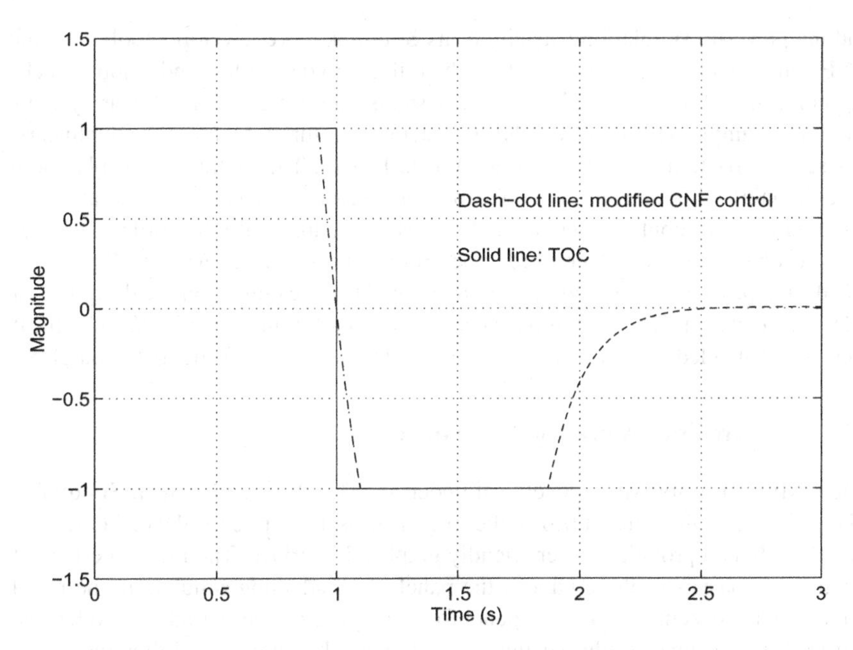

Figure 5.7. Control signals of TOC and modified CNF control

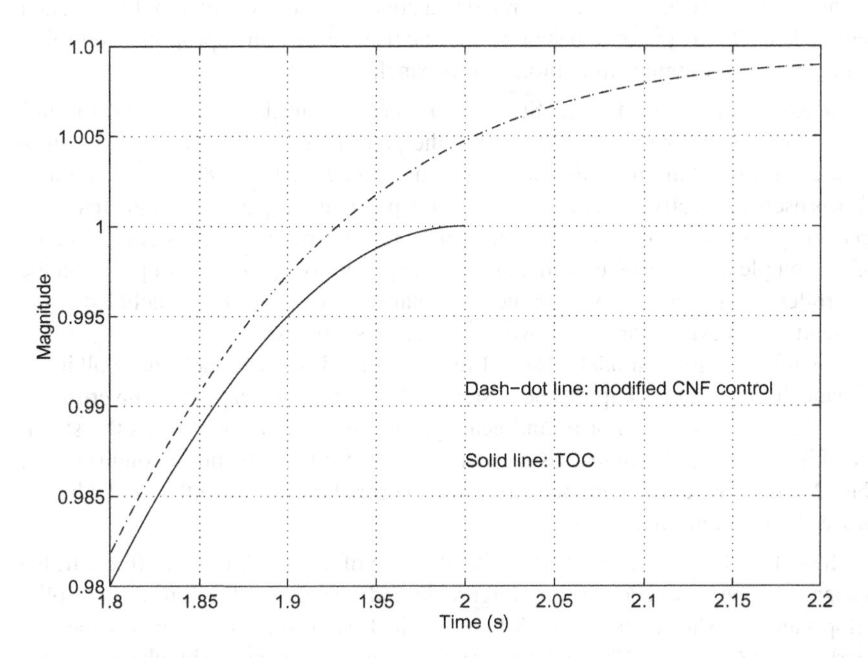

Figure 5.8. Controlled output responses around the target reference

and the powerful simulation capacity of its Simulink®, we develop a toolkit for the CNF control technique. The toolkit can be utilized to design fast and smooth tracking controllers for general SISO linear systems and a class of nonlinear systems (see, for example, [136] for details) with actuator saturation and other nonlinearities such as friction, as well as external disturbances. The toolkit can display both time-domain and frequency-domain responses on its main panel, and generate three different types of control laws, namely, the state feedback, the full-order measurement feedback and the reduced-order measurement feedback controllers. The usage and design procedure of the toolkit are illustrated by an example on the design of an HDD servo system. The toolkit has been reported earlier in Cheng *et al.* [54, 55]. It can be downloaded at the website `http://hdd.ece.nus.edu.sg/~bmchen`.

5.5.1 Software Framework and User Guide

The CNF control toolkit is developed under MATLAB® together with Simulink®. The software toolkit fully utilizes the graphical user interface (GUI) resources of MATLAB® and provides a user-friendly graphical interface. The main interface of the toolkit consists of three panels, the panel for conducting simulation, the panel for setting up system data and the panel for specifying an appropriate controller. We illustrate the design procedure using our CNF control toolkit in the following:

STAGE 1: INITIALIZATION. Once the toolkit is properly executed, a main panel as shown in Figure 5.9 will be generated in a popup window. Users have to first enter required data for a system to be controlled and then specify an appropriate controller structure before running simulation on this panel.

STAGE 2: PLANT MODEL SETUP. To enter system data, users need to click on the box labeled with `PLANT` to open the plant model setup panel as shown in Figure 5.10. In addition to the state-space model of Equation 5.89, the toolkit also allows users to specify resonance modes of the plant on this panel. We note that high-frequency resonance modes are existent in almost all mechanical systems. Because of the complexity of resonance modes, they are generally ignored or simplified in the controller design stage. However, these resonance modes have to be included in the simulation and evaluation of the overall control system design.

Each time a plant model is keyed in or modified on the panel, the toolkit automatically runs a checkup on the system stabilizability, detectability, invertibility, and other requirements. For a nonlinear system, the toolkit also checks the stability of its nonlinear dynamics. Users will be warned if the solvability conditions for the CNF tracking control are not satisfied and users have to revise the model before proceeding to controller design.

STAGE 3: CONTROLLER SETUP. As the core of this toolkit, the CNF controller design is to be proceeded in a configurable and convenient fashion. A controller setup panel as shown in Figure 5.11 is opened when the user activates the box marked with `CONTROLLER` in the main panel. This panel carries a block diagram for an adjustable controller configuration, which automatically refreshes when the user makes any change or reselection on the controller structure. Users need to decide

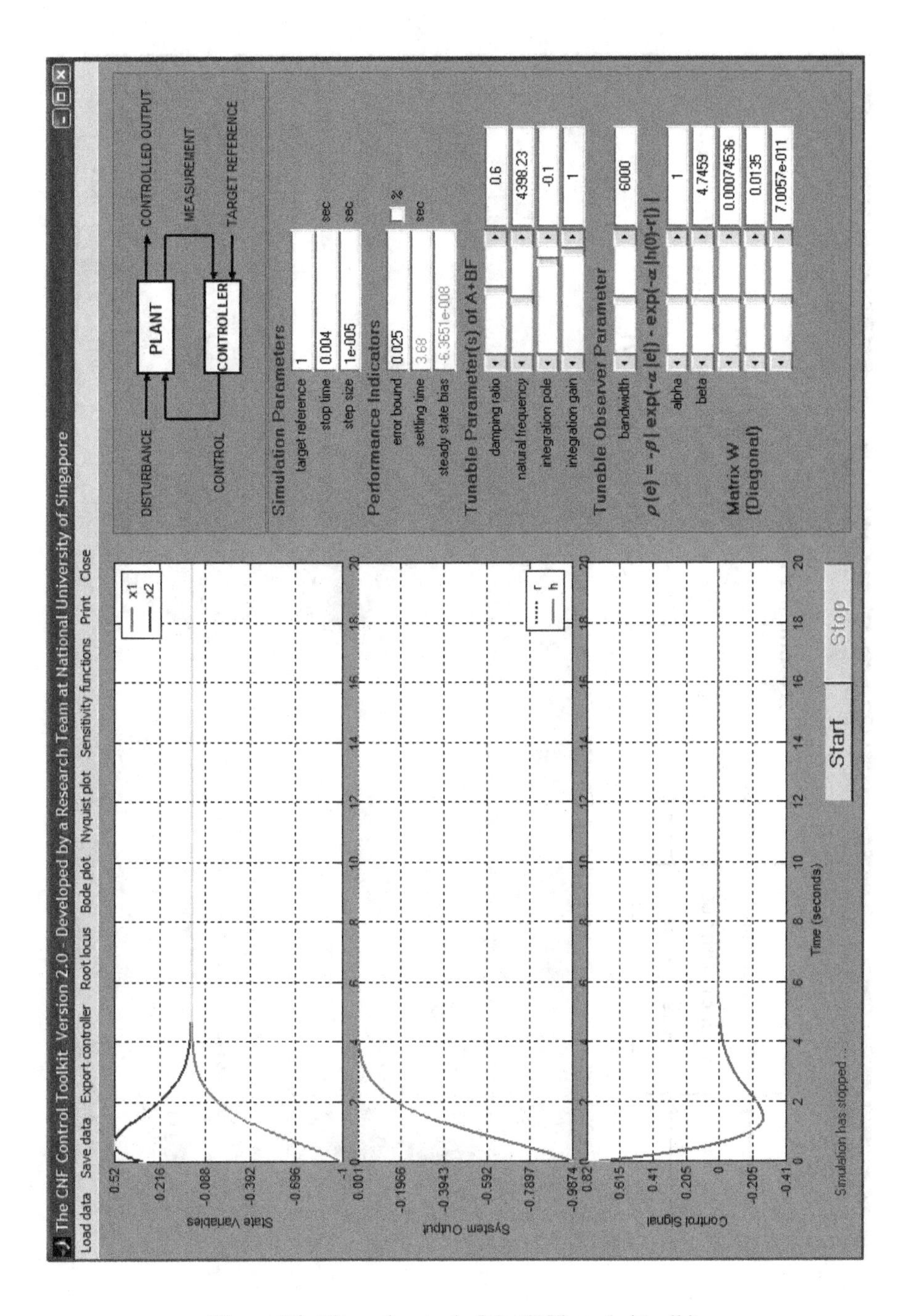

Figure 5.9. The main panel of the CNF control toolkit

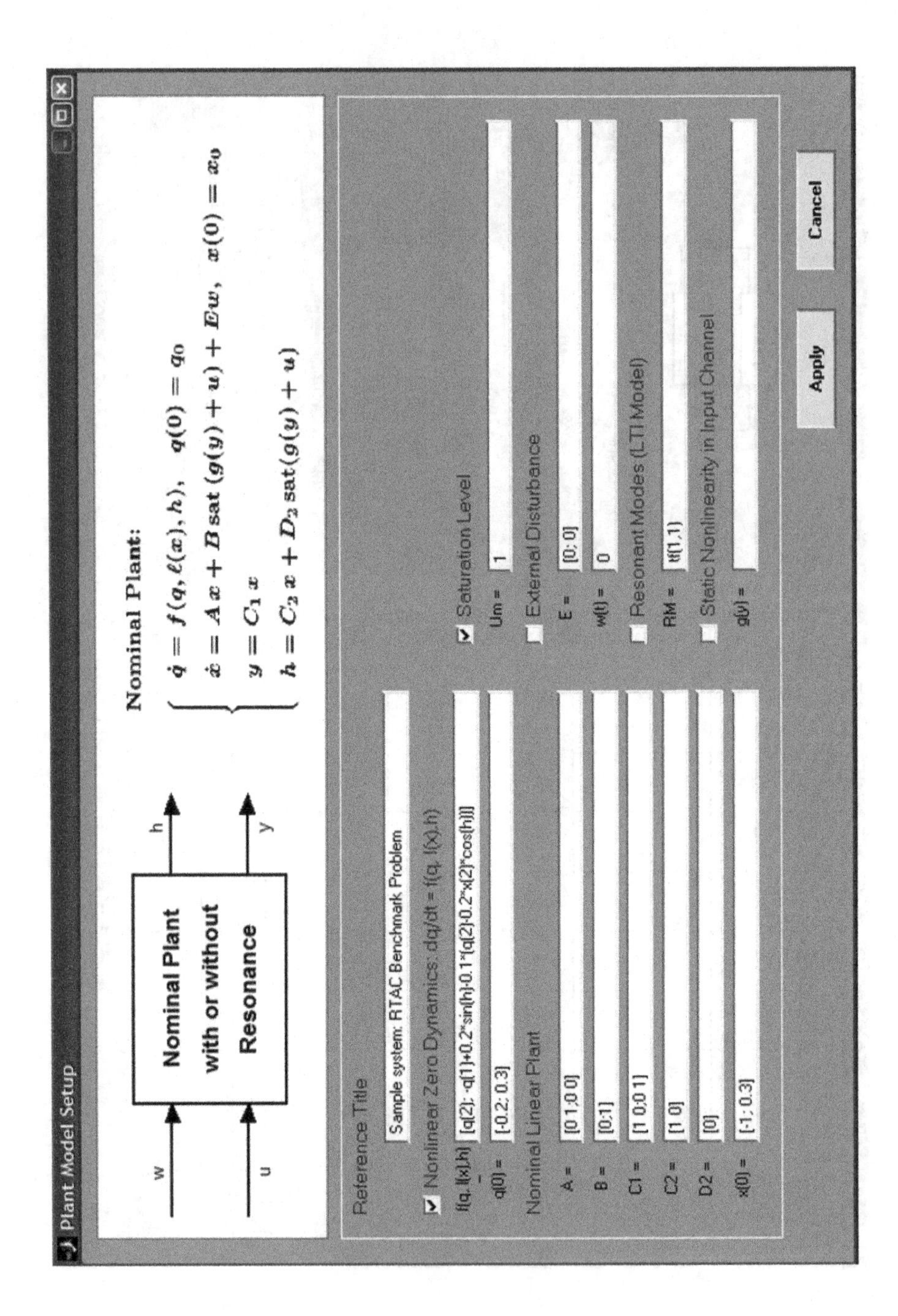

Figure 5.10. The panel for the plant model setup

a controller structure first before proceeding to specify the corresponding controller parameters. The following options are available for the controller in the toolkit:

1. If the given plant in Equation 5.89 has some nonlinearities, users can choose the precompensation option and enter an appropriate nonlinear function to cancel as many nonlinearities as possible.
2. If the given plant in Equation 5.89 has some unknown constant disturbances or other types of disturbances, users can select a controller structure with integrator to remove the steady-state bias.
3. If the given plant in Equation 5.89 is the nominal model of a noisy plant, which has high-frequency resonance modes, users might enter a predesigned lowpass or notch filter to minimize their effects on the overall performance.
4. Based on the properties of the given plant in Equation 5.89 and personal interest, users can then choose a controller with either one of the following options:
   a) State feedback;
   b) Full-order measurement feedback;
   c) Reduced-order measurement feedback.

When it comes to the design of the state feedback gain F, users have three choices. Users can either specify an explicit matrix (obtained through any other design methodologies such as H_∞ control) for F, or employ the H_2 control technique (provided on the panel), or use the pole-placement method characterized by the damping ratio and natural frequency of the dominant poles as well as integration pole and gain (if applicable). For the pole-placement method, the remaining closed-loop poles are placed three times faster than the dominant pair with a Butterworth pattern.

In the case of measurement feedback control, users also have three choices for the design of the observer gain K. Users can directly enter a predesigned solution for K, or design it online using the H_2 control technique, or using the pole-placement method to organize the observer poles into a Butterworth pattern with an appropriate bandwidth. In any case, the corresponding parameters can be further tuned on the main panel to obtain a satisfactory performance for the overall system.

STAGE 4: DESIGN AND SIMULATION. Once the plant model setup and the controller setup are completed, users can then specify directly on the main panel the simulation parameters, such as the setpoint for the target reference, the duration of simulation and the step size. Users can also define the tracking performance indicator and obtain the result for settling time and steady-state bias of the controlled output response. For example, in Figure 5.9, the settling time is defined as the time when the controlled output of the closed-loop system enters the neighborhood of ± 0.025 of the target reference. Alternatively, one can define such a neighborhood in terms of the percentage of final target instead of the absolute error bound.

As mentioned in the previous section, the CNF controller consists of two parts, a linear part and a nonlinear part. On the main panel, users are able to tune the properties of the linear part by selecting appropriate values of the damping ratio and natural frequency of the dominant modes of the linear state feedback dynamical matrix $\bar{A} + \bar{B}F$. The remaining eigenvalues of $\bar{A} + \bar{B}F$ are placed three times faster

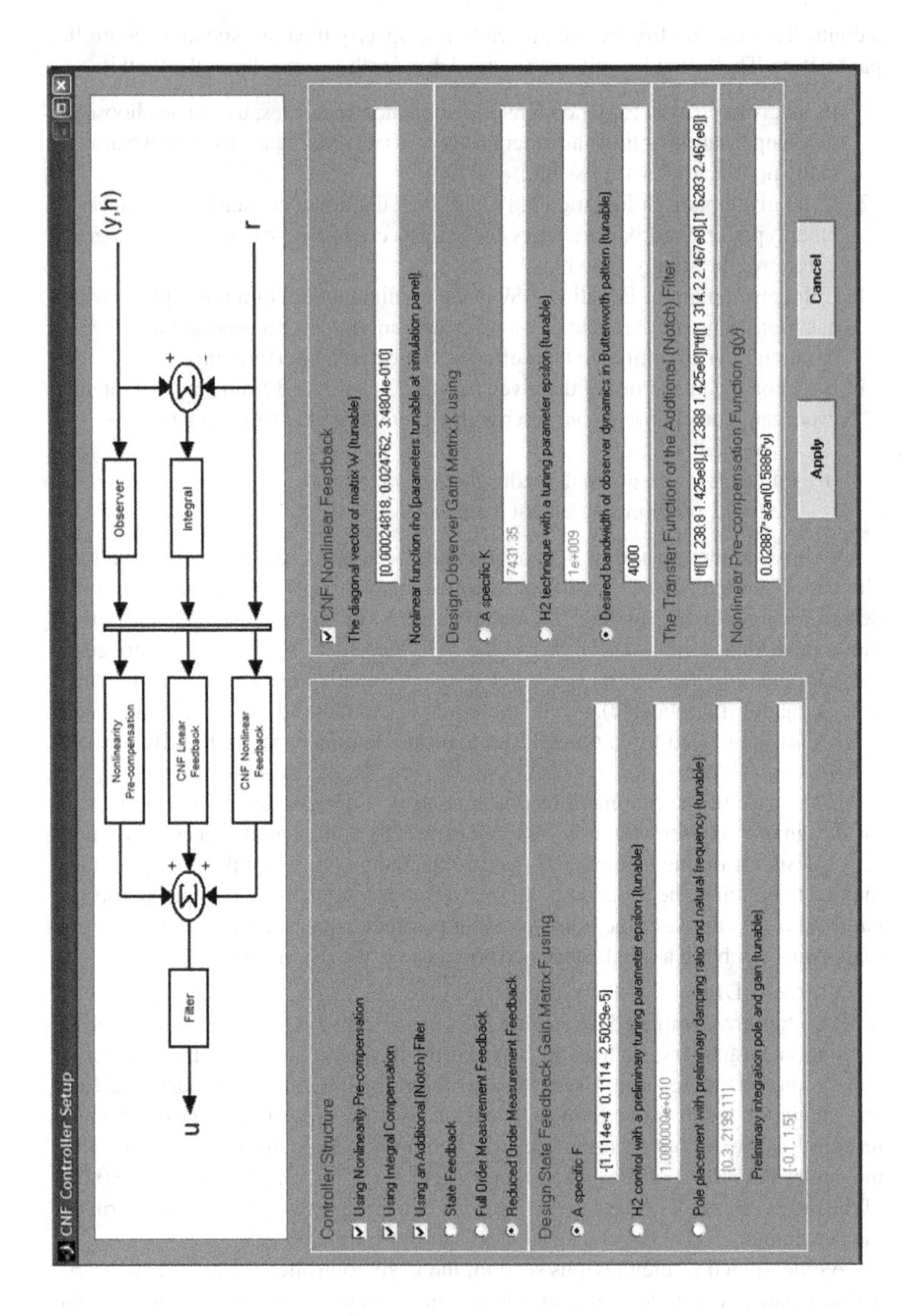

Figure 5.11. The panel for the CNF controller setup

than the dominant modes with a Butterworth pattern. Such an arrangement is purely for the simplification of the control system design. The pole corresponding to the integration part, if applicable, can be tuned on the main panel as well. Alternatively, users are allowed to apply the H_2 design with a tunable parameter ε, or directly specify a state feedback gain matrix (obtained using any design method) in the controller setup panel without any tunable parameters.

Design parameters W and $\rho(e)$ for the nonlinear part of the CNF controller can also be tuned online on the main panel. In particular, the parameters α and β for the nonlinear function $\rho(e)$ of Equation 5.128 can be easily adjusted using the computer mouse or by directly keying their values onto the spaces provided. In the current version of the CNF control toolkit, the design parameter W is restricted to a diagonal matrix for the simplicity of software implementation.

There are three windows on the main panel for displaying the system state variables, the controlled output response and the control input signal, together with a block diagram showing the structure of the overall control system. Using the right button of the computer mouse to click on the window displaying the state variables, output response and control signal, users are prompted by a small text window showing options to redraw the plots on a new popup window or export the simulation data to the MATLAB® workspace.

Finally, the following commands and functions are also implemented on the main panel for saving and loading data as well as for evaluating the frequency-domain properties of the overall control system:

1. Load data: This function is used to load data previously saved in the toolkit.
2. Save data: This function is to save the system and controller data for future use.
3. Export controller: This function is to export the data of the CNF controller obtained to the MATLAB® workspace. The controller data are given by

$$\left.\begin{aligned} \dot{z} &= \kappa_{\rm i}(h - r) \\ \dot{x}_{\rm v} &= A_{\rm cmp}\, x_{\rm v} + B_{\rm ycmp}\, y + B_{\rm ucmp}\, \mathrm{sat}(\bar{u}) \\ \hat{x} &= C_{\rm cmp}\, x_{\rm v} + D_{\rm cmp}\, y \\ \bar{u} &= F \begin{pmatrix} z \\ \hat{x} \end{pmatrix} + G\, r + \rho(e) F_{\rm n} \left[\begin{pmatrix} z \\ \hat{x} \end{pmatrix} - G_{\rm e}\, r \right] \\ u &= H(s)\bar{u} - g(y) \end{aligned}\right\} \tag{5.258}$$

   where $A_{\rm cmp}$, $B_{\rm ycmp}$, $B_{\rm ucmp}$, $C_{\rm cmp}$, $D_{\rm cmp}$, F, $F_{\rm n}$, G, $G_{\rm e}$ are constant vectors or matrices, and $\rho(e)$ and $g(y)$ are scalar functions. If a filter is used to reduce the effects of noise or high-frequency resonance modes of a physical plant, $H(s)$ will represent the transfer function of such a filter. All these parameters can be saved under a structured workspace variable (specified by the user) in the command window.
4. Root locus: This function is to generate the root locus of the control system with the CNF controller with respect to the change of the nonlinear function $\rho(e)$.
5. Bode plot: This function is to generate the Bode magnitude and phase responses of the open-loop system comprising the plant and the controller in the steady-

state situation when the nonlinear gain is converging to a constant. This function can be used to evaluate the frequency-domain properties of the control system, such as gain and phase margins.
6. Nyquist plot: Similar to the Bode plot given above. Note that for both the Bode and Nyquist plots, users are allowed to specify a frequency range of interest.
7. Sensitivity functions: This function is to plot the sensitivity and complementary sensitivity functions of the overall design with a prespecified frequency range.
8. Print: This function is to print the items shown on the main panel.
9. Close: This command is to close up the CNF control toolkit.

Although the CNF control toolkit is meant to design a composite nonlinear feedback controller, users have the option to choose only the linear portion of the CNF control law for their design. This option is particularly useful for the comparison of performances of the CNF controller and the linear controller.

5.5.2 An Illustrative Example

We illustrate the CNF control toolkit using a practical example on a micro hard disk drive (HDD) servo system design. The following dynamical model of an IBM microdrive (DMDM-10340) has been recently reported in Peng *et al.* [138] and will be further studied in Chapter 9:

$$\begin{cases} \dot{x} = \begin{bmatrix} 0 & 1 \\ 0 & 0 \end{bmatrix} x + \begin{bmatrix} 0 \\ 2.35 \times 10^8 \end{bmatrix} \left[\text{sat}\Big(g(y) + u\Big) + w \right] \\ y = h = [1 \quad 0]\, x \end{cases} \tag{5.259}$$

where the state variable x, consists of the displacement and velocity of the read/write (R/W) head of the microdrive; y and h are simply the displacement (in μm) of the R/W head of the drive; w represents friction and some unknown disturbances of the system, and is set to be -1 mV for this demonstration; u is the control input, limited by ± 3 V; and finally the nonlinearity $g(y) = -0.02887\text{artan}(0.5886y)$ is generated by the data flex cable in the drive. The high-frequency resonance modes of the microdrive as given in Chapter 9 are included in simulation.

For hard disk drive servo systems, the task of the controller is to move the actuator R/W head to a desired track as fast as possible and to maintain the R/W head as close as possible to the target track center when data are being read or written. To ensure reliable data reading and writing, it is required that the deviation of the R/W head from the target track center should not exceed 5% of the track pitch. In this demonstration, we focus our attention on designing a servo system that yields an optimal performance for short-distance tracking. In particular, we set the target reference $r = 1$ μm. Thus, the settling time is defined as the total duration for the R/W head to enter the ± 0.05 μm region of the target track.

Using the CNF control toolkit of [55] with few online adjustments on the design parameters, we manage to obtain a reduced-order CNF controller that yields a good performance. The dynamics equations of the controller is given by:

$$\begin{pmatrix} \dot{x}_{\rm i} \\ \dot{x}_{\rm v} \end{pmatrix} = \begin{bmatrix} 0 & 0 \\ 0 & -6000 \end{bmatrix} \begin{pmatrix} x_{\rm i} \\ x_{\rm v} \end{pmatrix} + \begin{bmatrix} 0 \\ 2.35 \times 10^8 \end{bmatrix} \text{sat}(u) - \begin{bmatrix} -10 \\ 3.6 \times 10^7 \end{bmatrix} y - \begin{bmatrix} 10 \\ 0 \end{bmatrix} r \tag{5.260}$$

$$\begin{aligned} u = \rho(e) \, [0.02398 \;\; 0.02646 \;\; 2.5472 \times 10^{-5}] & \begin{pmatrix} x_{\rm i} \\ y - r \\ x_{\rm v} + 6000y \end{pmatrix} \\ - [2.0851 \;\; 0.20911 \;\; 6.3830 \times 10^{-6}] & \begin{pmatrix} x_{\rm i} \\ y - r \\ x_{\rm v} + 6000y \end{pmatrix} - g(y), \end{aligned} \tag{5.261}$$

with $\rho(e) = -4.4 \left| {\rm e}^{-4|e|} - {\rm e}^{-4} \right|$. Figure 5.12 shows the controlled output response of the servo system with the obtained CNF controller. The resulting 2.5% settling time is about 0.27 ms. The frequency-domain properties of the CNF control servo system are, respectively, shown in Figures 5.13 to 5.15.

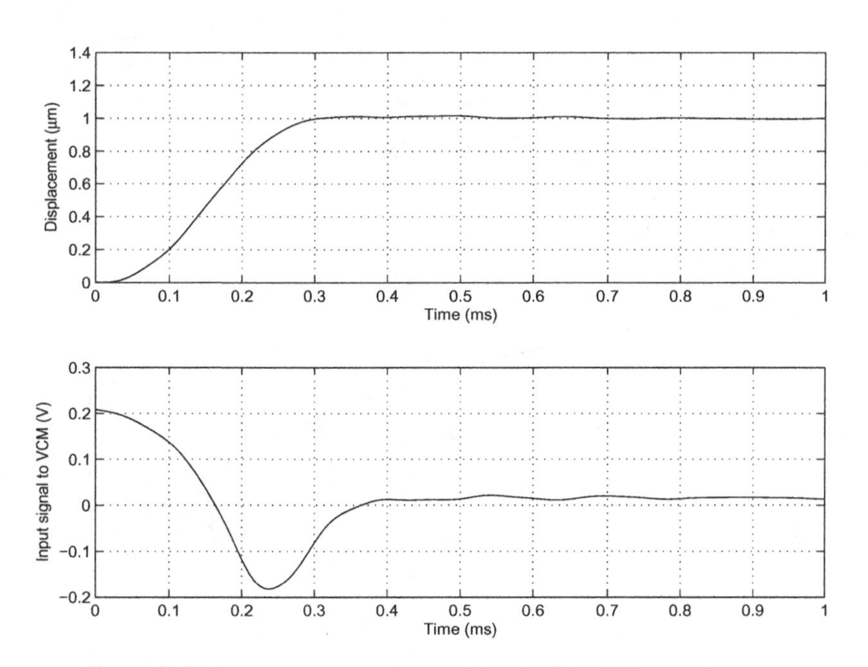

Figure 5.12. Output response and control signal of the CNF control system

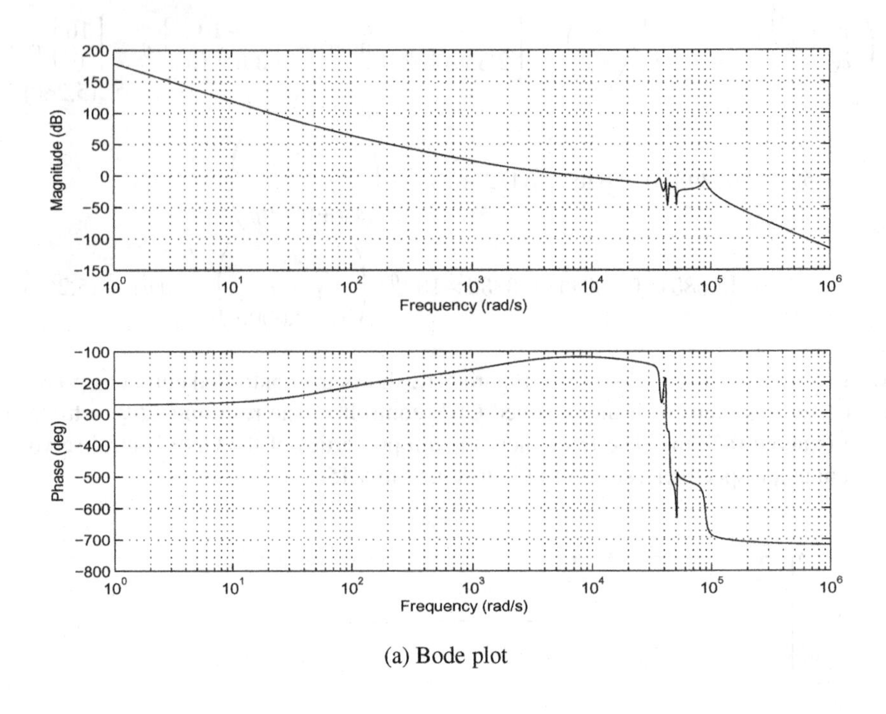

(a) Bode plot

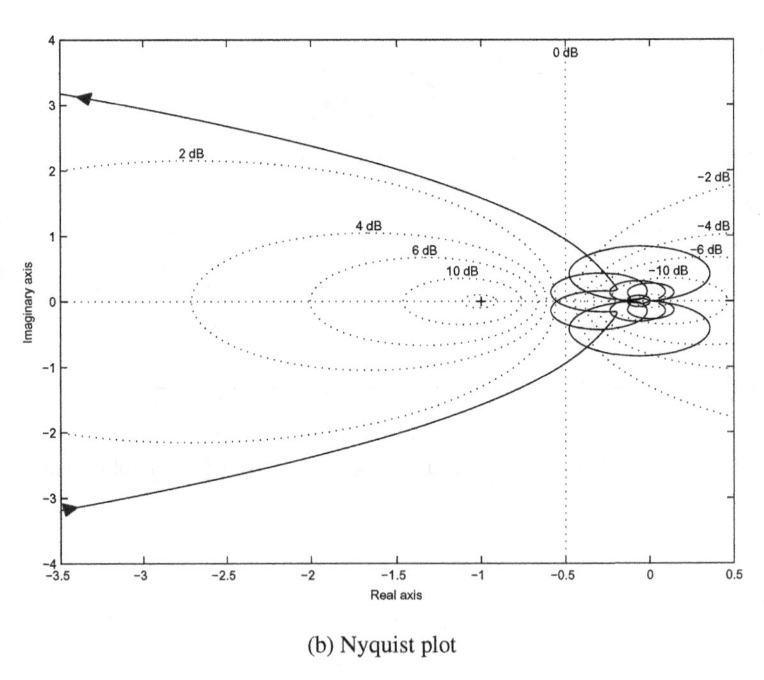

(b) Nyquist plot

Figure 5.13. Bode and Nyquist plots of the CNF control system

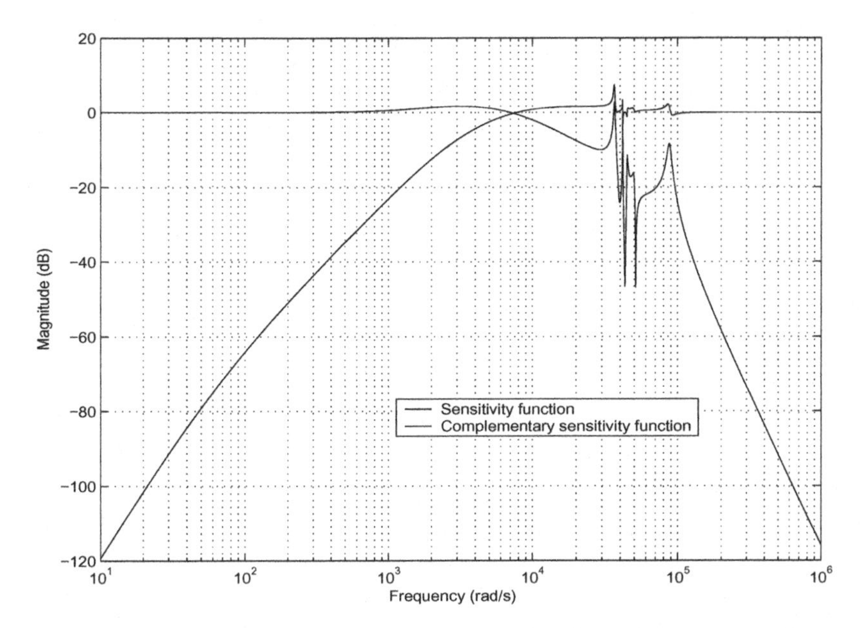

Figure 5.14. Sensitivity and complementary sensitivity functions of the CNF control system

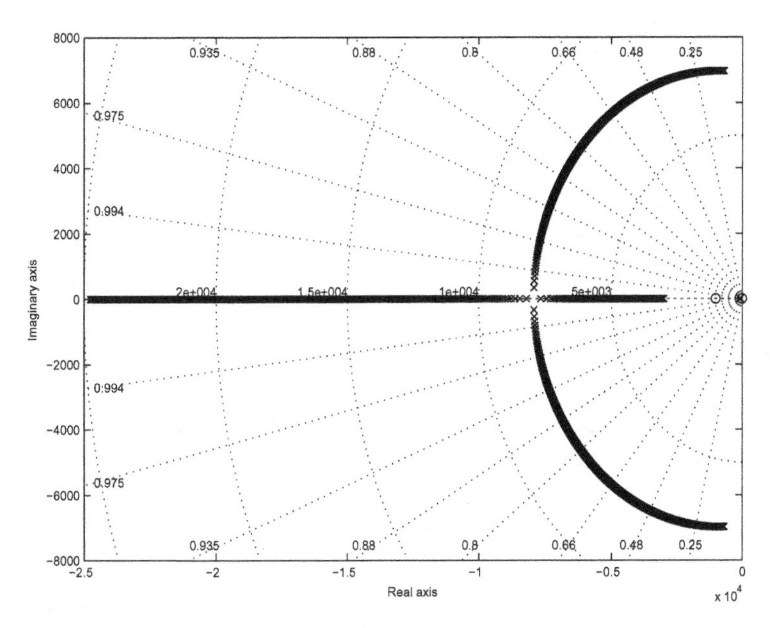

Figure 5.15. Root locus of the closed-loop system versus the nonlinear function $\rho(e)$

Part II

HDD Servo Systems Design

6

Track Following of a Single-stage Actuator

6.1 Introduction

The prevalent trend in hard disk design is towards smaller hard disks with increasingly larger capacities. This implies that the track width has to be smaller, leading to lower error tolerance in the positioning of the head. The controller for track following has to achieve tighter regulation in the control of the servomechanism. Current HDDs use a combination of classical control techniques, such as lead-lag compensators, PID compensators, and notch filters. These classical methods can no longer meet the demand for HDDs of higher performance. Thus, many control approaches have been tried, such as LQG and/or LTR approach (see, *e.g.*, [18, 19]), and adaptive control (see, *e.g.*, [28]) and so on.

The purpose of this chapter is to use the results of the RPT control method of Chapter 3 and the CNF control technique of Chapter 5 to carry out the design of track-following controllers for an HDD with a single VCM actuator. We first obtain a model of the VCM actuator and then cast the overall track-following control system design into an RPT design framework. A second-order dynamic measurement feedback controller is then designed to achieve robust and perfect tracking for any step reference. Our controller is theoretically capable of making the L_p-norm of the resulting tracking error arbitrarily small in faces of external disturbances and initial conditions. Some tradeoffs are then made in order for the RPT controller to be implementable using the existing hardware setup and to meet physical constraints such as sampling rates and the limit of control of the system. The CNF control technique and the classical PID control technique is also employed to design track-following controllers for the same drive. The simulation and implementation results of these controllers are done in both time domain and frequency domain and compared. The result on the RPT design presented in this chapter are rooted in earlier work reported in [3] (see also [74]). However, it is obtained using a new HDD, namely a Maxtor (Model 51536U3) HDD. Furthermore, we introduce a notch filter to minimize the effect of the resonance modes and follow the development of a recent work by Cheng *et al.* [140] to add an additional integrator to the RPT controller. The new design is capable of eliminating steady-state bias.

6.2 VCM Actuator Model

In this section, we present the modeling of the VCM actuator that is well known in the research community of the HDD servo systems to have a characteristic of a double integrator cascaded with some high-frequency resonance, which can reduce the system stability if neglected. There are some bias forces in the HDD system that cause steady-state errors in tracking performance. Moreover, there are also some nonlinearities in the system at low frequencies, which are primarily due to the pivot-bearing friction. All these factors have to be taken into consideration when considering the design of a controller for the VCM actuator. For the purpose of developing a model, we have to compromise between accuracy and simplicity. In this section, a relatively simplified model of the VCM actuator is identified and presented.

The dynamics of an ideal VCM actuator can be formulated as a second-order state-space model as follows:

$$\begin{pmatrix} \dot{y} \\ \dot{v} \end{pmatrix} = \begin{bmatrix} 0 & k_{\mathrm{y}} \\ 0 & 0 \end{bmatrix} \begin{pmatrix} y \\ v \end{pmatrix} + \begin{pmatrix} 0 \\ k_{\mathrm{v}} \end{pmatrix} u \tag{6.1}$$

where u is the actuator input (in volts), y and v are the position (in tracks) and the velocity of the R/W head, k_{y} is the position measurement gain and $k_{\mathrm{v}} = k_{\mathrm{t}}/m$, with k_{t} being the current–force conversion coefficient and m being the mass of the VCM actuator. Thus, the transfer function of an ideal VCM actuator model appears to be a double integrator, *i.e.*

$$G_{\mathrm{v1}}(s) = \frac{k_{\mathrm{v}} k_{\mathrm{y}}}{s^2} \tag{6.2}$$

However, if we also consider the high-frequency resonance modes, a more realistic model for the VCM actuator will be

$$G_{\mathrm{v}}(s) = \frac{k_{\mathrm{v}} k_{\mathrm{y}}}{s^2} \prod_{i=1}^{N} G_{\mathrm{r},i}(s) \tag{6.3}$$

where N is the number of significant resonance modes in the frequency range of interest, and $G_{\mathrm{r},i}(s)$, $i = 1, 2 \ldots, N$, are the transfer functions for the resonance modes. The frequency characteristics of the Maxtor (Model 51536U3) HDD have been obtained using an LDV and an HP Dynamic Signal Analyzer. The actual frequency response is shown in Figure 6.1. Applying the least squares estimation identification method given in Chapter 2 (see also [13, 59]) to the measured data from the actual system, we obtain a tenth-order model for the actuator:

$$G_{\mathrm{v}}(s) = \frac{6.4013 \times 10^7}{s^2} \prod_{i=1}^{4} G_{\mathrm{r},i}(s) \tag{6.4}$$

with

$$G_{\mathrm{r},1}(s) = \frac{0.912s^2 + 457.4s + 1.433 \times 10^8}{s^2 + 359.2s + 1.433 \times 10^8} \tag{6.5}$$

$$G_{r,2}(s) = \frac{0.7586s^2 + 962.2s + 2.491 \times 10^8}{s^2 + 789.1s + 2.491 \times 10^8} \tag{6.6}$$

$$G_{r,3}(s) = \frac{9.917 \times 10^8}{s^2 + 1575s + 9.917 \times 10^8} \tag{6.7}$$

and

$$G_{r,4}(s) = \frac{2.731 \times 10^9}{s^2 + 2613s + 2.731 \times 10^9} \tag{6.8}$$

Figure 6.1 shows that the frequency response of the identified model matches the measured data very well for the frequency range from 0 to 1.8 kHz, which far exceeds the working range of the VCM actuator.

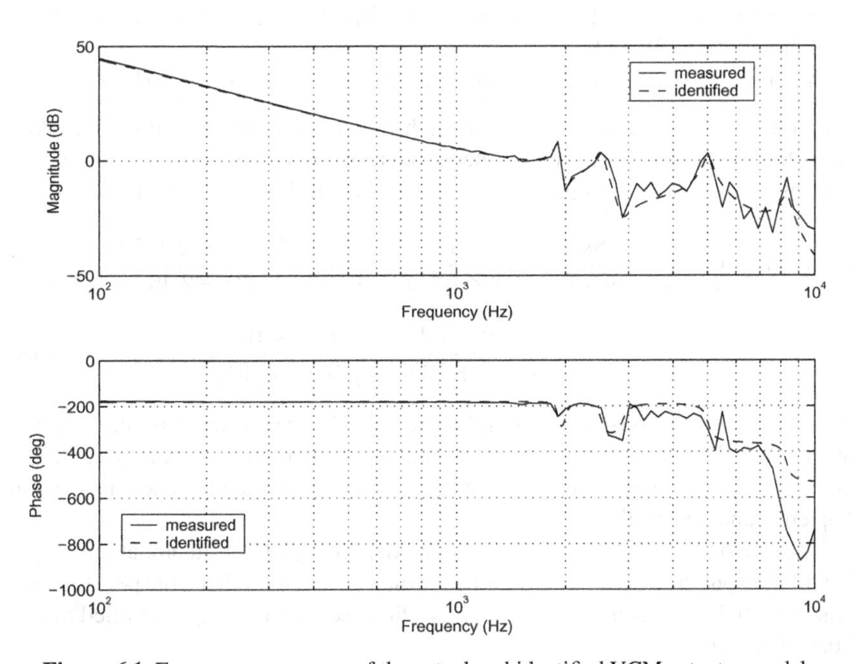

Figure 6.1. Frequency responses of the actual and identified VCM actuator models

6.3 Track-following Controller Design

We now present the control system design for the actuator identified in the previous section. Basically, the majority of commercially available HDD servo systems to date are designed using a conventional PID approach. For drives with a single VCM actuator, designers would encounter problems if they wished to push up the track-following speed. Usually, there are some huge overshoot peaks in step response. Thus, in practice, one would have to make tradeoffs between the track-following speed and the overshoot by selecting appropriate PID controller gains. These drawbacks can be overcome by using our newly developed control techniques.

We design an HDD servo system that meets the following design constraints and specifications:

1. the control input does not exceed ± 3 V owing to physical constraints on the actual VCM actuator;
2. the overshoot and undershoot of the step response are kept to less than 5% as the R/W head can start to read or write within $\pm 5\%$ of the target;
3. the 5% settling time in the step response is as short as possible;
4. the gain margin and phase margin of the overall design are, respectively, greater than 6 dB and 30°;
5. the maximum peaks of the sensitivity and complementary sensitivity functions are less than 6 dB; and
6. the sampling frequency in implementing the actual controller is 20 kHz.

In order to minimize the effect of the high-frequency resonance modes, we add a notch filter to the plant to cancel as much as possible of the unwanted responses. For the plant considered in this chapter, we introduce the following notch filter

$$G_{\text{notch}}(s) = \left(\frac{s^2 + 238.8s + 1.425 \times 10^8}{s^2 + 2388s + 1.425 \times 10^8}\right) \times \left(\frac{s^2 + 314.2s + 2.467 \times 10^8}{s^2 + 3142s + 2.467 \times 10^8}\right) \times \left(\frac{s^2 + 628.3s + 9.87 \times 10^8}{s^2 + 12570s + 9.87 \times 10^8}\right) \tag{6.9}$$

whose frequency response is given in Figure 6.2. The overall response of the plant together with the notch filter is given in Figure 6.3, which shows that the effect of the first two resonance modes are indeed reduced. This same notch filter will be used in Chapters 7 and 8 as well.

Thus, we only consider a second-order model for the VCM actuator at this stage. We will then put the resonance modes back when we are to evaluate the performance of the overall design. Thus, in our design, we first use the following simplified model of the VCM actuator:

$$\dot{x} = Ax + Bu = \begin{bmatrix} 0 & 1 \\ 0 & 0 \end{bmatrix} x + \begin{bmatrix} 0 \\ 6.4013 \times 10^7 \end{bmatrix} u \tag{6.10}$$

and

$$y = C_1 x = [1 \quad 0]\, x \tag{6.11}$$

Next, we define the output to be controlled as

$$h = y = C_2 x + D_2 u = [1 \quad 0]\, x \tag{6.12}$$

The overall control system is depicted as in Figure 6.4.

RPT Control System Design. Consider the reference $r(t)$ to be a step function with a magnitude α, *i.e.* $r(t) = \alpha \cdot 1(t)$, where $1(t)$ is a unit step function. Then, we have

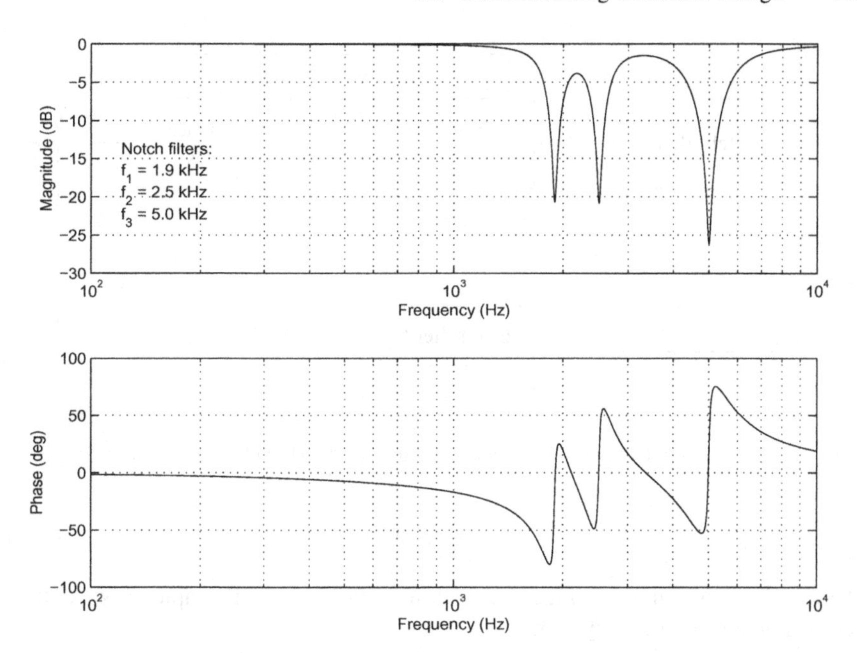

Figure 6.2. Frequency responses of the notch filter

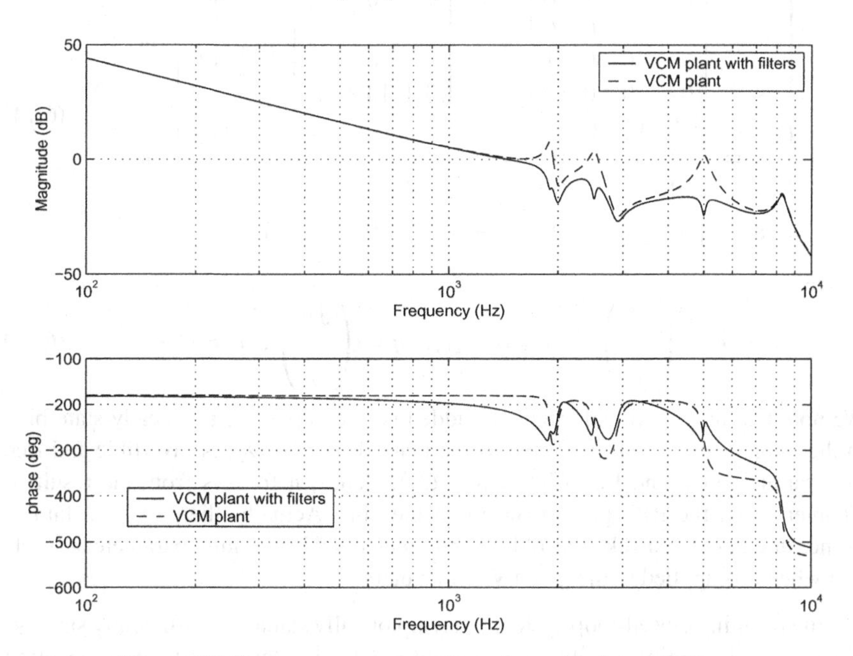

Figure 6.3. Frequency responses of the plant with the notch filter

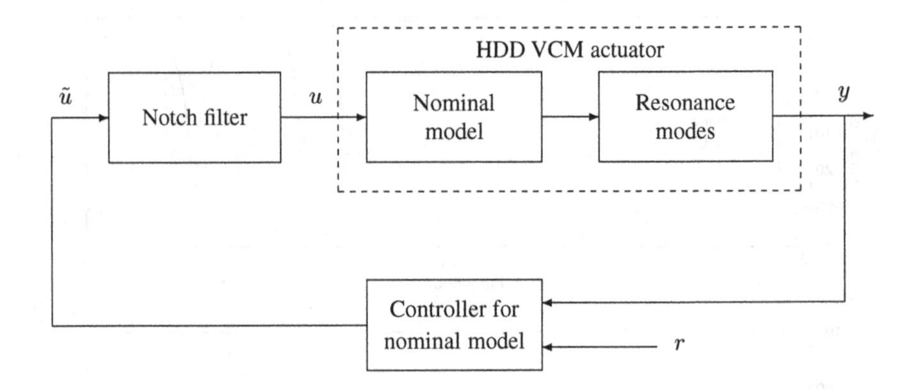

Figure 6.4. Control system configuration for the HDD VCM actuator

$$\dot{r}(t) = \alpha \cdot \delta(t) \tag{6.13}$$

where $\delta(t)$ is a unit impulse function. Following the results of Chapter 3, we obtain a corresponding auxiliary system:

$$\begin{cases} \dot{\boldsymbol{x}} = \begin{bmatrix} 0 & -1 & 1 & 0 \\ 0 & 0 & 0 & 0 \\ 0 & 0 & 0 & 1 \\ 0 & 0 & 0 & 0 \end{bmatrix} \boldsymbol{x} + \begin{bmatrix} 0 \\ 0 \\ 0 \\ 6.4013 \times 10^7 \end{bmatrix} u + \begin{bmatrix} 0 \\ 1 \\ 0 \\ 0 \end{bmatrix} \boldsymbol{w} \\ \boldsymbol{y} = \begin{bmatrix} 1 & 0 & 0 & 0 \\ 0 & 1 & 0 & 0 \\ 0 & 0 & 1 & 0 \end{bmatrix} \boldsymbol{x} + \begin{bmatrix} 0 \\ 0 \\ 0 \end{bmatrix} \boldsymbol{w} \\ e = \begin{bmatrix} 0 & -1 & 1 & 0 \end{bmatrix} \boldsymbol{x} + 0 \quad u \end{cases} \tag{6.14}$$

where

$$\boldsymbol{x} = \begin{pmatrix} \int e \\ r \\ x \end{pmatrix}, \quad \boldsymbol{w} = \alpha \cdot \delta(t), \quad \boldsymbol{y} = \begin{pmatrix} \int e \\ r \\ y \end{pmatrix}, \quad e = h - r \tag{6.15}$$

We note that the additional integrator added is to compensate the steady-state bias in the implementation. It is simple to see that (A, B, C_2, D_2) is invertible and free of invariant zeros, and Ker (C_1) = Ker (C_2). Hence, it follows from the result of Chapter 3 that the RPT performance is achievable. Actually, there exists a family of measurement feedback control laws, parameterized by a tuning parameter ε, such that when it is applied to the given VCM actuator:

1. the resulting closed-loop system is asymptotically stable for sufficiently small ε;
2. for any given initial condition x_0 and any $p \in [1, \infty)$, the l_p-norm of the resulting tracking error e has the property $\|e\|_p \to 0$, as $\varepsilon \to 0$.

Following the procedure for the reduced-order RPT controller given in Chapter 3, we obtain a parameterized measurement feedback control law of the form

$$\dot{x}_{\rm v} = A_{\rm RC}(\varepsilon)x_{\rm v} + B_{\rm RC}(\varepsilon)\begin{pmatrix} r \\ y \end{pmatrix}, \quad \tilde{u} = C_{\rm RC}(\varepsilon)x_{\rm v} + D_{\rm RC}(\varepsilon)\begin{pmatrix} r \\ y \end{pmatrix} \tag{6.16}$$

with

$$\left.\begin{aligned} A_{\rm RC}(\varepsilon) &= \begin{bmatrix} 0 & 0 \\ -696.4/\varepsilon^3 & -7229.5/\varepsilon \end{bmatrix} \\ B_{\rm RC}(\varepsilon) &= \begin{bmatrix} -1 & 1 \\ 69.962/\varepsilon^2 & -359.14/\varepsilon^2 \end{bmatrix} \\ C_{\rm RC}(\varepsilon) &= [\,-1.0879/\varepsilon^3 \quad -5.0451/\varepsilon\,] \\ D_{\rm RC}(\varepsilon) &= [\,0.10929/\varepsilon^2 \quad -0.3111/\varepsilon^2\,] \end{aligned}\right\} \tag{6.17}$$

The actual control input to the HDD system is then given by

$$u = G_{\rm notch}(s)\tilde{u} \tag{6.18}$$

where $G_{\rm notch}(s)$ is as given in Equation 6.9. Figure 6.5 clearly shows that the RPT problem is solved as we tune the tuning parameter ε to be smaller and smaller. Unfortunately, owing to the constraints of the physical system, *i.e.* the limits in control inputs and sampling rates, as well as resonance modes, it is impossible to implement a controller that tracks the reference with zero time. We would thus have to make some compromises in the track-following speed because of these limitations. After several trials, we found that the controller parameters of Equation 6.17 with $\varepsilon = 1$ would give us a satisfactory performance.

We then discretize it using a ZOH transformation with a sampling frequency of 20 kHz. The discretized controller is given by

$$\begin{cases} x_{\rm v}(k+1) = \begin{bmatrix} 1 & 0 \\ -0.14611 & 0.69665 \end{bmatrix} x_{\rm v}(k) + \begin{bmatrix} -1 & 1 \\ 293.64 & -1507 \end{bmatrix} \times 10^{-5} \begin{pmatrix} r(k) \\ y(k) \end{pmatrix} \\ \tilde{u}(k) \quad = [\,-5.4395 \quad -5.0451\,]\, x_{\rm v}(k) + [\,0.10929 \quad -0.3111\,] \begin{pmatrix} r(k) \\ y(k) \end{pmatrix} \end{cases} \tag{6.19}$$

and

$$u(k) = G_{\rm notch}(s)\Big|_{\rm ZOH} \cdot \tilde{u}(k) \tag{6.20}$$

CNF Control System Design. Following the procedure given in Chapter 5 for the CNF design for systems with disturbances, in which an integrator is added to the controller to eliminate steady-state bias, we choose an integration gain $\kappa_{\rm i} = 1$ and obtain a corresponding augmented plant:

$$\begin{cases} \dot{\bar{x}} = \begin{bmatrix} 0 & 1 & 0 \\ 0 & 0 & 1 \\ 0 & 0 & 0 \end{bmatrix} \bar{x} + \begin{bmatrix} 0 \\ 0 \\ 6.4013 \times 10^7 \end{bmatrix} {\rm sat}(u) + \begin{bmatrix} -1 \\ 0 \\ 0 \end{bmatrix} r \\ y = \begin{bmatrix} 1 & 0 & 0 \\ 0 & 1 & 0 \end{bmatrix} \bar{x} \\ h = [\,0 \quad 1 \quad 0\,]\, \bar{x} \end{cases} \tag{6.21}$$

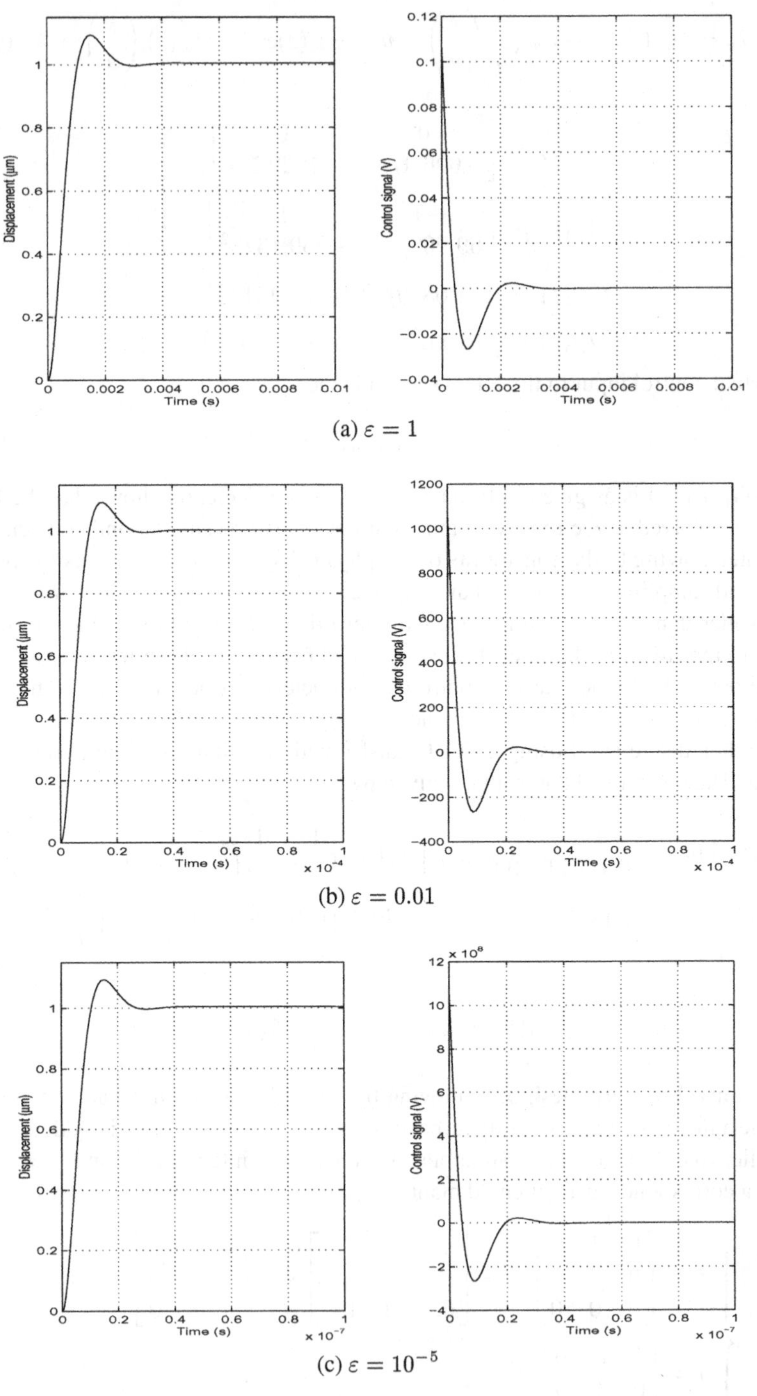

(a) $\varepsilon = 1$

(b) $\varepsilon = 0.01$

(c) $\varepsilon = 10^{-5}$

Figure 6.5. Responses of the closed-loop systems with RPT controller

For track following, $r = 1$. Following a few simulation tries, we obtain a state feedback gain matrix,

$$F = -\left[\,9.8676\times10^{-5} \quad 0.098676 \quad 2.3557\times10^{-5}\,\right] \tag{6.22}$$

which places the closed-loop poles at $-801.11 \pm j2547.4$ and -0.001. Next, we choose W to be a diagonal matrix with diagonal elements being 1.95×10^{-4}, 0.0195 and 3×10^{-10}, respectively. Solving the Lyapunov equation in Equation 5.98, we obtain

$$P = \begin{bmatrix} 0.09751 & 2.3284\times10^{-5} & 1.5436\times10^{-8} \\ 2.3284\times10^{-5} & 9.4272\times10^{-6} & 1.5473\times10^{-9} \\ 1.5436\times10^{-8} & 1.5473\times10^{-9} & 1.1255\times10^{-12} \end{bmatrix} > 0 \tag{6.23}$$

The reduced-order observer gain matrix is selected as $K_{\rm R} = -4000$, which places the observer pole at -4000, and the nonlinear gain function is selected as follows:

$$\rho(e) = -1.5029\left|{\rm e}^{-|e|} - 0.3679\right| \tag{6.24}$$

Finally, the reduced-order enhanced CNF control law for the servo system is given by

$$\begin{pmatrix} \dot{x}_{\rm i} \\ \dot{x}_{\rm v} \end{pmatrix} = \begin{bmatrix} 0 & 0 \\ 0 & -4000 \end{bmatrix}\begin{pmatrix} x_{\rm i} \\ x_{\rm v} \end{pmatrix} - \begin{bmatrix} -1 \\ 1.6\times10^{7} \end{bmatrix} y + \begin{bmatrix} 0 \\ 6.4013\times10^{7} \end{bmatrix}\tilde{u} - \begin{pmatrix} r \\ 0 \end{pmatrix} \tag{6.25}$$

$$\begin{aligned} \tilde{u} = {}& \rho(e)\left[\,0.98808 \quad 0.099044 \quad 7.2048\times10^{-5}\,\right]\begin{pmatrix} x_{\rm i} \\ y-r \\ x_{\rm v}+4000y \end{pmatrix} \\ & - \left[\,9.8676\times10^{-5} \quad 0.098676 \quad 2.3557\times10^{-5}\,\right]\begin{pmatrix} x_{\rm i} \\ y-r \\ x_{\rm v}+4000y \end{pmatrix} \end{aligned} \tag{6.26}$$

and

$$u = G_{\rm notch}(s)\cdot\tilde{u} \tag{6.27}$$

The above CNF controller is discretized using the ZOH method with a sampling frequency of 20 kHz when implemented onto the actual HDD hardware.

PID Control System Design. Following the result of Chapter 3, we obtain a finely tuned PID controller, which has the following transfer function:

$$\tilde{u}(k) = \frac{0.50495z^2 - 0.99147z + 0.48653}{z^2 - 1.09091z + 0.09091}\cdot\frac{0.2696}{z - 0.7304}\cdot\Big[r(k) - y(k)\Big] \tag{6.28}$$

where the second term is a low-pass filter added to improve the implementation performance of the control system. In our experimental tests, we find that it is necessary to add the low-pass filter to the PID control law in order to obtain a meaningful experimental performance. The above discrete-time PID controller is obtained from a continuous-time counterpart using the ZOH method with a sampling frequency of 20 kHz. Once again, we note that the above PID controller is tuned to meet the requirements on the gain and phase margins, and the design specifications on the sensitivity and complementary sensitivity functions. Although the PID control has the simplest structure, its dynamical order, which is 3, is higher than that of the RPT and CNF controllers. As expected, the complete control input is given by

$$u(k) = G_{\text{notch}}(s)\Big|_{\text{ZOH}} \cdot \tilde{u}(k) \tag{6.29}$$

6.4 Simulation and Implementation Results

In this section we present the simulation and actual implementation results of our designs and their comparison. The following tests are presented: i) the track-following test of the closed-loop systems, ii) the frequency-domain test including the Bode and Nyquist plots as well as the plots of the resulting sensitivity and complementary sensitivity functions, iii) the runout disturbance test, and lastly iv) the PES test. Our controller was implemented on an open HDD with a sampling rate of 20 kHz. Closed-loop actuation tests were performed using an LDV to measure the R/W head position. The resolution used for LDV was 2 μm/V. This displacement output is then fed into the DSP, which would then generate the necessary control signal to the VCM actuator. The actual implementation setup is as depicted in Figure 1.7.

6.4.1 Track-following Test

The simulation result and actual implementation result of the closed-loop responses for the control systems are, respectively, shown in Figures 6.6 and 6.7. It is noted that the PID control generates large overshoots in both simulation and implementation, while the systems with the RPT and CNF control have very little overshoot. We summarize the resulting 5% settling time, which is commonly used in the HDD research community, in Table 6.1. Clearly, the CNF control gives the best performance in the time domain compared to those of the other two systems.

Table 6.1. Performances of the track-following controllers

	Settling time (ms)		
	PID control	RPT control	CNF control
Simulation	3.10	0.95	0.80
Implementation	2.65	1.05	0.85

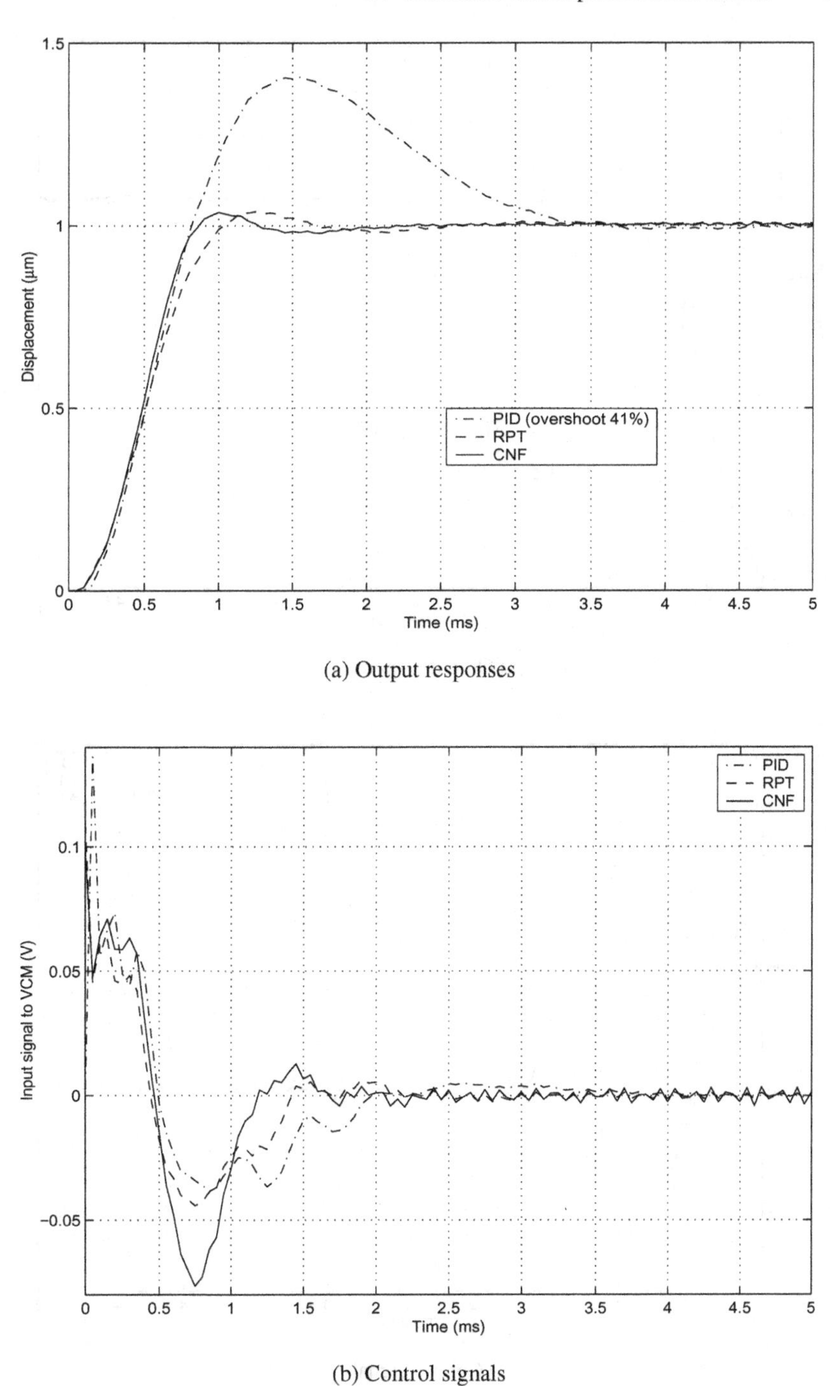

(a) Output responses

(b) Control signals

Figure 6.6. Simulation result: step responses with PID, RPT and CNF control

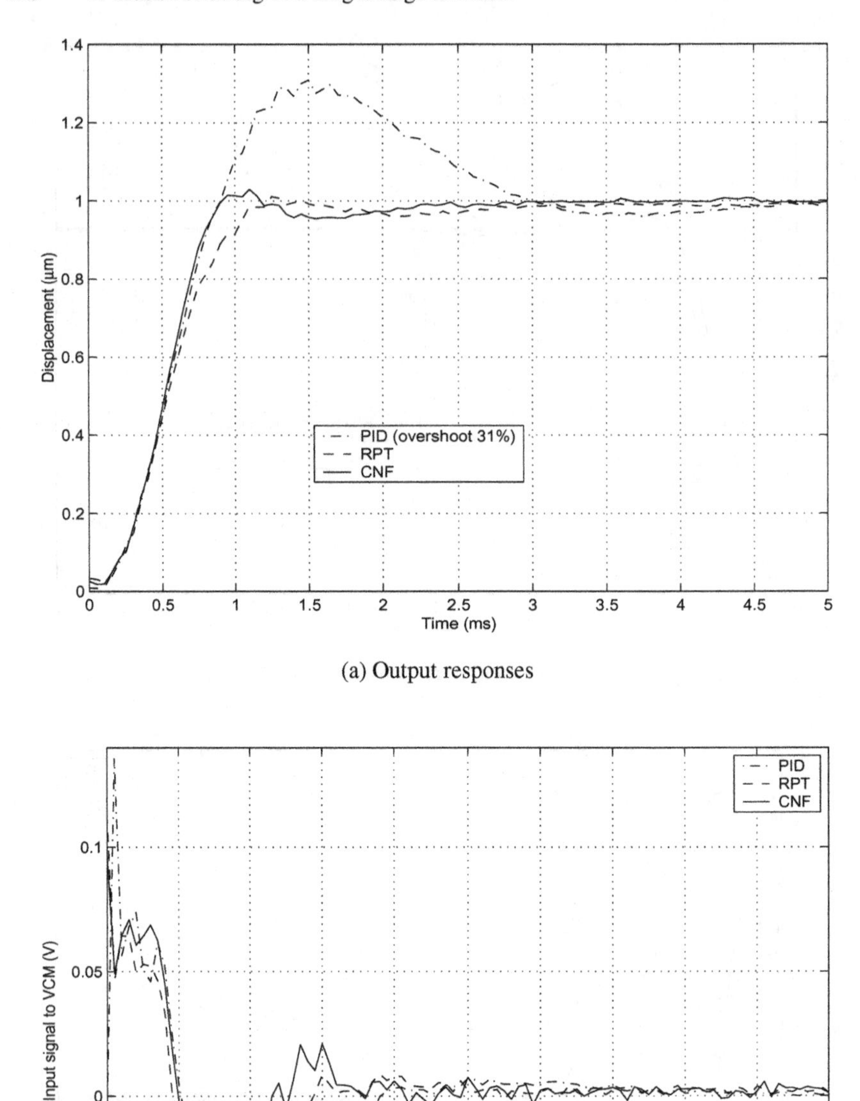

(a) Output responses

(b) Control signals

Figure 6.7. Implementation result: step responses with PID, RPT and CNF control

We believe that the shortcoming of the PID control is mainly due to its structure, *i.e.* it only feeds in the error signal, $y - r$, instead of feeding in both y and r independently. We trust that the same problem might be present in other control methods if the only signal fed is $y - r$. The PID control structure might well be as simple as most researchers and engineers have claimed. However, the RPT controller is even simpler, but it and the CNF controller have fully utilized all available information associated with the actual system.

Unfortunately, we could not compare our results with those in the literature. Most of the references we found in the open literature contained only simulation results in this regard. Some of the implementation results we found were, however, very different in nature. For example, Hanselmann and Engelke [18] reported an implementation result of a disk drive control system design using the LQG approach with a sampling frequency of 34 kHz. The overall step response in [18] with a higher-order LQG controller and higher sampling frequency is worse than that of ours.

6.4.2 Frequency-domain Test

For practical consideration, it is important and necessary to examine the frequency-domain properties of control system design, which include the results of gain and phase margins and the plots of sensitivity and complementary sensitivity functions. Traditionally, gain and phase margins can be obtained through the Bode plot of the open-loop transfer function comprising the given plant and the controller. However, for the HDD system considered in our design, which has additional high-frequency resonance modes, the corresponding Bode plots might have more than one gain and/or phase crossover frequencies. Thus, it is important to verify the stability margins obtained from the associated Nyquist plots. Figures 6.8 to 6.13, respectively, show the Bode plot, the Nyquist plot, and the sensitivity and complementary sensitivity functions, as well as the closed-loop transfer functions (from the reference input r to the controlled output $h = y$) of the resulting control systems. For the CNF design, which is a nonlinear controller, its frequency-domain functions are calculated at the steady-state situation for which the nonlinear gain function $\rho(e)$ has approached its final constant value. The results show that all these designs meet the frequency-domain specifications and have about the same closed-loop bandwidth.

6.4.3 Runout Disturbance Test

Although we do not consider the effects of runout disturbances in our problem formulation, it turns out that our controllers are capable of rejecting the repeatable runout disturbances, which are mainly due to the imperfectness of the data tracks and the spindle motor speeds, and commonly have frequencies at the multiples of the spindle speed, which is about 55 Hz. We simulate these runout effects by injecting a sinusoidal signal into the measurement output, *i.e.* the new measurement output is the sum of the actuator output and the runout disturbance. Figure 6.14 shows the implementation result of the output responses of the overall control system comprising the tenth-order model of the VCM actuator model and the controllers together

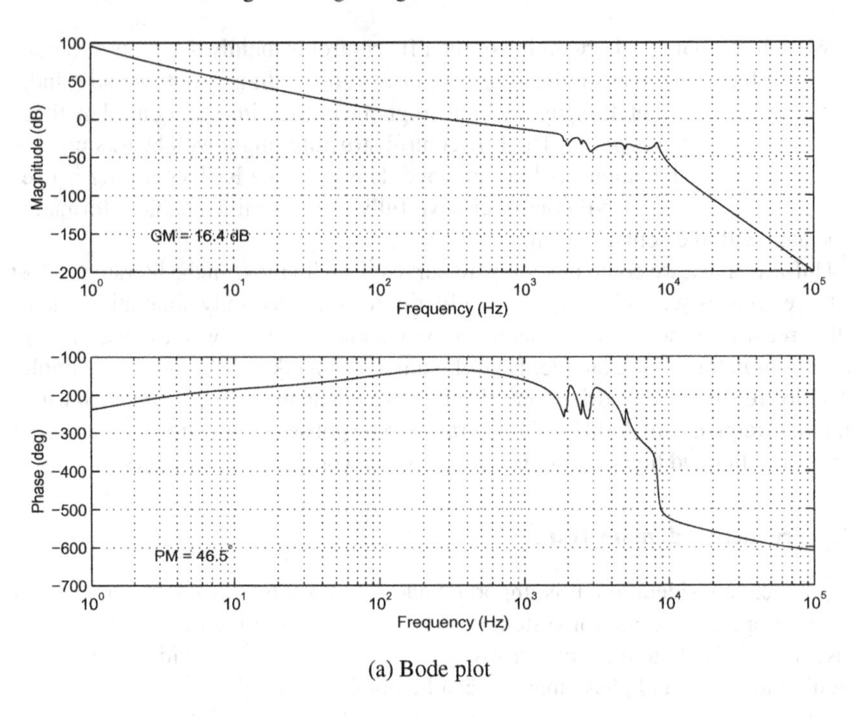

(a) Bode plot

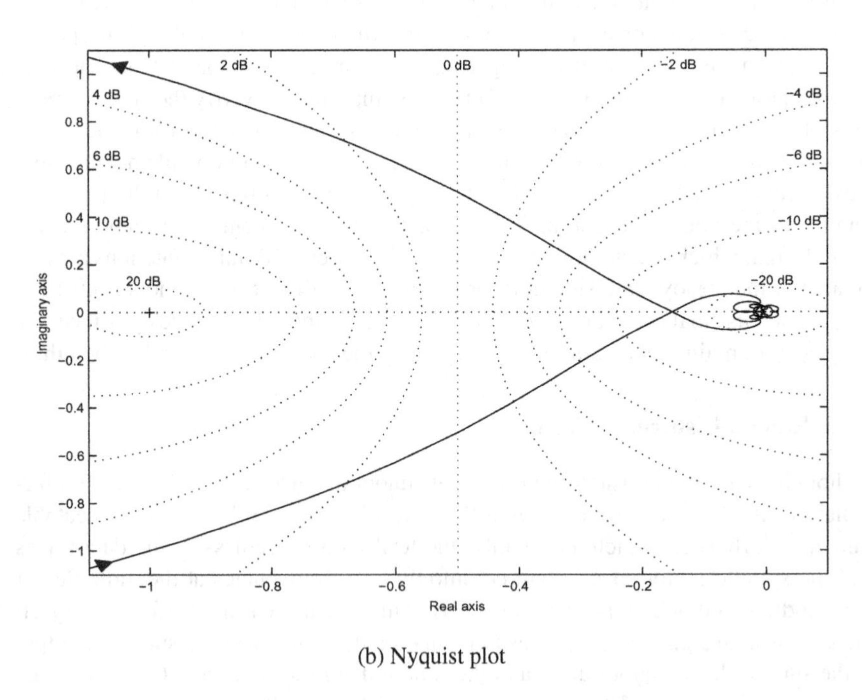

(b) Nyquist plot

Figure 6.8. Bode and Nyquist plots of the PID control system

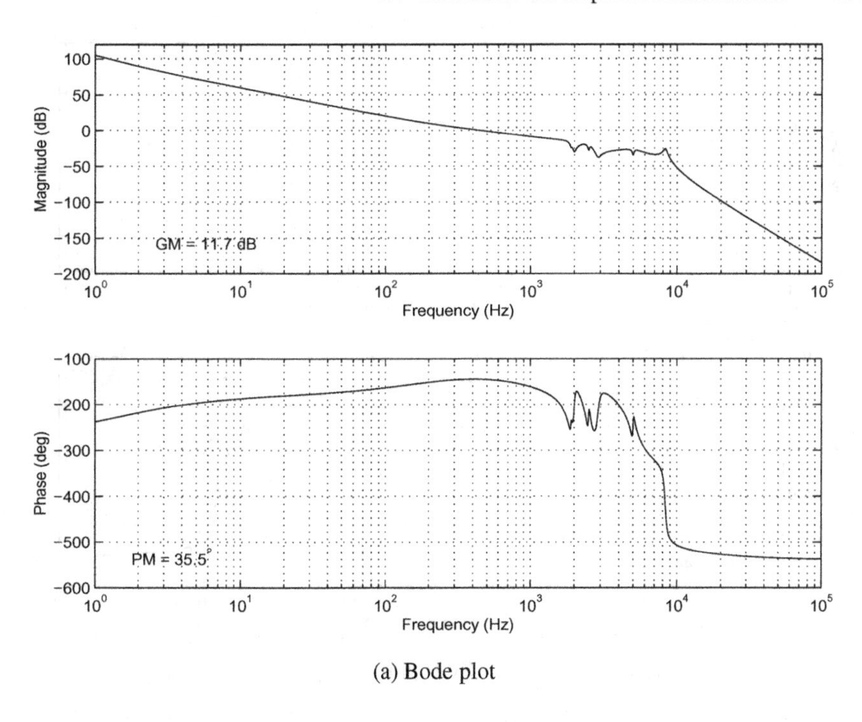

(a) Bode plot

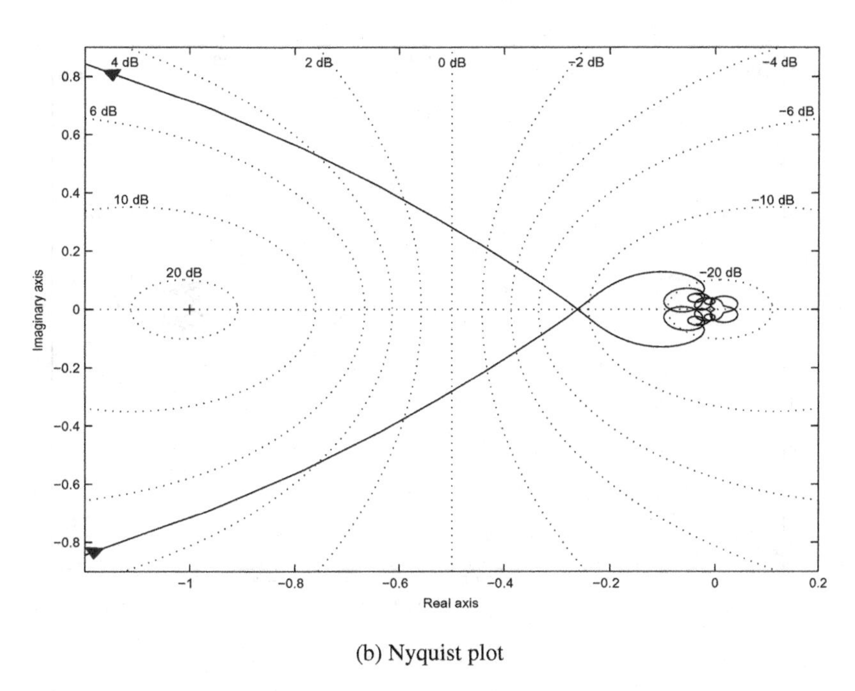

(b) Nyquist plot

Figure 6.9. Bode and Nyquist plots of the RPT control system

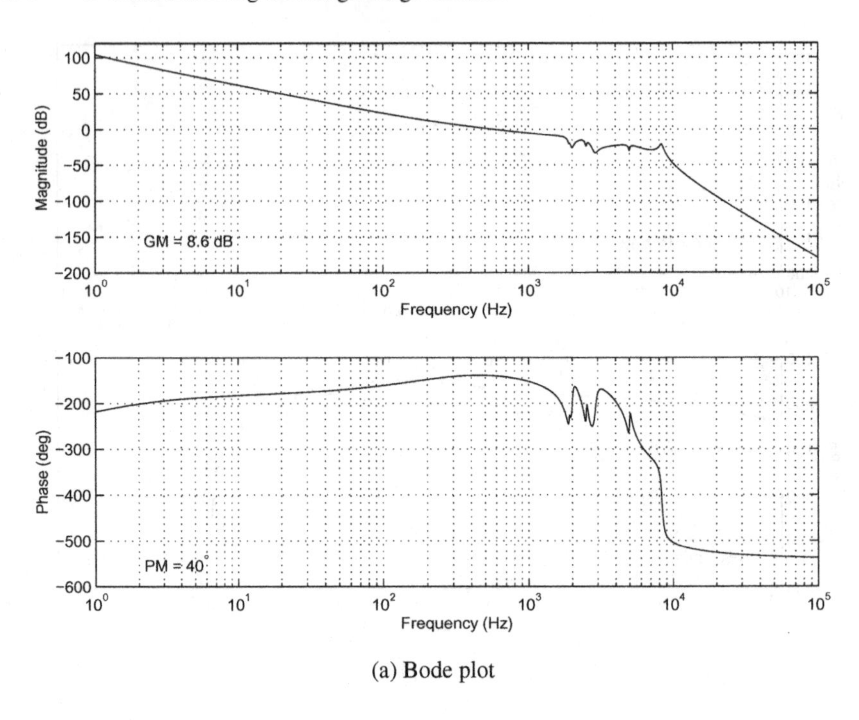

(a) Bode plot

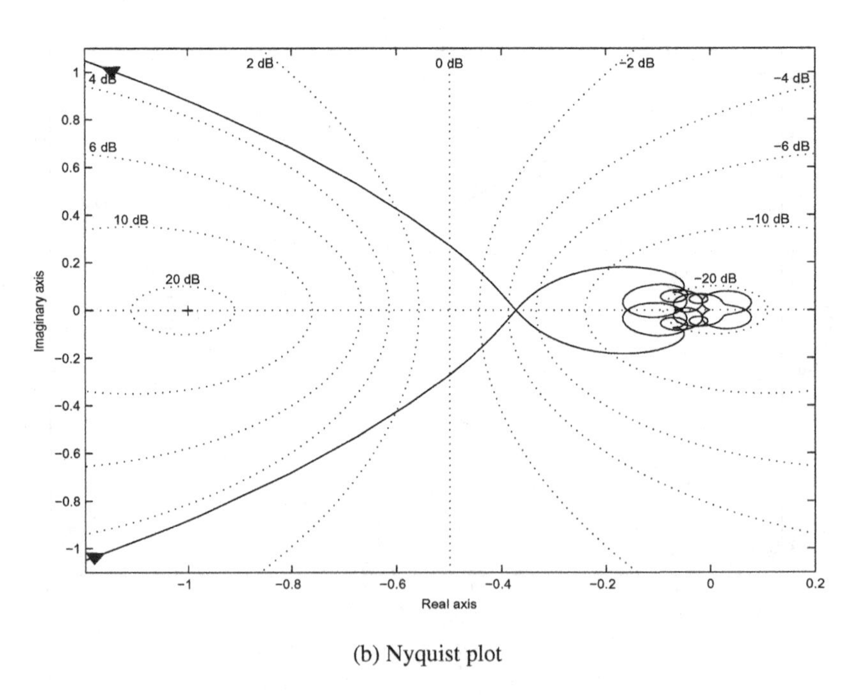

(b) Nyquist plot

Figure 6.10. Bode and Nyquist plots of the CNF control system

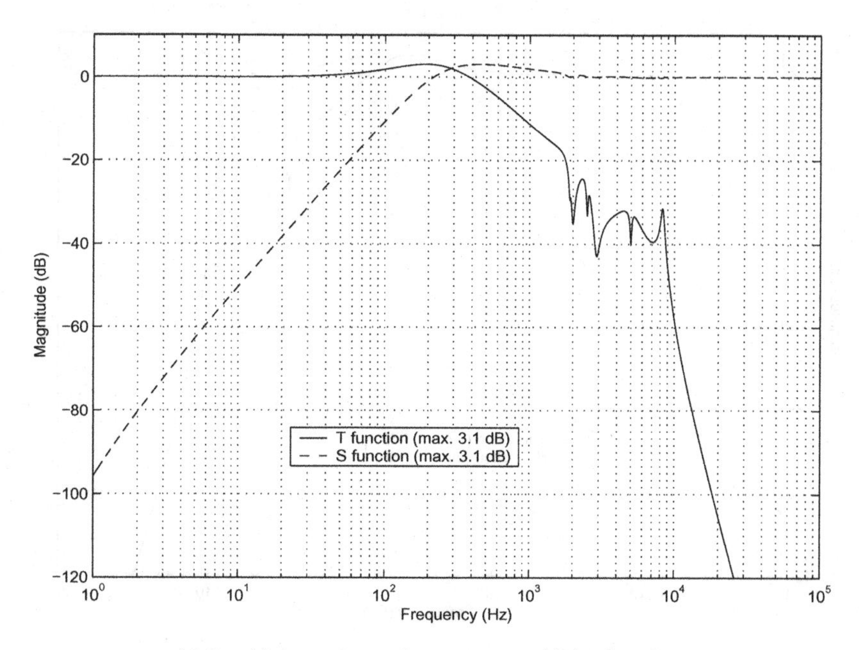

(a) Sensitivity and complementary sensitivity functions

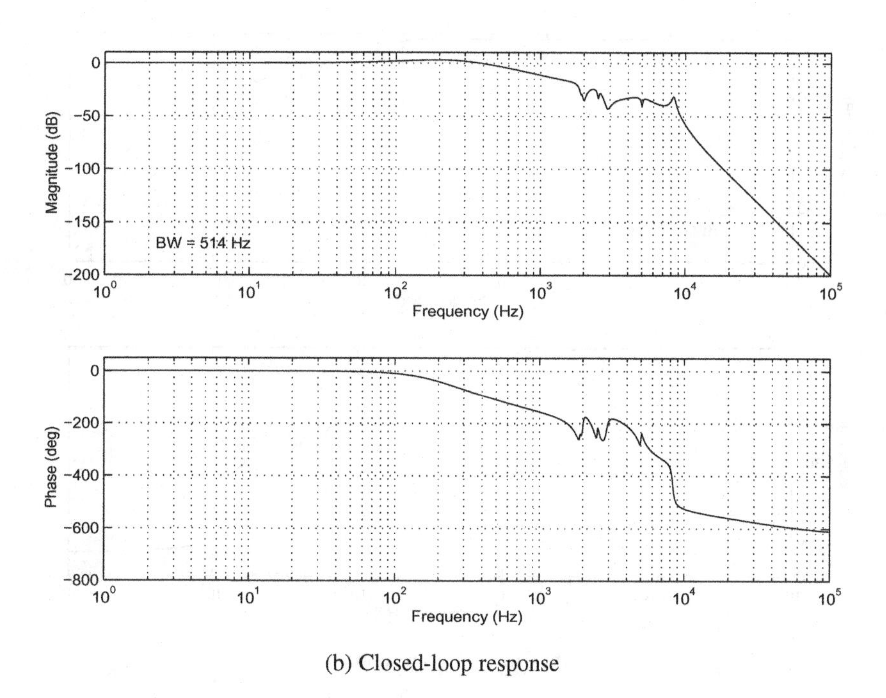

(b) Closed-loop response

Figure 6.11. Sensitivity functions and closed-loop transfer function of the PID control system

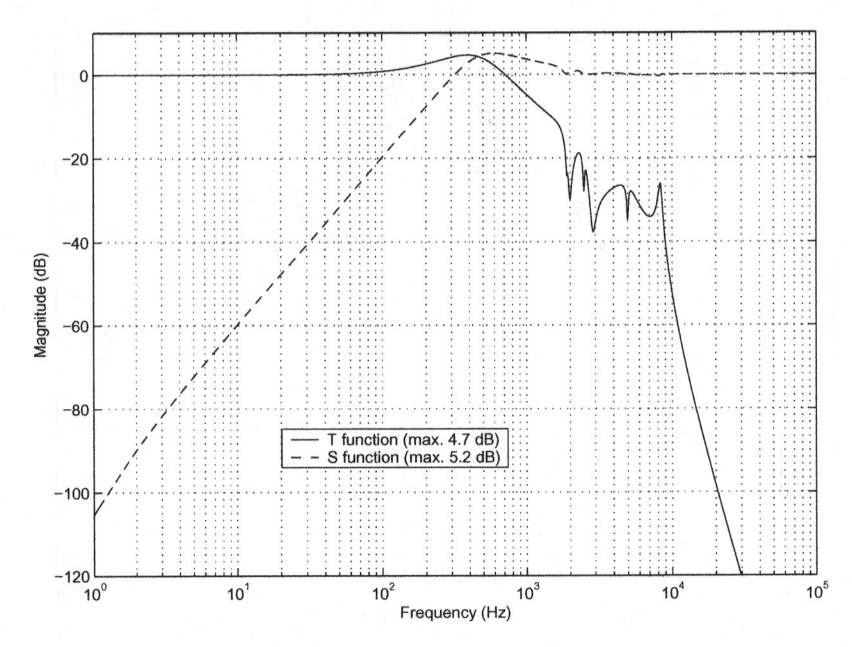

(a) Sensitivity and complementary sensitivity functions

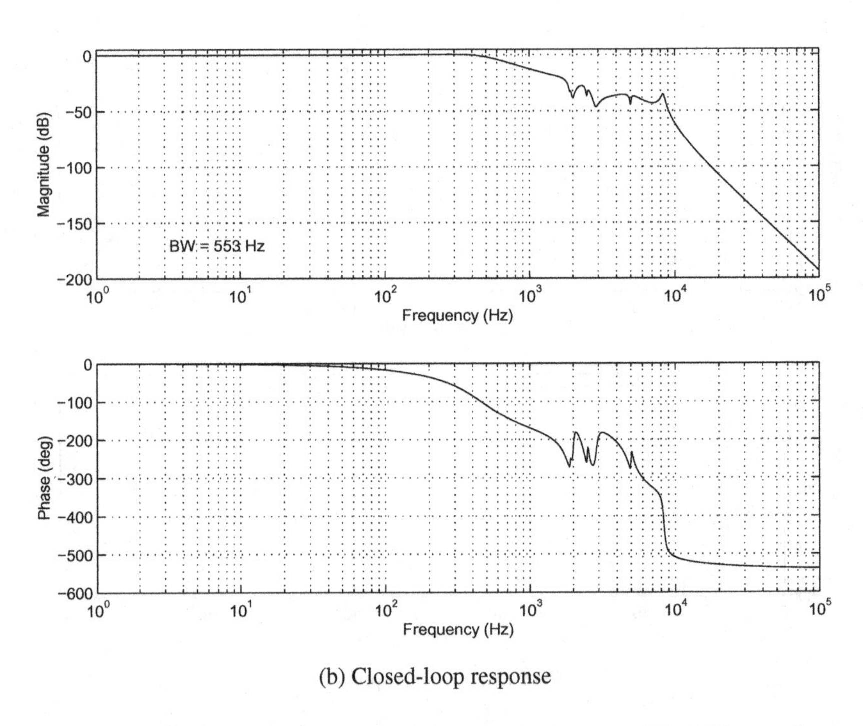

(b) Closed-loop response

Figure 6.12. Sensitivity functions and closed-loop transfer function of the RPT control system

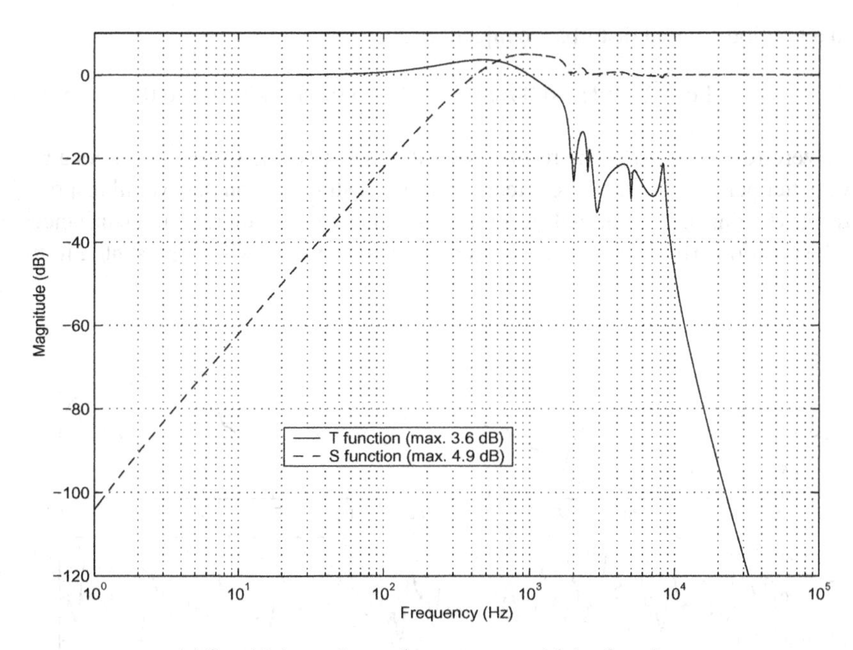

(a) Sensitivity and complementary sensitivity functions

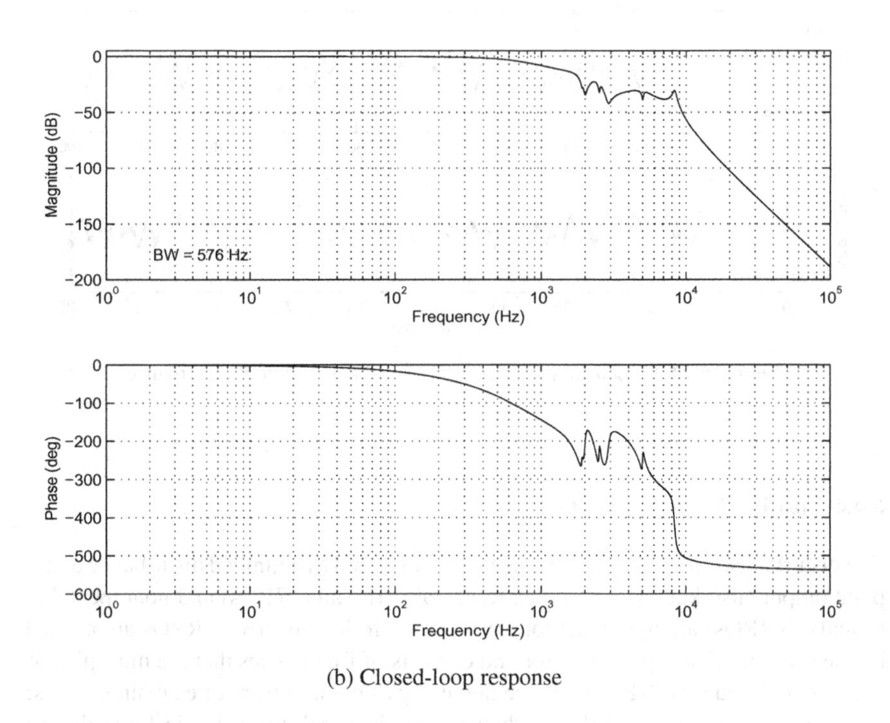

(b) Closed-loop response

Figure 6.13. Sensitivity functions and closed-loop transfer function of the CNF control system

with a fictitious runout disturbance injection

$$\tilde{w}(t) = 0.5+0.1\cos(110\pi t)+0.05\sin(220\pi t)+0.02\sin(440\pi t)+0.01\sin(880\pi t) \tag{6.30}$$

and a zero reference $r(t)$. The result shows that the RPT and CNF controllers again have better performance and the effects of such a disturbance on the overall response under CNF control are minimal. A more comprehensive test on runout disturbances, *i.e.* the position error signal (PES) test on the actual system will be presented in the next section.

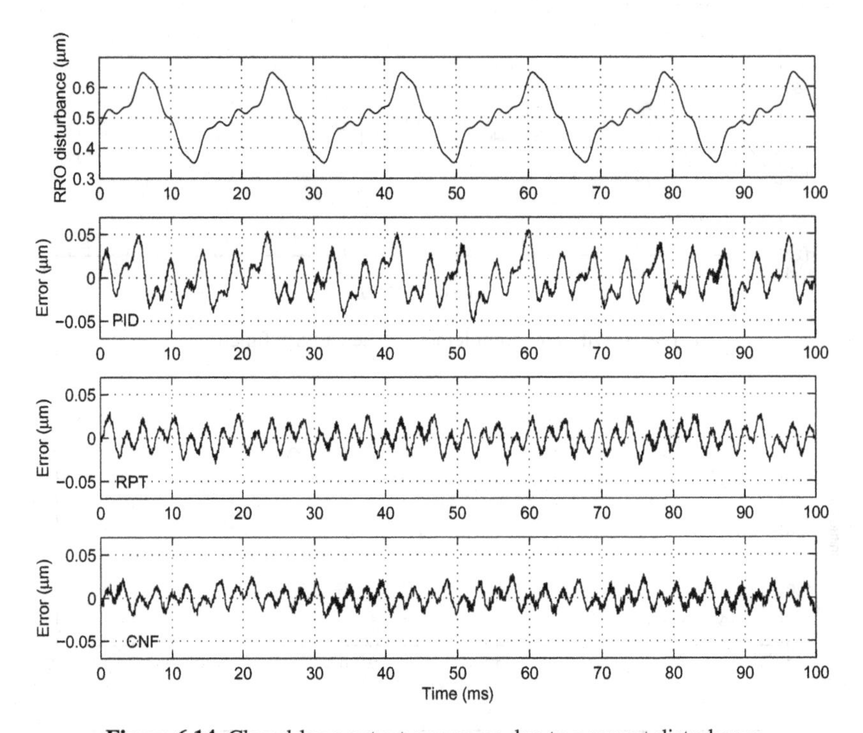

Figure 6.14. Closed-loop output responses due to a runout disturbance

6.4.4 Position Error Signal Test

The disturbances in a real HDD are usually considered as a lumped disturbance at the plant output, also known as runouts. Repeatable runouts (RROs) and nonrepeatable runouts (NRROs) are the major sources of track-following errors. RROs are caused by the rotation of the spindle motor and consists of frequencies that are multiples of the spindle frequency. NRROs can be perceived as coming from three main sources: vibration shocks, mechanical disturbance and electrical noise. Static force due to flex cable bias, pivot-bearing friction and windage are all components of the vibration shock disturbance. Mechanical disturbances include spindle motor variations, disk flutter and slider vibrations. Electrical noises include quantization errors, media noise, servo demodulator noise and power amplifier noise. NRROs are usually random and unpredictable by nature, unlike repeatable runouts. They are also of a lower magnitude (see, *e.g.*, [1]). A perfect servo system for HDDs has to reject both the RROs and NRROs.

In our experiment, we have simplified the system somewhat by removing many sources of disturbances, especially that of the spinning magnetic disk. Therefore, we actually have to add the runouts and other disturbances into the system manually. Based on previous experiments, we know that the runouts in real disk drives are composed mainly of RROs, which are basically sinusoidal with a frequency of about 55 Hz, equivalent to the spin rate of the spindle motor. By manually adding this "noise" to the output while keeping the reference signal at zero, we can then read off the subsequent position signal as the expected PES in the presence of runouts. For actual drives, prewritten PES data might be estimated at high sampling rates using servo sector measurements (see, for example, [141]). In disk drive applications, the variation in the position of the R/W head from the center of the track during track following, which can be directly read off as the PES, is very important. Track-following servo systems have to ensure that the PES is kept to a minimum. Having deviations that are above the tolerance of the disk drive would result in too many read or write errors, making the disk drive unusable. A suitable measure is the standard deviation of the readings, $\sigma_{\rm pes}$. A useful guideline is to make the $3\sigma_{\rm pes}$ value less than 10% of the track pitch, which is about 0.1 μm for a track density of 25 kTPI.

Figure 6.15 shows the histograms of the tracking errors of the respective control systems under the disturbance of the runouts. The $3\sigma_{\rm pes}$ values of the PES test are summarized in Table 6.2. Again, the CNF control yields the best performance in the PES test.

Table 6.2. The $3\sigma_{\rm pes}$ values of the PES test

	PID control	RPT control	CNF control
$3\sigma_{\rm pes}$ (μm)	0.0615	0.0375	0.0288

In conclusion, the RPT and CNF controllers have much better performance in track following and in the PES tests compared with that of the PID controller. We note that the results can be further improved if we used a better VCM actuator and arm assembly (such as those used in minidrives and microdrives) with a higher resonance frequency. We will carry out a detailed study on the servo system of a microdrive later in Chapter 9.

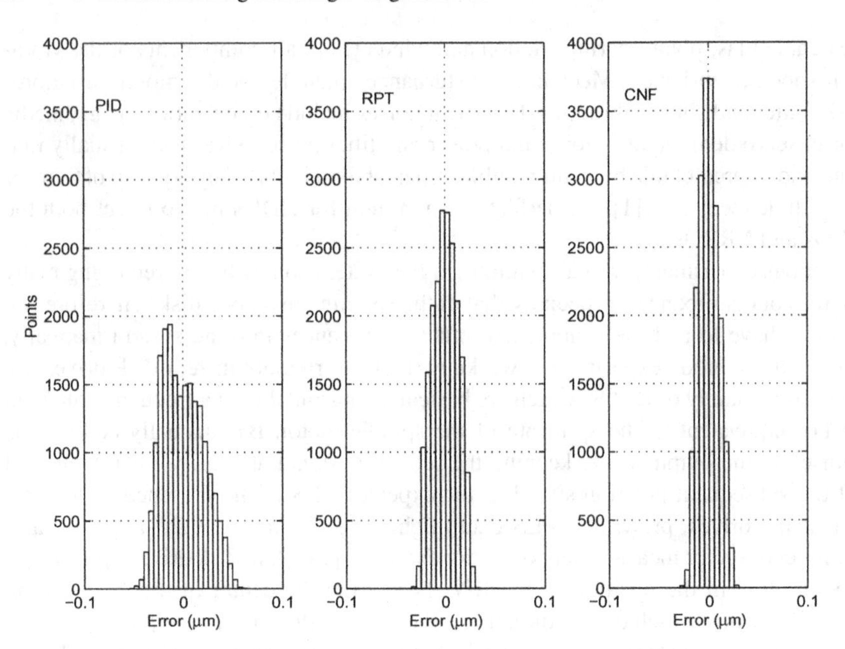

Figure 6.15. Implementation result: histograms of the PES tests

7

Track Seeking of a Single-stage Actuator

7.1 Introduction

In this chapter, we proceed to design track-seeking controllers for a single-stage actuated HDD that would give high-speed seeking performance. We utilize the nonlinear control techniques reported in Chapters 4 and 5 as well as the linear techniques reported in Chapter 3 to carry out the design of three different types of track-seeking controllers for a Maxtor HDD with a single VCM actuator. More specifically, we design the servo systems using the conventional PTOS approach, the CNF control technique, and the MSC system with PTOS and RPT controllers.

As in Chapter 6, a Maxtor (Model 51536U3) HDD is used to implement our design. The actual frequency response and the identified model are given Figure 6.1. The frequency-domain model has been identified earlier in Chapter 6 and is given in Equations 6.4–6.8. The same notch filter as in Equation 6.9 is again utilized for track seeking. With such a formulation, it is safe to approximate the VCM actuator model with the notch filter as a double integrator with an appropriate gain. Such an approximation simplifies the overall design procedure a great deal. Most importantly, it works very well. However, in order to make our design more realistic, all our simulation results are done using the tenth-order model. The final implementation is, of course, to be carried out on the actual system.

The following state-space model is then used throughout our design of track-seeking controllers:

$$\dot{x} = \begin{bmatrix} 0 & 1 \\ 0 & 0 \end{bmatrix} x + \begin{bmatrix} 0 \\ 6.4013 \times 10^7 \end{bmatrix} \mathrm{sat}(u), \quad x = \begin{pmatrix} y \\ v \end{pmatrix} \tag{7.1}$$

where y and v are, respectively, the position of the VCM actuator head in micrometers and velocity in micrometers per second, and u is the control input in volts. In general, the velocity of the VCM actuator in the actual system is not available, and thus y is the only measurable state variable. For this particular system, the controlled output is also the measurement output, *i.e.*

$$h = y = [1 \quad 0]\, x \tag{7.2}$$

Our objective is to design a servo controller that meets the following physical constraints and design specifications:

1. the control input does not exceed ± 3 V owing to physical constraints on the actual VCM actuator;
2. the overshoot and undershoot of track seeking are kept to less than 0.5 μm, the limit of our measurement device for large displacement. As such, the settling time used in this chapter is defined as the time taken for the R/W head to reach the ± 0.5 μm of the target track from its initial point.
3. the gain margin and phase margin of the overall design are, respectively, greater than 6 dB and 30°.

As mentioned earlier, three different approaches, namely the PTOS method, the MSC method and the CNF method, are presented in the following to design appropriate servo systems for the given HDD. We carry out control system design for each method first. All simulation and implementation results, as well as their comparison are to be discussed in the last section.

7.2 Track Seeking with PTOS Control

We present in this section the design and implementation of an HDD servo system using the PTOS approach (see Chapter 4). The first step is to find the state feedback gains k_1 and k_2 in the PTOS control law based on the design specifications. To get specifications in terms of required closed-loop poles we need the natural frequency ω_n and the damping ratio ζ. Let us choose the natural frequency to be 3141.6 rad/s, *i.e.* 500 Hz, and the damping factor to be 0.7255 so as to have an acceleration discount factor α of 0.95, which yields a reasonably good performance for seek lengths up to 300 μm. It follows from Equation 4.32 that a PTOS control law with such a discount factor only increases the total tracking time by about 1.6% from that required in the TOC. Clearly, the performance of the PTOS control is pushed very closely to its limit, the TOC. Interested readers are referred to [30, 142, 143] for detailed information on the selection of these parameters. Note that the relation between the damping ratio ζ and the acceleration discount factor in PTOS control law is given by (see [30])

$$\alpha = \frac{1}{2\zeta^2} \tag{7.3}$$

Then, the corresponding s-plane closed-loop poles are

$$s_{1,2} = -2279.2 \pm j2162.2 \tag{7.4}$$

Using the m-function `acker` in MATLAB®, we obtain the following feedback gains

$$k_1 = 0.15418 \quad \text{and} \quad k_2 = 7.1209 \times 10^{-5} \tag{7.5}$$

and the length of the linear region in PTOS can be found from Equation 4.31 and is given by $y_\ell = 19.458$ μm. Thus, the PTOS control law for our disk drive is as follows:

$$\tilde{u}_{\mathrm{P}} = u_{\max} \cdot \operatorname{sat}\left(\frac{k_2\left[f_{\mathrm{P}}(e) - v\right]}{u_{\max}}\right) \tag{7.6}$$

where $e = r - y$ with r being the target reference, and

$$f_{\mathrm{P}}(e) = \begin{cases} \dfrac{k_1}{k_2}e, & \text{for } |e| \le y_\ell \\ \operatorname{sgn}(e)\left[\sqrt{2u_{\max}a\alpha \mid e \mid} - \dfrac{u_{\max}}{k_2}\right], & \text{for } |e| > y_\ell \end{cases} \tag{7.7}$$

with

$$a = 6.4013 \times 10^7, \quad k_1 = 0.15418, \quad k_2 = 7.1209 \times 10^{-5} \tag{7.8}$$

and

$$y_\ell = 19.458, \quad \alpha = 0.95 \tag{7.9}$$

The advantage of this control scheme is that it is quite simple to understand. The implementation of such a controller requires an estimation of the VCM actuator velocity (with the estimator pole being placed at -4000). More precisely, the following velocity estimator is used:

$$\dot{x}_{\mathrm{c}} = -4000x_{\mathrm{c}} + 6.4013 \times 10^7\tilde{u}_{\mathrm{P}} - 1.6 \times 10^7 y \tag{7.10}$$

and

$$\hat{v} = x_{\mathrm{c}} + 4000y \tag{7.11}$$

In the actual simulation and implementation of the PTOS controller of Equation 7.6, v is replaced with $\hat{v}$ of Equation 7.11 and

$$u = G_{\mathrm{notch}}(s) \cdot \tilde{u}_{\mathrm{P}} \tag{7.12}$$

where $G_{\mathrm{notch}}(s)$ is as given in Equation 6.9. The simulation and implementation results of the above design will be given later in Section 7.5.

7.3 Track-seeking with MSC

In this section, we apply the MSC method of Chapter 4 to the disk drive given earlier. The MSC scheme uses the proximate time-optimal controller in the track-seeking mode, and the RPT controller in the track-following mode. We note that in MSC, initially, the plant is controlled by the seeking controller and at the end of the seeking mode a switch changes it to a track-following controller. In [127], the mode switching was done after finding the optimal mode-switching conditions such that the impact of the initial values on settling performance was minimized. But the impact of the resulting control signal on the resonance modes was not considered. It has been shown [74, 106] that the RPT controller is independent of these initial values. The optimal mode-switching conditions in our scheme can just be set such that the control signal is small enough so as not to excite the resonance vibrations. The problem of unmodeled mechanical resonance can be treated more rigorously either by minimizing the jerk as defined by $\|du/dt\|_2$ as reported in [144] or by using a method developed in the frequency domain in [145]. However, by utilizing the features of RPT control, such as it works for a wide range of resonance frequencies (see Chapter 6), the mode-switching conditions can be determined in a very simple way (see Chapter 4).

We now move to present an MSC controller for the HDD with a single VCM actuator. The control law in track-seeking mode (here we label its control signal as u_P) is given in Equation 7.6, as this mode uses the PTOS control. The control law in the track-following mode, *i.e.* the reduced-order measurement feedback RPT control law, is given by

$$\begin{pmatrix} \dot{x}_i \\ \dot{x}_c \end{pmatrix} = A_{RC} \begin{pmatrix} x_i \\ x_c \end{pmatrix} + B_{RC} \begin{pmatrix} r \\ y \end{pmatrix}, \quad \tilde{u}_R = C_{RC} \begin{pmatrix} x_i \\ x_c \end{pmatrix} + D_{RC} \begin{pmatrix} r \\ y \end{pmatrix} \tag{7.13}$$

with r being the target reference and

$$\left.\begin{aligned} A_{RC} &= \begin{bmatrix} 0 & 0 \\ -1.0268 \times 10^3 & -8.8166 \times 10^3 \end{bmatrix} \\ B_{RC} &= \begin{bmatrix} -1 & 1 \\ 1.0316 \times 10^2 & -4.5583 \times 10^2 \end{bmatrix} \\ C_{RC} &= \begin{bmatrix} -1.6041 & -7.5245 \end{bmatrix} \\ D_{RC} &= \begin{bmatrix} 0.16116 & -0.46214 \end{bmatrix} \end{aligned}\right\} \tag{7.14}$$

Next we find the mode-switching conditions as defined in Equation 4.62. Using the RPT controller parameters, and following the results of Chapter 4, the mode-switching conditions can be determined as $|e(t_1)| \leq 2$ µm $< y_\ell = 19.458$ µm and $|v(t_1)| \leq 5000$ µm/s. We select the MSC law

$$\tilde{u} = \begin{cases} \tilde{u}_P, & t < t_1 \\ \tilde{u}_R, & t \geq t_1 \end{cases} \tag{7.15}$$

in which $\tilde{u}_P$ is as given in Equation 7.6 and t_1 is chosen such that

$$|\, y(t_1) - r \,| = 2 \text{ µm} \quad \text{and} \quad |\, \hat{v}(t_1) \,| \leq 5000 \text{ µm/s} \tag{7.16}$$

As in the PTOS case, the actual control signal is generated by

$$u = G_{\text{notch}}(s) \cdot \tilde{u} \tag{7.17}$$

The overall closed-loop system comprising the given VCM actuated HDD and the MSC control law is asymptotically stable. For easy comparison, the simulation and implementation of the overall system with the MSC control law will again be presented in Section 7.5.

7.4 Track Seeking with CNF Control

We now move to the design of a reduced-order continuous-time composite nonlinear control law as given by Equations 5.79 and 5.80 for the commercial hard disk model shown in Figure 6.1. As the CNF control law depends on the size of the step command input, we derive, for our HDD model given in Equation 7.1, the following parameterized state feedback gain $F(\varepsilon)$:

$$F(\varepsilon) = -\frac{1}{6.4013\times 10^7}\begin{bmatrix} \frac{4\pi^2 f^2 s_0}{\kappa_i \varepsilon^3} & \frac{4\pi^2 f^2 + 4\pi f \zeta s_0}{\varepsilon^2} & \frac{4\pi f \zeta + s_0}{\varepsilon} \end{bmatrix} \tag{7.18}$$

which places the eigenvalues of $\bar{A} + \bar{B}F(\varepsilon)$ exactly at $(-\zeta \pm j\sqrt{1-\zeta^2})2\pi f/\varepsilon$ and $-s_0/\varepsilon$. The latter is the pole associated with the integration dynamics. Following the design procedure given in Chapter 5 and the physical properties of the given system, for $r = 300$ μm, which is to be used in the next section for simulation and implementation, we choose a damping ratio of $\zeta = 0.52$ and $f = 200$ Hz, which corresponds roughly to the normal working frequency range of the linear part of the CNF control law with $\varepsilon = 1$. Selecting $\kappa_i = 1$, $s_0 = 0.001$ and

$$W = \begin{bmatrix} 1.2\times 10^{-5} & 0 & 0 \\ 0 & 10^{-3} & 0 \\ 0 & 0 & 10^{-20} \end{bmatrix} \tag{7.19}$$

we obtain

$$P = \begin{bmatrix} 6.0005\times 10^{-3} & 4.9660\times 10^{-6} & 3.7995\times 10^{-9} \\ 4.9660\times 10^{-6} & 8.0005\times 10^{-7} & 3.1977\times 10^{-10} \\ 3.7995\times 10^{-9} & 3.1977\times 10^{-10} & 2.4468\times 10^{-13} \end{bmatrix} \tag{7.20}$$

and a reduced-order CNF control law as follows:

$$\dot{x}_i = y - r \tag{7.21}$$

with r being the target reference,

$$\dot{x}_c = -4000x_c + 6.4013\times 10^7\tilde{u} - 1.6\times 10^7 y \tag{7.22}$$

and

$$\tilde{u} = \text{sat}\left\{ \begin{bmatrix} K_1 + \rho(e)K_2 \end{bmatrix} \begin{pmatrix} x_i \\ y - r \\ x_c + 4000y \end{pmatrix} \right\} \tag{7.23}$$

where

$$K_1 = [\,-2.4669\times 10^{-5} \quad -2.4669\times 10^{-2} \quad -2.0416\times 10^{-5}\,] \tag{7.24}$$

$$K_2 = [\,0.24322 \quad 2.0470\times 10^{-2} \quad 1.5663\times 10^{-5}\,] \tag{7.25}$$

and

$$\rho(e) = -2.4287 \left| e^{-2|e|/r} - e^{-2} \right| \quad (7.26)$$

Again, the actual control signal is generated as follows:

$$u = G_{\text{notch}}(s) \cdot \tilde{u} \quad (7.27)$$

Note that in both simulation and implementation, the initial condition of x_{i} is set to zero at $t = 0$ and reset to zero when the R/W head of the actuator reaches the point that is 2 μm from the target to reinforce the integration action. Again, the simulation and implementation results of the servo system with the CNF control law will be presented in the next section for an easy comparison.

7.5 Simulation and Implementation Results

Now, we are ready to present the simulation and implementation results for all three servo systems discussed in the previous sections and do a full-scale comparison on the performances of these methods. In particular, we study the following tests:

1. track-seeking test;
2. frequency-domain test;
3. runout disturbance test; and
4. PES test.

All simulation results presented in this section have been obtained using Simulink® in MATLAB® and all implementation results are carried out using our own experimental setup as described in Chapter 1. The sampling frequency for actual implementation is chosen as 20 kHz. Here, we note that all our controllers are discretized using the ZOH technique.

7.5.1 Track-seeking Test

In our simulation and implementation, we use a track pitch of 1 μm for the HDD. In what follows, we present results for a track seek length of 300 μm. Unfortunately, owing to the capacity of the LDV that has been used to measure the displacement of the R/W head of the VCM actuator, the absolute errors of our implementation results given below are 0.5 μm. As such, the settling time for both implementation and simulation results is defined as the total time required for the R/W head to move from its initial position to the entrance of the region of the final target with plus and minus the absolute error. This is the best we can do with our current experimental setup. Nonetheless, the results we obtain here are sufficient to illustrate our design ideas and philosophy. The simulation and implementation results of the track-seeking performances of the obtained servo systems are, respectively, shown in Figures 7.1 and 7.2. We summarize the overall results on settling times in Table 7.1.

Clearly, the simulation and implementation results show that the servo system with the CNF controller has the best performance. We believe that this is due to the fact that the CNF control law unifies the nonlinear and linear components without switching, whereas the other two servo systems involve switching elements between the nonlinear and linear parts, which degrades the overall performance.

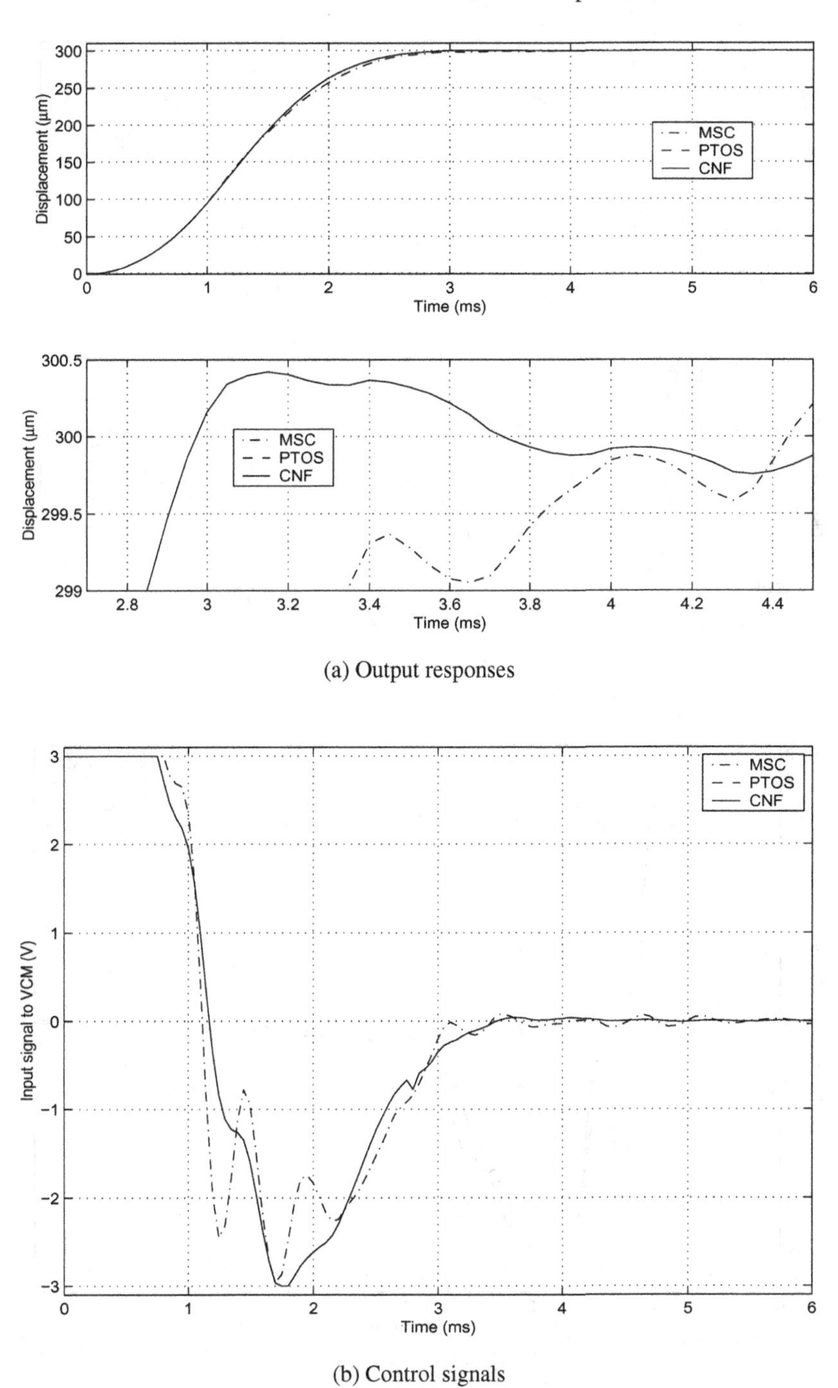

Figure 7.1. Simulation result: response and control of the track-seeking systems

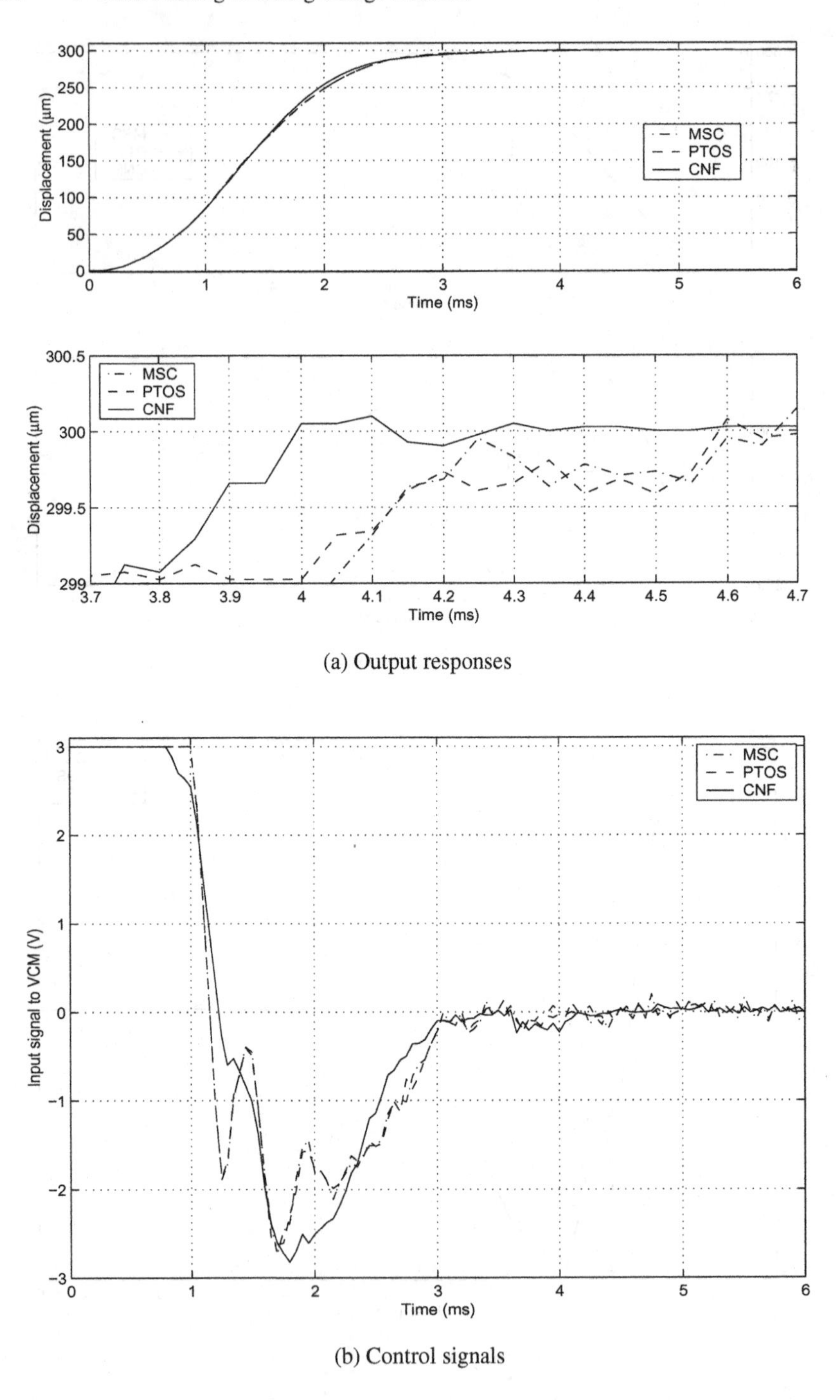

(a) Output responses

(b) Control signals

Figure 7.2. Implementation result: response and control of the track-seeking systems

Table 7.1. Simulation and implementation: settling time of the track-seeking systems

Settling time (ms)					
Simulation			Implementation		
PTOS	MSC	CNF	PTOS	MSC	CNF
3.85	3.85	2.95	4.15	4.15	3.90

7.5.2 Frequency-domain Test

As seen in the previous sections, all three track-seeking controllers that we have designed are nonlinear in nature. The frequency-domain properties of nonlinear track-seeking controllers are not well defined and in fact not as important as those of track-following controllers. For completeness and for comparison, we define the frequency responses of the open and closed-loop systems with a nonlinear controller as those corresponding to the steady-state situation when the nonlinear action of the controller has become a constant or vanished. The Bode plot, Nyquist plot, sensitivity and complementary sensitivity functions as well as the closed-loop frequency responses for the system with each control technique are, respectively, shown in Figures 7.3 to 7.8. All three control systems meet the design specifications in the frequency domain. Finally, we note that for track-seeking controllers, the PES test is not necessary.

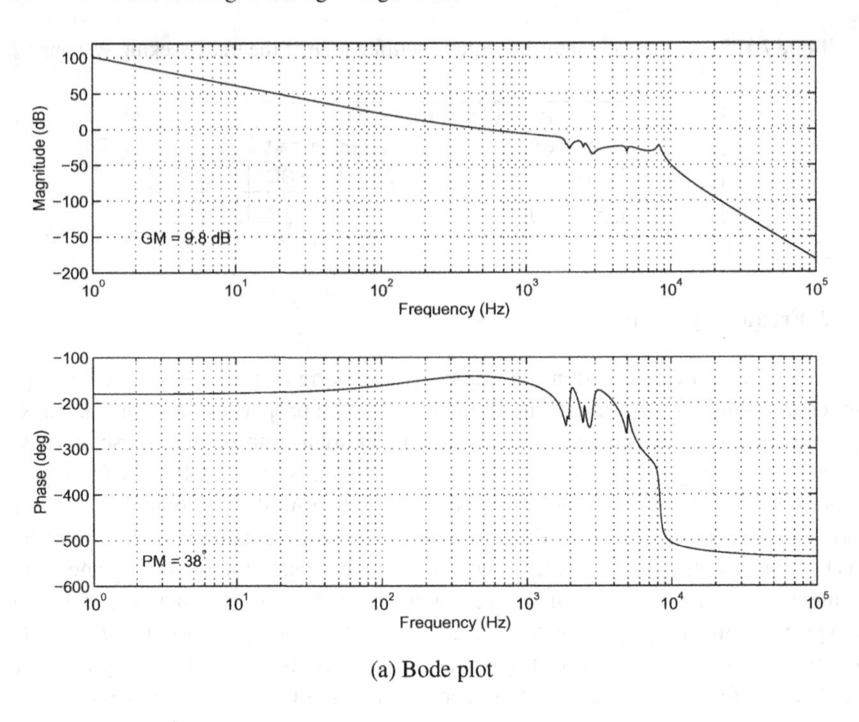

(a) Bode plot

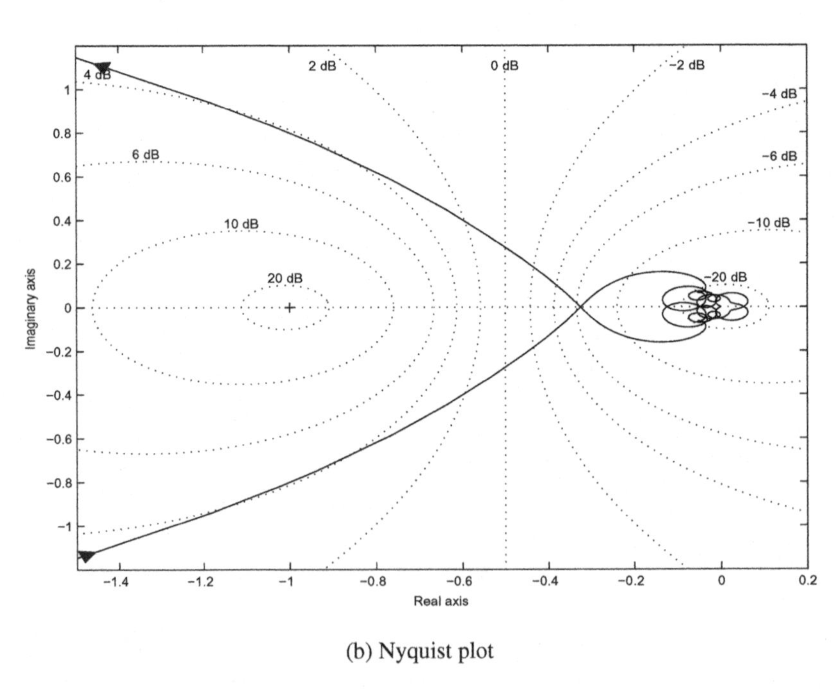

(b) Nyquist plot

Figure 7.3. Bode and Nyquist plots of the PTOS control system

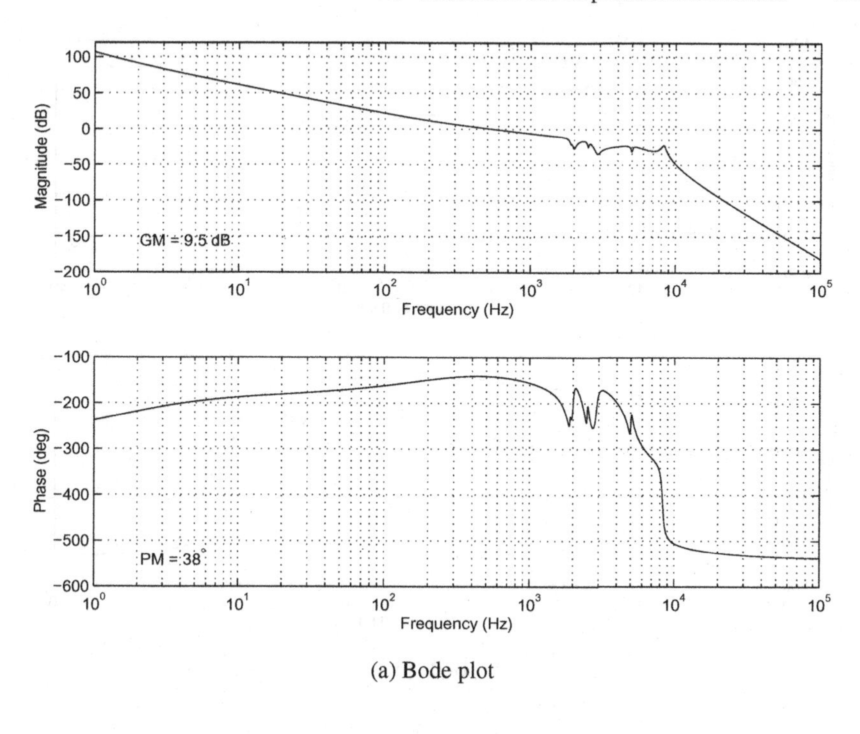

(a) Bode plot

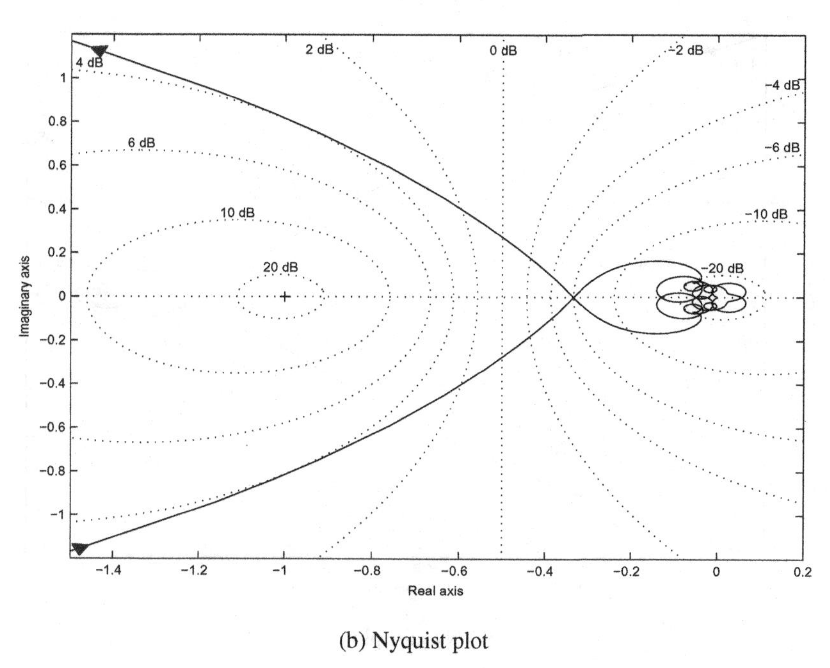

(b) Nyquist plot

Figure 7.4. Bode and Nyquist plots of the MSC control system

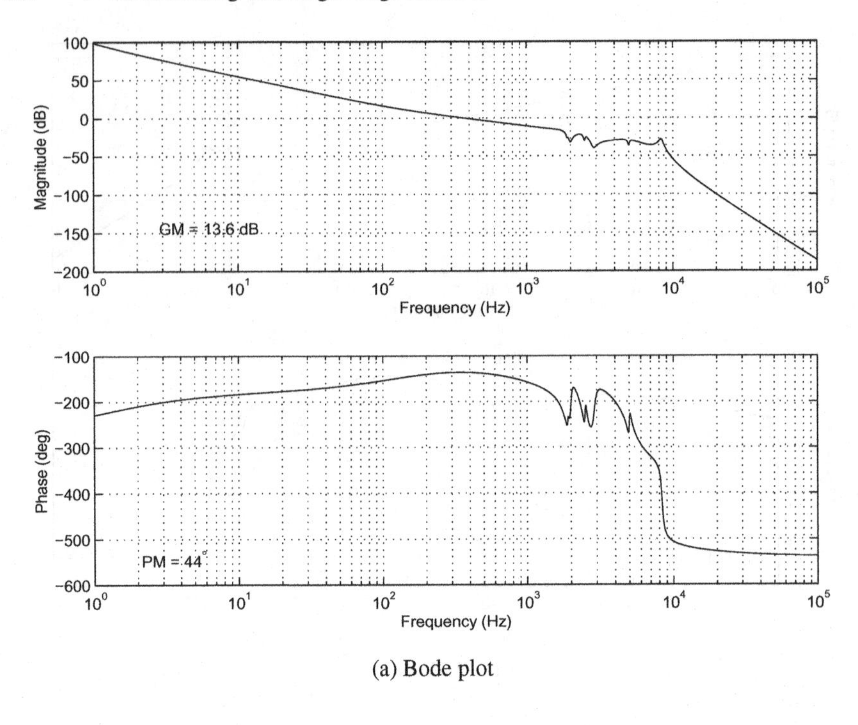

(a) Bode plot

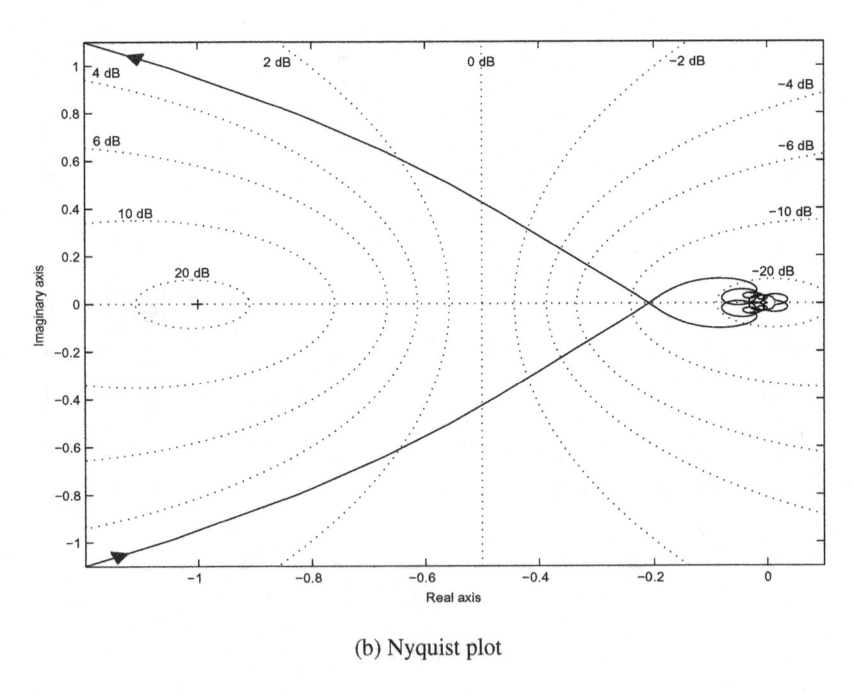

(b) Nyquist plot

Figure 7.5. Bode and Nyquist plots of the CNF control system

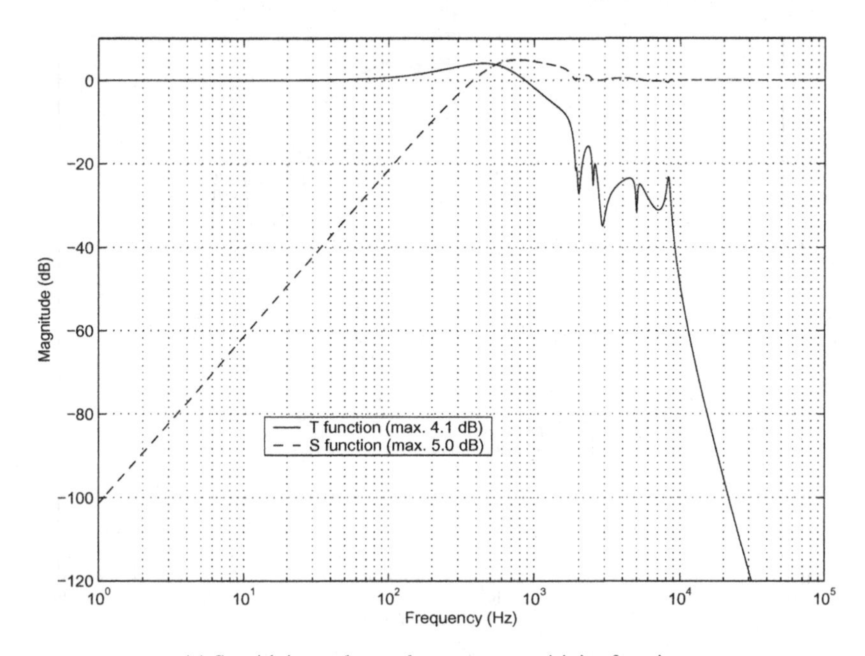

(a) Sensitivity and complementary sensitivity functions

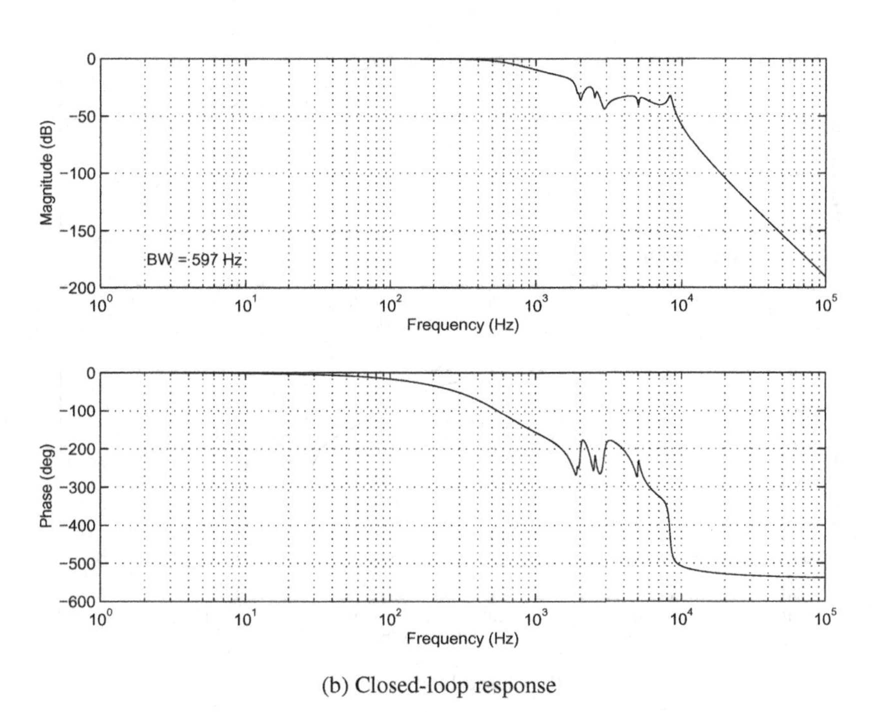

(b) Closed-loop response

Figure 7.6. Sensitivity functions and closed-loop transfer function of the PTOS control system

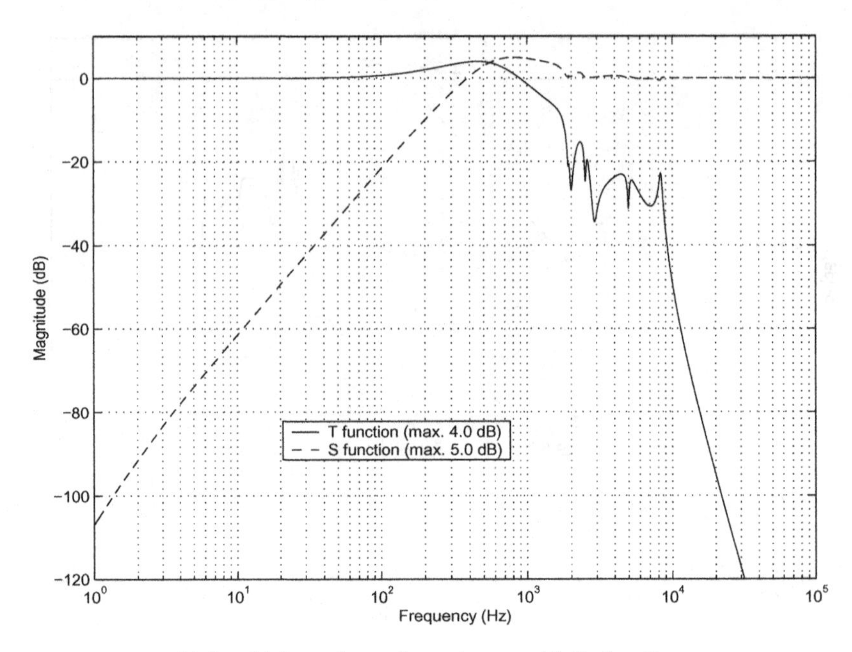

(a) Sensitivity and complementary sensitivity functions

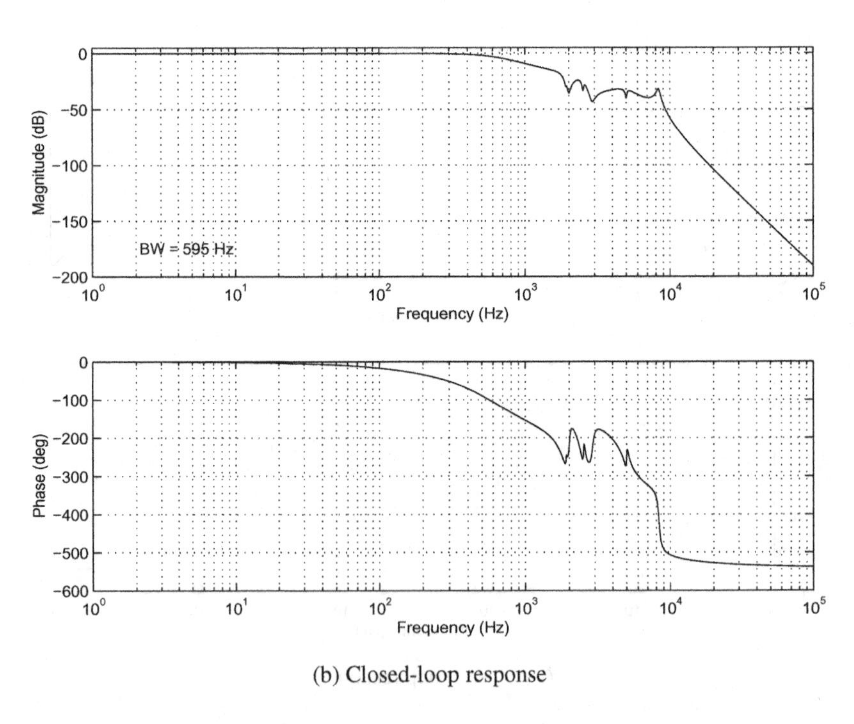

(b) Closed-loop response

Figure 7.7. Sensitivity functions and closed-loop transfer function of the MSC control system

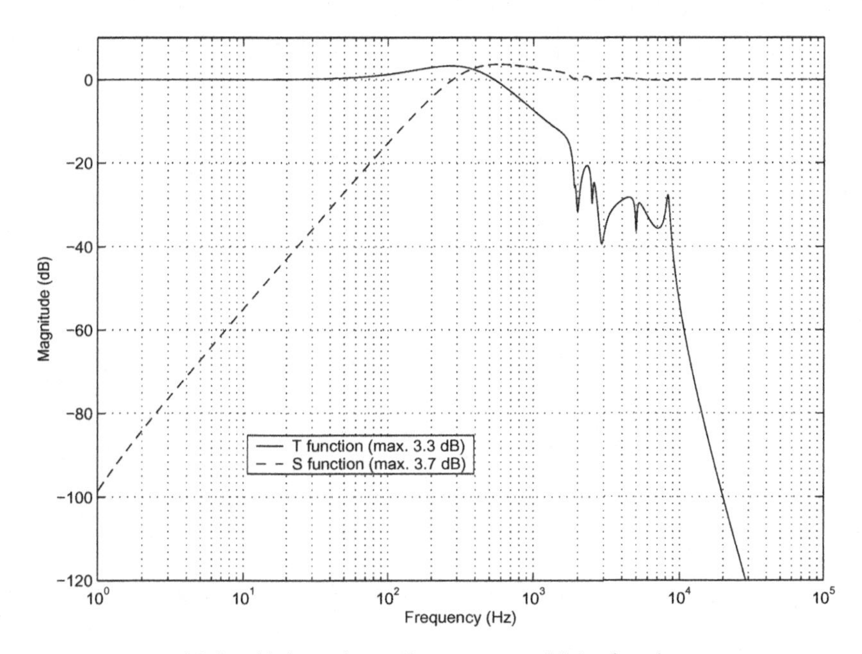

(a) Sensitivity and complementary sensitivity functions

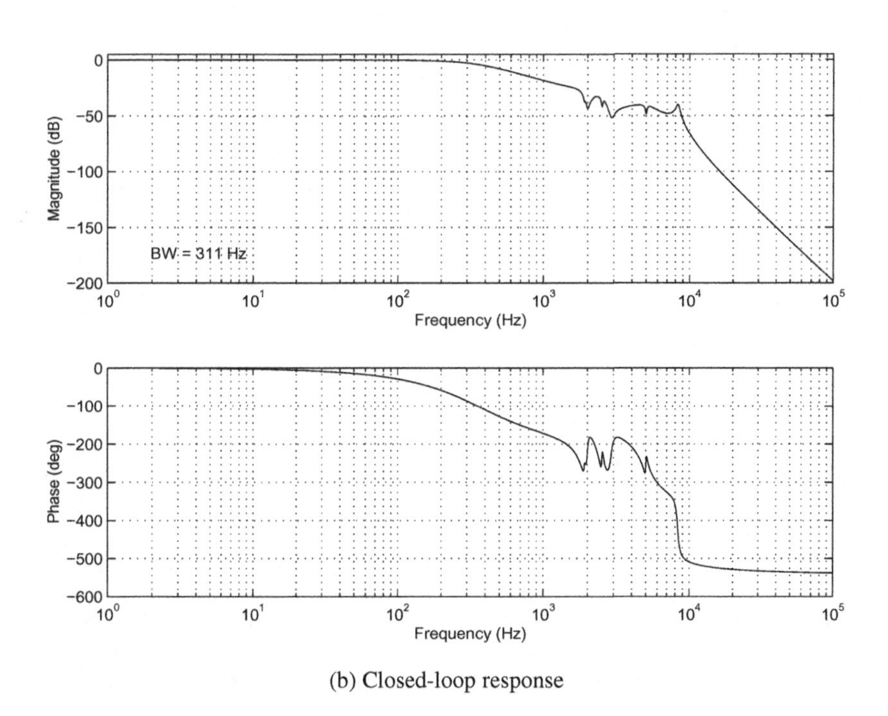

(b) Closed-loop response

Figure 7.8. Sensitivity functions and closed-loop transfer function of the CNF control system

(a) [illegible]

(b) [illegible]

Figure 7.8. [illegible]

8

Dual-stage Actuated Servo Systems

8.1 Introduction

The present demand for large-capacity disk drives is leading to an increase in areal density at a rate of 100% per year. This requires a positioning accuracy of the order of a few nanometers. The servo bandwidth of the current disk drive actuators makes it very hard to achieve this. The VCM actuator used in conventional disk drives has hundreds of flexible resonances at high frequencies (see, *e.g.*, [1, 30]), which limits the increase of bandwidth and hence the positioning accuracy. In order to develop high bandwidth (track-following) servo systems, dual-stage actuation has been proposed as a possible solution. Dual-stage actuator refers to the fact that there is a microactuator mounted on a large conventional VCM actuator. A dual-stage actuated HDD was successfully demonstrated by Tsuchiura *et al.* [146] of Hitachi. In [146], a fine positioner based on a piezoelectric structure was mounted at the end of a primary VCM stage to form the dual actuator. The higher bandwidth of the fine positioner allowed the R/W heads to be positioned accurately. There have been other instances where electromagnetic [147] and electrostatic [148] microactuators have been used for fine positioning of R/W heads. The two most fundamental choices in a dual-stage system are the actuator configuration and the control algorithm. There have been proposals for electromagnetic, electrostatic, piezoelectric, shape memory and rubber microactuators, *etc.*, each with their own advantages and disadvantages. Many research studies have been done and reported in the literature (see, *e.g.*, [20, 148–167] just to name a few). In this chapter, we focus on the design of complete HDD servo systems with a dual-stage actuator with a piezoelectric actuator in its second stage (see Figure 8.1).

Diverse control strategies and methods have been reported in the design of HDD servo systems with a dual-stage actuator (see, *e.g.*, [150, 154, 156, 157, 159, 161–163]. Guo *et al.* [154] have proposed four control strategies to design the dual-stage actuator control system: the so-called parallel loop, master–slave loop, dual feedback loop and master-slave with decoupling methods. Hu *et al.* [159] and Guo *et al.* [156] have also utilized the well-known LQG/LTR method to design the dual-stage actuator control system. These studies have accelerated the progress to improve HDD

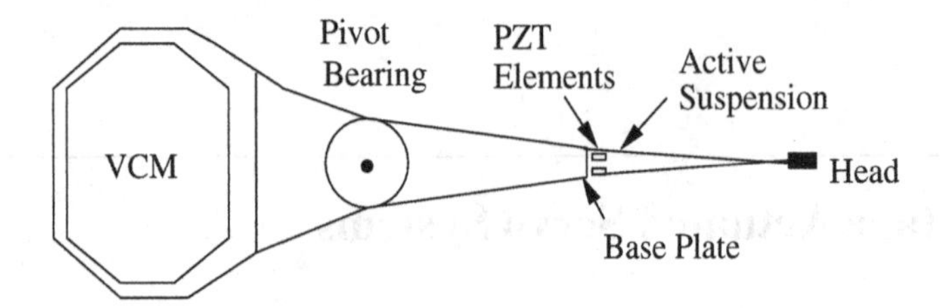

Figure 8.1. A dual-stage HDD actuator

servo system performances, but more studies need to be done before such dual-stage actuated HDDs can be considered for commercialization.

We present in this chapter the design of track-following controllers for the dual-stage actuated HDD with three different approaches, *i.e.* the PID control, the RPT control method, and the CNF control technique. In each design, the VCM actuator is controlled by a control law obtained using one of these three approaches and the microactuator is controlled by a proportional gain together with an appropriate filter.

8.2 Modeling of a Dual-stage Actuator

In this section we develop a model for the dual-stage actuator. Since the VCM actuator and the microactuator are decoupled, we need to identify two separate models for the VCM actuator and for the microactuator. A Maxtor dual-stage HDD is used in our design and implementation. It has a fine positioner based on a piezoelectric suspension mounted at the end of a primary VCM arm (see Figure 8.1), and the microactuators produce the relative motion of the R/W head along the radial direction. Here we note that only the displacement of the R/W head is available as the measurement output. Also, the VCM arm in this HDD is quite similar to that in the HDD studied in Chapters 6 and 7. Figures 8.2 and 8.3 respectively show the frequency-response characteristics of the VCM actuator and the microactuator.

Using the data measured from the actual system, and the identification algorithm given in Chapter 2 (see also [13, 59]), we obtain a tenth-order model for the actuator (see also Chapter 6):

$$G_{\rm v}(s) = \frac{6.4013 \times 10^7}{s^2} \prod_{i=1}^{4} G_{{\rm v,r},i}(s) \tag{8.1}$$

with

$$G_{\rm v,r,1}(s) = \frac{0.912s^2 + 457.4s + 1.433 \times 10^8}{s^2 + 359.2s + 1.433 \times 10^8} \tag{8.2}$$

$$G_{\rm v,r,2}(s) = \frac{0.7586s^2 + 962.2s + 2.491 \times 10^8}{s^2 + 789.1s + 2.491 \times 10^8} \tag{8.3}$$

$$G_{\rm v,r,3}(s) = \frac{9.917 \times 10^8}{s^2 + 1575s + 9.917 \times 10^8} \tag{8.4}$$

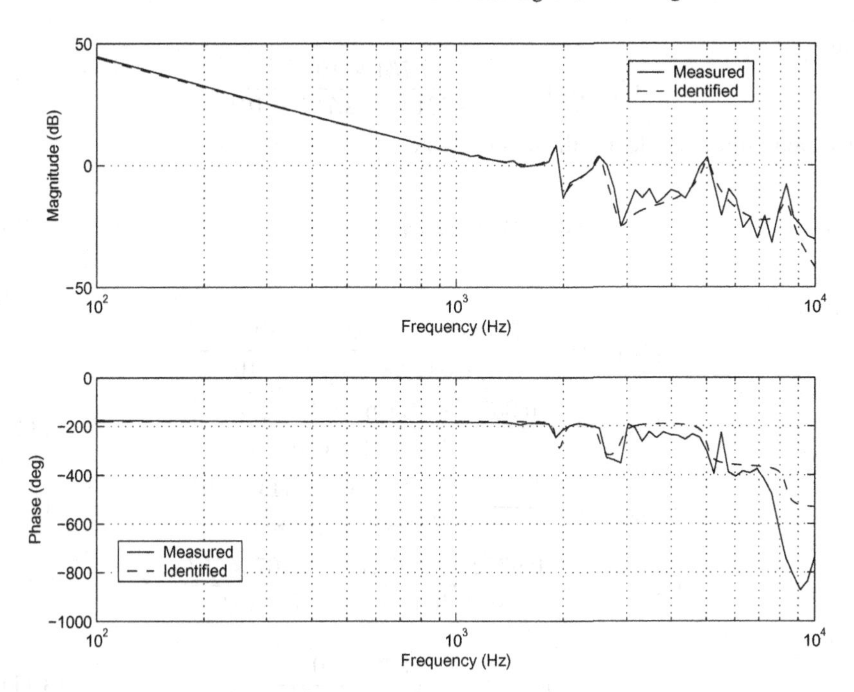

Figure 8.2. Frequency responses of the actual and identified VCM actuator models

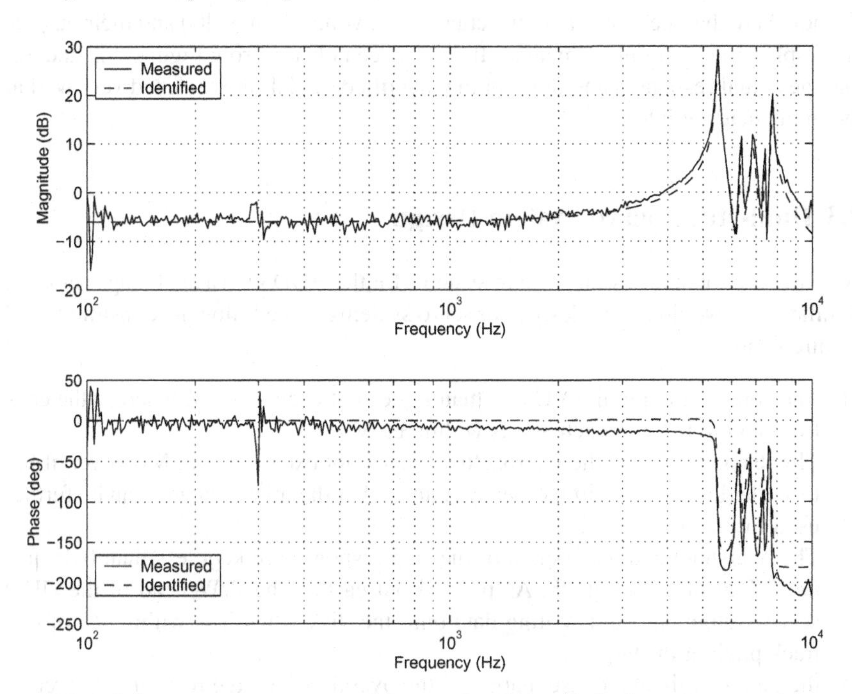

Figure 8.3. Frequency responses of the actual and identified microactuator models

and

$$G_{\mathrm{v,r,4}}(s) = \frac{2.731 \times 10^9}{s^2 + 2613s + 2.731 \times 10^9} \tag{8.5}$$

and a tenth-order model for the microactuator,

$$G_{\mathrm{m}}(s) = 0.5 \prod_{i=1}^{5} G_{\mathrm{m,r},i}(s) \tag{8.6}$$

with

$$G_{\mathrm{m,r,1}}(s) = \frac{0.7938s^2 + 767.9s + 1.189 \times 10^9}{s^2 + 344.8s + 1.189 \times 10^9} \tag{8.7}$$

$$G_{\mathrm{m,r,2}}(s) = \frac{0.955s^2 + 978.6s + 1.605 \times 10^9}{s^2 + 400.6s + 1.605 \times 10^9} \tag{8.8}$$

$$G_{\mathrm{m,r,3}}(s) = \frac{0.8912s^2 + 1013s + 1.843 \times 10^9}{s^2 + 1073s + 1.843 \times 10^9} \tag{8.9}$$

$$G_{\mathrm{m,r,4}}(s) = \frac{0.9772s^2 + 460.1s + 2.167 \times 10^9}{s^2 + 465.5s + 2.167 \times 10^9} \tag{8.10}$$

and

$$G_{\mathrm{m,r,5}}(s) = \frac{2.376 \times 10^9}{s^2 + 487.4s + 2.376 \times 10^9} \tag{8.11}$$

We note here that the inputs to both actuators are voltages (in volts) and their outputs are displacements (in micrometers). It can be seen clearly from Figures 8.2 and 8.3 that the actual responses and those of the identified models are matched very well at the frequencies of interest.

8.3 Dual-stage Servo System Design

We now carry out the design of servo systems for the HDD with a dual-stage actuator. Similarly, we would like to design our servo systems to the following constraints and requirements.

1. The control input to the VCM actuator does not exceed ±3 V, whereas the control input to the microactuator is within ±2 V.
2. The displacement of the microactuator does not exceed 1 μm. Moreover, it has to settle down to zero in the steady state so that the microactuator can be further used for the next move.
3. The overshoot and undershoot of the step response are kept less than 0.05 μm, *i.e.* 5% of one track pitch. As pointed out earlier, the R/W head of the HDD servo system can start writing data onto the disk when it is within 5% of one track pitch of the target.
4. the gain margin and phase margin of the overall design are respectively greater than 6 dB and 30°;

5. the maximum peaks of the sensitivity and complementary sensitivity functions are less than 6 dB; and
6. the sampling frequency in implementing the actual controller is 20 kHz.

Unfortunately, the only available measurement in the dual-stage actuated HDD is the displacement of the R/W head, which is a combination of the displacement of the VCM actuator and that of the microactuator. Practically, we have to control both actuators using the same measurement, which makes the servo system design very difficult. Observing that the frequency response of the microactuator is as in Figure 8.3, we find that it is nothing more than a constant gain at low frequencies with a gain of 0.5. This property is valid so long as we do not push the speed of the microactuator too high. As such, we propose in Figure 8.4 a control configuration for the dual-stage actuated servo system. We note that the filter in the feedforward path of the microactuator is the combination of an appropriate notch filter, which is to attenuate the high-frequency resonance modes, and a lead-lag compensator, which is to compensate some phase losses resulted in the notch filter.

To be more specific, we estimate the displacement of the microactuator $\hat{y}_{\mathrm{m}}$ directly from its input u_{m} and then the estimation of the displacement of the VCM actuator can be obtained as $y - \hat{y}_{\mathrm{m}}$. It can be observed from the configuration in Figure 8.4 that y_{m}, the displacement of the microactuator, settles down to zero as the tracking error approaches zero. As pointed out earlier, such a feature would enable the microactuator to be used for the next move. Since the microactuator can only produce a maximum displacement of 1 μm, it is helpful in track seeking. As such, we only focus on the design of controllers for track following, in which the microactuator is expected to contribute significantly.

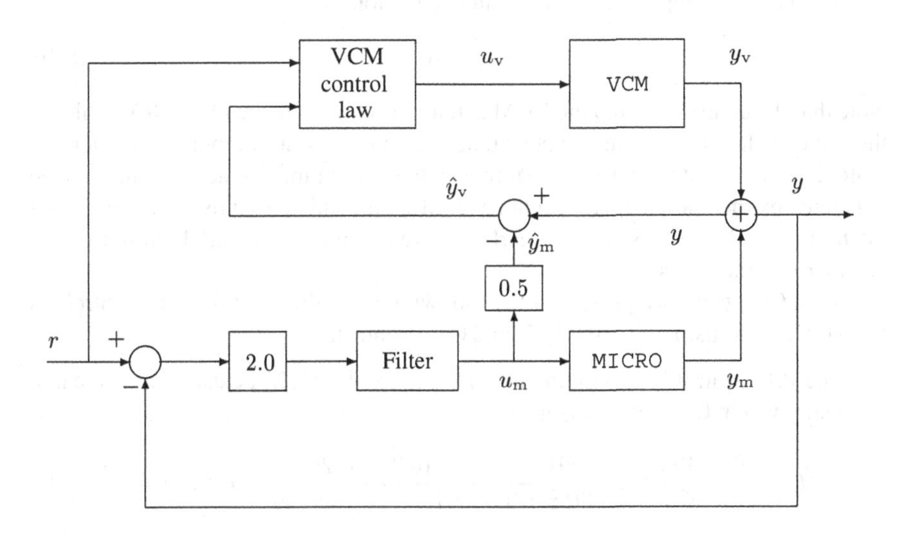

Figure 8.4. The schematic representation of a dual-stage actuator control

To reduce the effects of the resonance modes in the VCM actuator, as in Chapter 6, we introduce a notch filter whose transfer function is given as in Equation 6.9. We can then approximate the models of the VCM actuator as follows:

$$\Sigma_{\text{v}} : \begin{cases} \dot{x}_{\text{v}} = \begin{bmatrix} 0 & 1 \\ 0 & 0 \end{bmatrix} x_{\text{v}} + \begin{bmatrix} 0 \\ 6.4013 \times 10^{7} \end{bmatrix} u_{\text{v}} \\ y_{\text{v}} = [1 \quad 0] \; x_{\text{v}} \end{cases} \tag{8.12}$$

In order to minimize the effects of the resonance modes in the microactuator, we introduce the following filter in the feedforward path:

$$G_{\text{filter}}(s) = \frac{N_{\text{filter}}(s)}{D_{\text{filter}}(s)} \tag{8.13}$$

where

$$\begin{aligned} N_{\text{filter}}(s) = {} & 1.164\times10^{5}s^{5} + 1.177\times10^{9}s^{4} + 3.99\times10^{1}4s^{3} \\ & + 3.681\times10^{18}s^{2} + 3.112\times10^{23}s + 2.617\times10^{27} \end{aligned} \tag{8.14}$$

and

$$\begin{aligned} D_{\text{filter}}(s) = {} & 0.0002449s^{7}+47.92s^{6}+3.99\times10^{6}s^{5}+1.953\times10^{11}s^{4} \\ & + 5.71\times10^{15}s^{3}+1.058\times10^{20}s^{2}+8.387\times10^{23}s+1.309\times10^{27} \end{aligned} \tag{8.15}$$

It is a combination of a notch filter and a low-pass filter. The frequency responses of the filter in Equations 8.13–8.15 and the compensated microactuator system are given, respectively, in Figures 8.5 and 8.6. Clearly, the dynamics of the microactuator can then be safely approximated by a static equation:

$$y_{\text{m}} = 0.5 \; u_{\text{m}} \tag{8.16}$$

Note that the control input to the VCM actuator is constrained within ±3 V, which is the same as the one used in Chapters 6 and 7, and the control input to the microactuator has to be kept within ±2 V. Moreover, the maximum displacement that can be generated by the microactuator cannot exceed 1 μm, which is equivalent to one track pitch of the HDD servo system. Clearly, we are having a plant with both sensor and actuator nonlinearities.

As in Chapter 6, we present in the following three different types of controllers for the VCM actuator, *i.e.* PID, RPT and CNF control.

1. The PID control law (with the notch filter) for the VCM actuator is identical to that given in Chapter 6. It is given by

$$\tilde{u}(k) = \frac{0.50495z^{2} - 0.99147z + 0.48653}{z^{2} - 1.09091z + 0.09091} \cdot \frac{0.2696}{z - 0.7304} \cdot \Big[r(k) - \hat{y}_{\text{v}}(k) \Big] \tag{8.17}$$

and

$$u_{\text{v}}(k) = G_{\text{notch}}(s)\Big|_{\text{ZOH}} \cdot \tilde{u}(k) \tag{8.18}$$

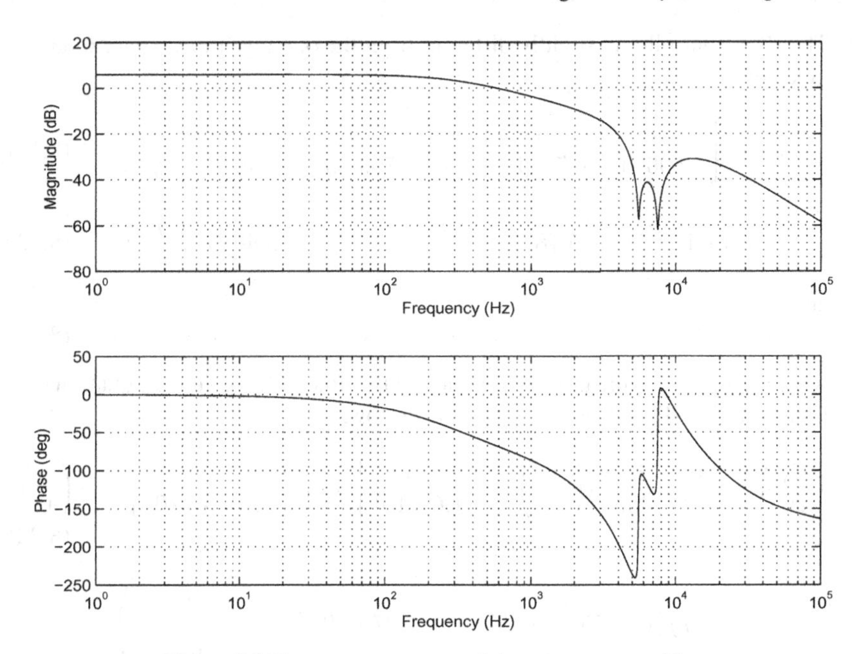

Figure 8.5. Frequency response of the microactuator filter

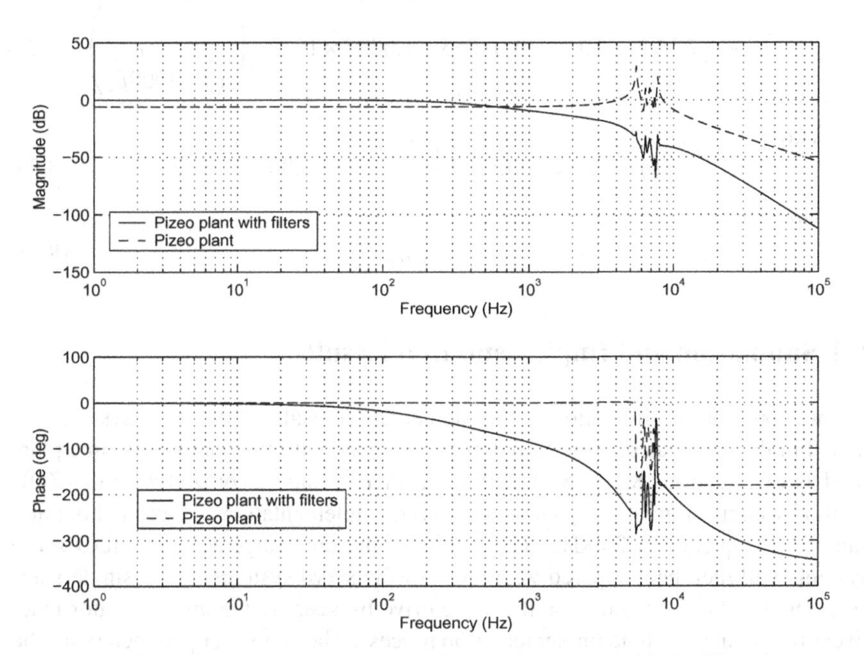

Figure 8.6. Frequency response of the microactuator with the filter

2. The RPT controller is slightly different from the one given in Chapter 6 and is given by

$$\dot{x}_{\rm v} = \begin{bmatrix} 0 & 0 \\ -696.40 & -7704.5 \end{bmatrix} x_{\rm v} + \begin{bmatrix} -1 & 1 \\ 70.009 & -378.19 \end{bmatrix} \begin{pmatrix} r \\ \hat{y}_{\rm v} \end{pmatrix} \tag{8.19}$$

$$\tilde{u} = [-1.0879 \quad -5.7871]\, x_{\rm v} + [0.10937 \quad -0.34085] \begin{pmatrix} r \\ \hat{y}_{\rm v} \end{pmatrix} \tag{8.20}$$

and

$$u_{\rm v} = G_{\rm notch}(s) \cdot \tilde{u} \tag{8.21}$$

where $G_{\rm notch}(s)$ is the transfer function of the notch filter given in Equation 6.9.
3. Finally, the CNF control law is given as follows:

$$\begin{pmatrix} \dot{x}_{\rm i} \\ \dot{x}_{\rm v} \end{pmatrix} = \begin{bmatrix} 0 & 0 \\ 0 & -4000 \end{bmatrix} \begin{pmatrix} x_{\rm i} \\ x_{\rm v} \end{pmatrix} - \begin{bmatrix} -1 \\ 1.6 \times 10^7 \end{bmatrix} \hat{y}_{\rm v} + \begin{bmatrix} 0 \\ 6.4013 \times 10^7 \end{bmatrix} \tilde{u} - \begin{pmatrix} r \\ 0 \end{pmatrix} \tag{8.22}$$

$$\begin{aligned} \tilde{u} = {} & \rho(\hat{y}_{\rm v} - r)\, [0.7555 \quad 0.07555 \quad 1.2597 \times 10^{-4}] \begin{pmatrix} x_{\rm i} \\ \hat{y}_{\rm v} - r \\ x_{\rm v} + 4000\hat{y}_{\rm v} \end{pmatrix} \\ & - [7.5549 \times 10^{-5} \quad 0.07555 \quad 1.0306 \times 10^{-5}] \begin{pmatrix} x_{\rm i} \\ \hat{y}_{\rm v} - r \\ x_{\rm v} + 4000\hat{y}_{\rm v} \end{pmatrix} \end{aligned} \tag{8.23}$$

with

$$\rho(\hat{y}_{\rm v} - r) = -0.0791 \left| {\rm e}^{-|\hat{y}_{\rm v} - r|} - 0.3679 \right| - 0.35 \tag{8.24}$$

and

$$u_{\rm v} = G_{\rm notch}(s) \cdot \tilde{u} \tag{8.25}$$

8.4 Simulation and Implementation Results

We now present the simulation and implementation results of the servo systems obtained in the previous section. Simulations are done with the continuous-time plant models of Equations 8.1 to 8.11 with controllers being discretized with the ZOH method with a sampling frequency of 20 kHz. Implementations are carried out at a sampling frequency of 20 kHz. The results of the dual-stage actuated HDD servo systems will then be compared with those of the servo systems with a single-stage actuator. The latter are done on the same drive by keeping the microactuator inactive throughout the whole implementation process. The controller parameters for the single-stage actuated systems are identical to those given in Chapter 6.

8.4.1 Track-following Test

The simulated output responses of the servo systems with $r = 1$ µm are given in Figure 8.7. Once again, the RPT and CNF control systems yield much better performances compared to that of the PID control system. For comparison, we present the responses and control signals of each individual approach for the dual-stage actuated servo system together with those of its single-stage counterpart in Figures 8.8 to 8.10, respectively. The implementation results of the corresponding servo systems are then presented in Figures 8.11 to 8.13. Finally, we summarize the resulting settling time of each control system in Table 8.1.

Table 8.1. Performances of the servo systems

(a) Simulation results

	Settling time (ms)		
	PID Control	RPT Control	CNF Control
Single-stage	3.10	0.95	0.80
Dual-stage	2.15	0.40	0.40

(b) Implementation results

	Settling time (ms)		
	PID Control	RPT Control	CNF Control
Single-stage	2.70	1.05	0.85
Dual-stage	1.80	0.30	0.30

8.4.2 Frequency-domain Test

To verify the frequency-domain properties of our designs, Figures 8.14 to 8.19, respectively, give the Bode plot, the Nyquist plot, and the sensitivity and complementary sensitivity functions, as well as the closed-loop transfer functions (from the reference input r to the controlled output $h = y$) of the resulting control systems. For the CNF design, once again, its frequency-domain functions are calculated at the steady-state situation for which the nonlinear gain function $\rho(e)$ has approached its final constant value. The results show that all these designs meet the frequency-domain design specifications.

8.4.3 Runout Disturbance Test

As in the previous chapters, we artificially add a runout disturbance

$$w(t) = 0.5 + 0.1\cos(100\pi t) + 0.05\sin(220\pi t) + 0.02\sin(440\pi t) + 0.01\sin(880\pi t) \tag{8.26}$$

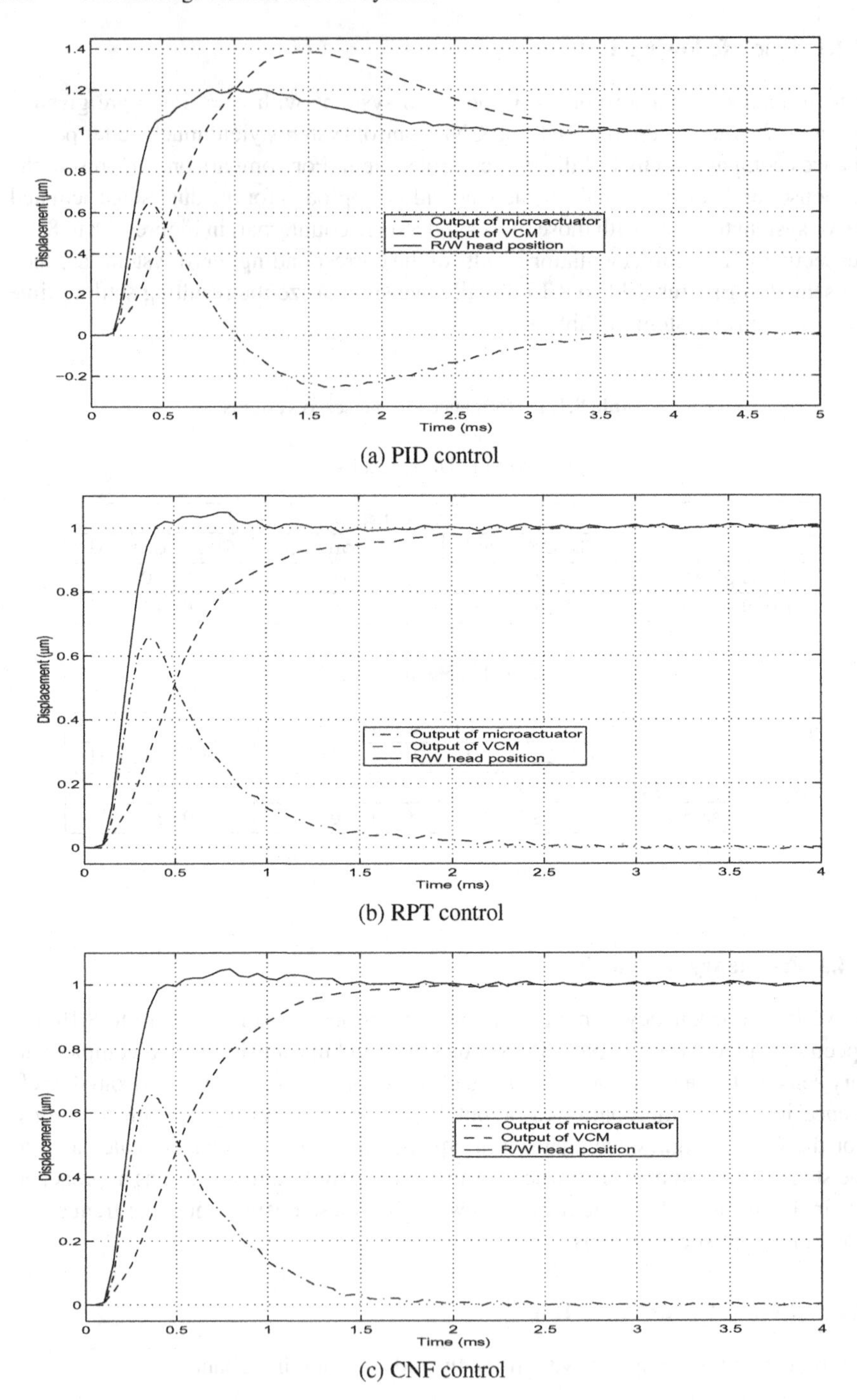

(a) PID control

(b) RPT control

(c) CNF control

Figure 8.7. Simulation: output responses of the dual-stage actuated systems

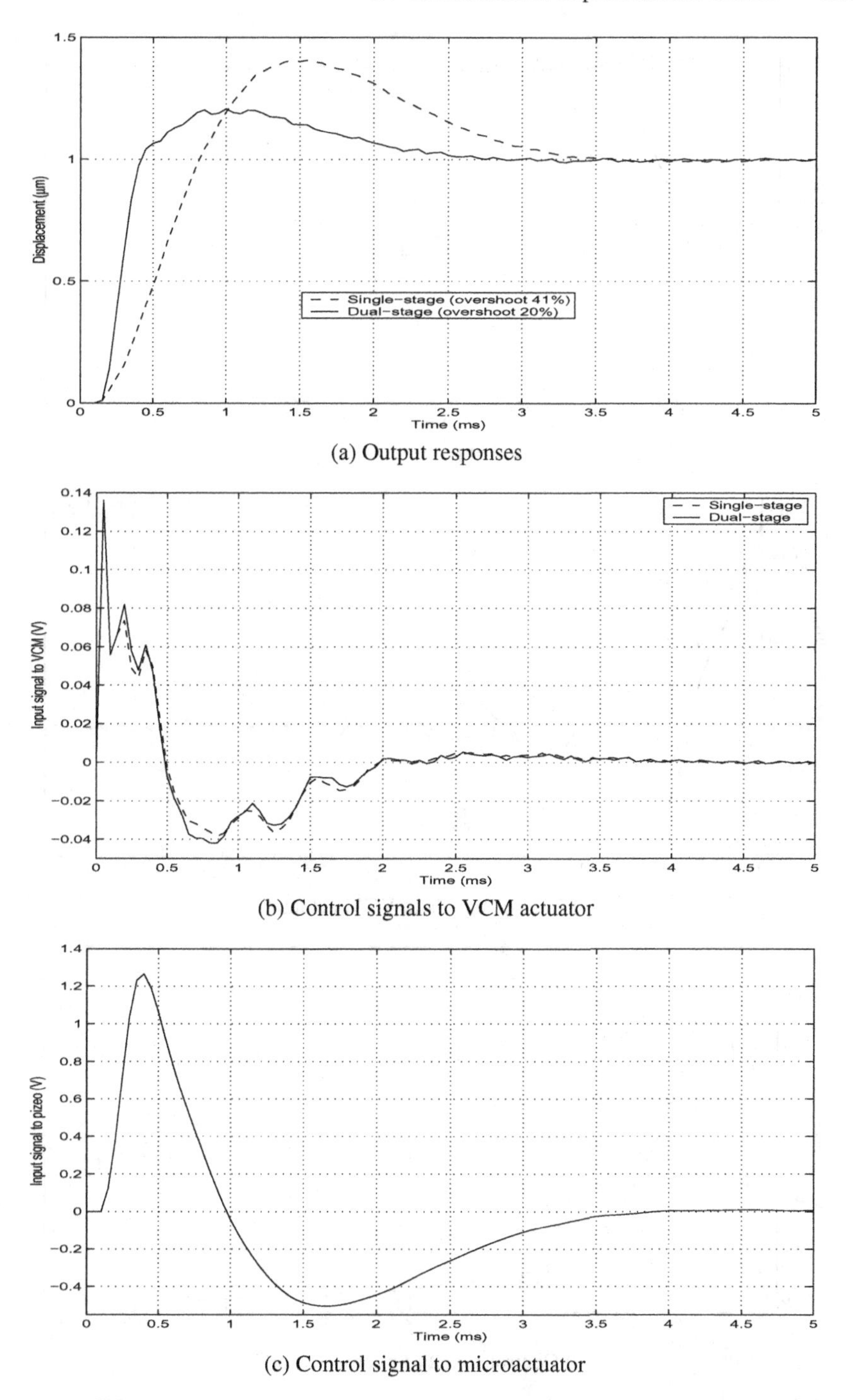

(a) Output responses

(b) Control signals to VCM actuator

(c) Control signal to microactuator

Figure 8.8. Simulation results of single- and dual-stage servo systems with PID control

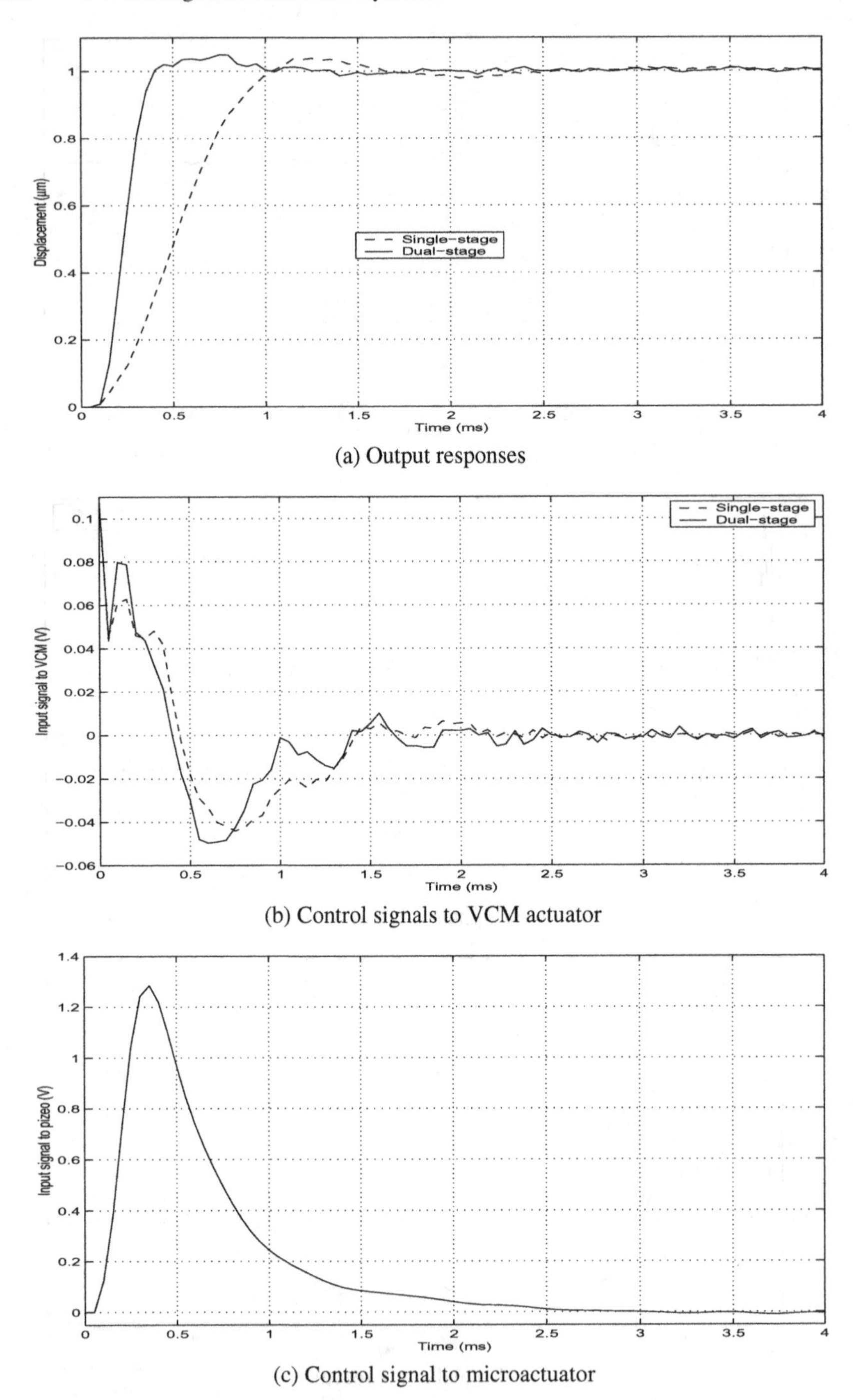

(a) Output responses

(b) Control signals to VCM actuator

(c) Control signal to microactuator

Figure 8.9. Simulation results of single- and dual-stage servo systems with RPT control

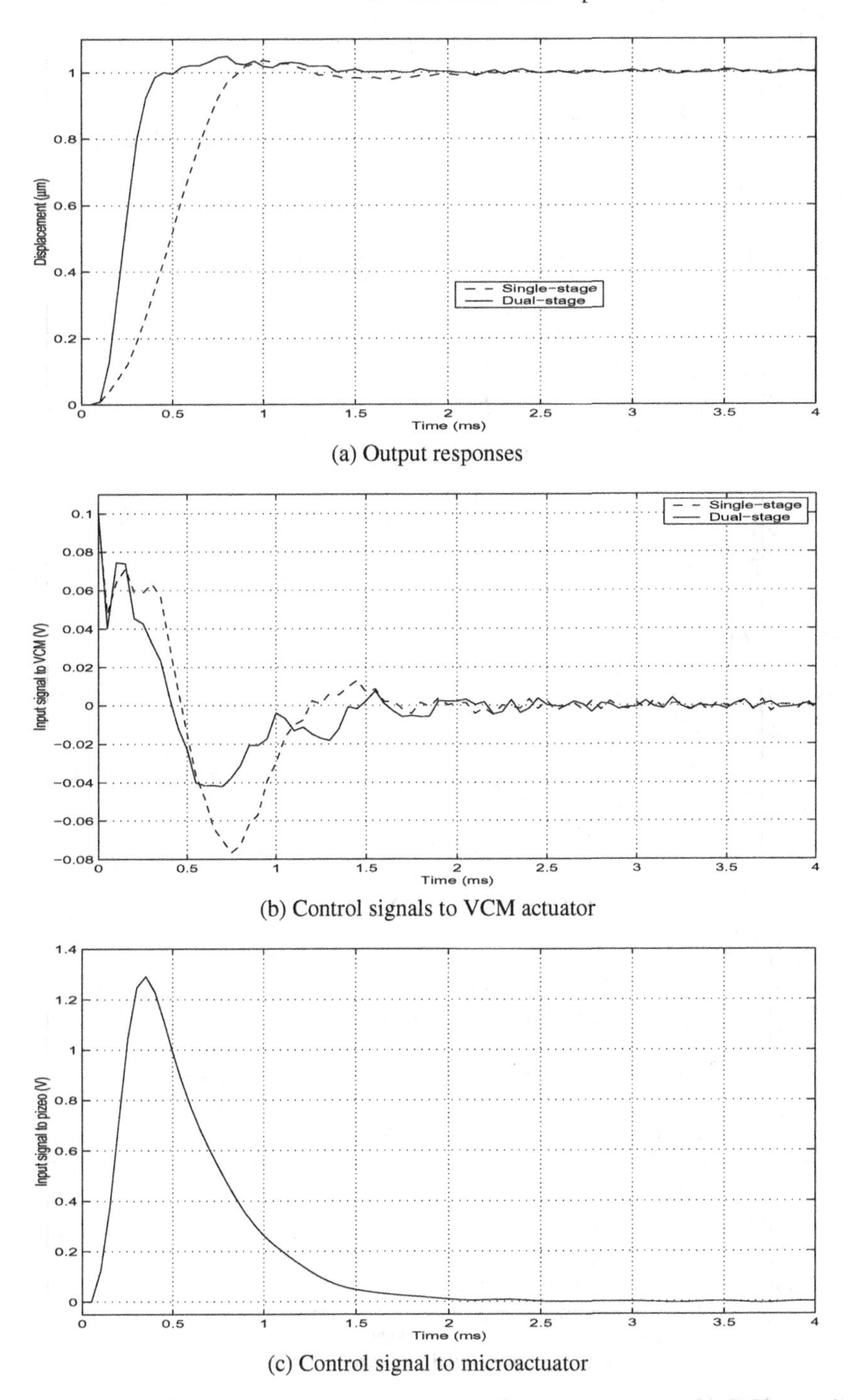

(a) Output responses

(b) Control signals to VCM actuator

(c) Control signal to microactuator

Figure 8.10. Simulation results of single- and dual-stage servo systems with CNF control

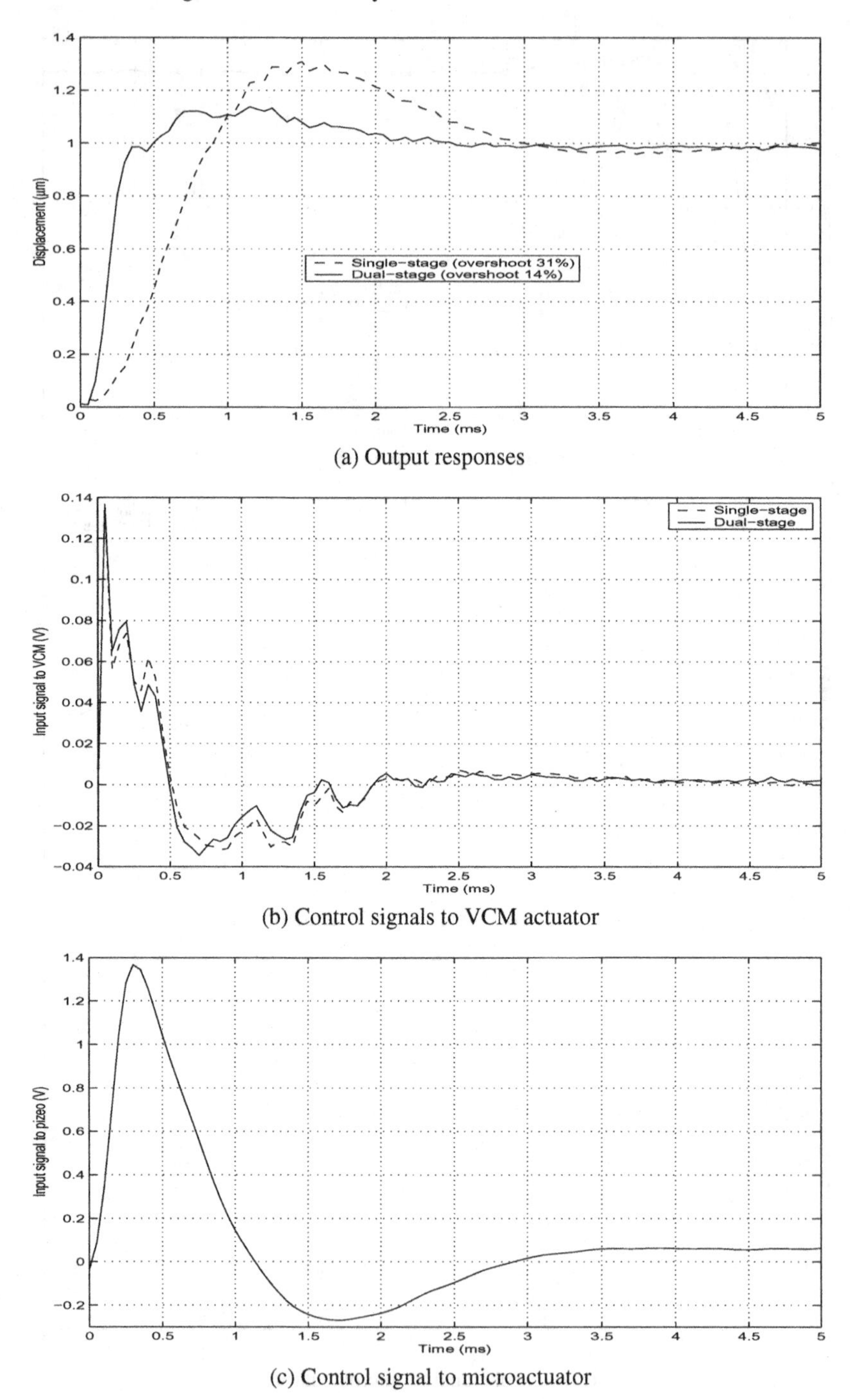

(a) Output responses

(b) Control signals to VCM actuator

(c) Control signal to microactuator

Figure 8.11. Implementation results of single- and dual-stage servo systems with PID control

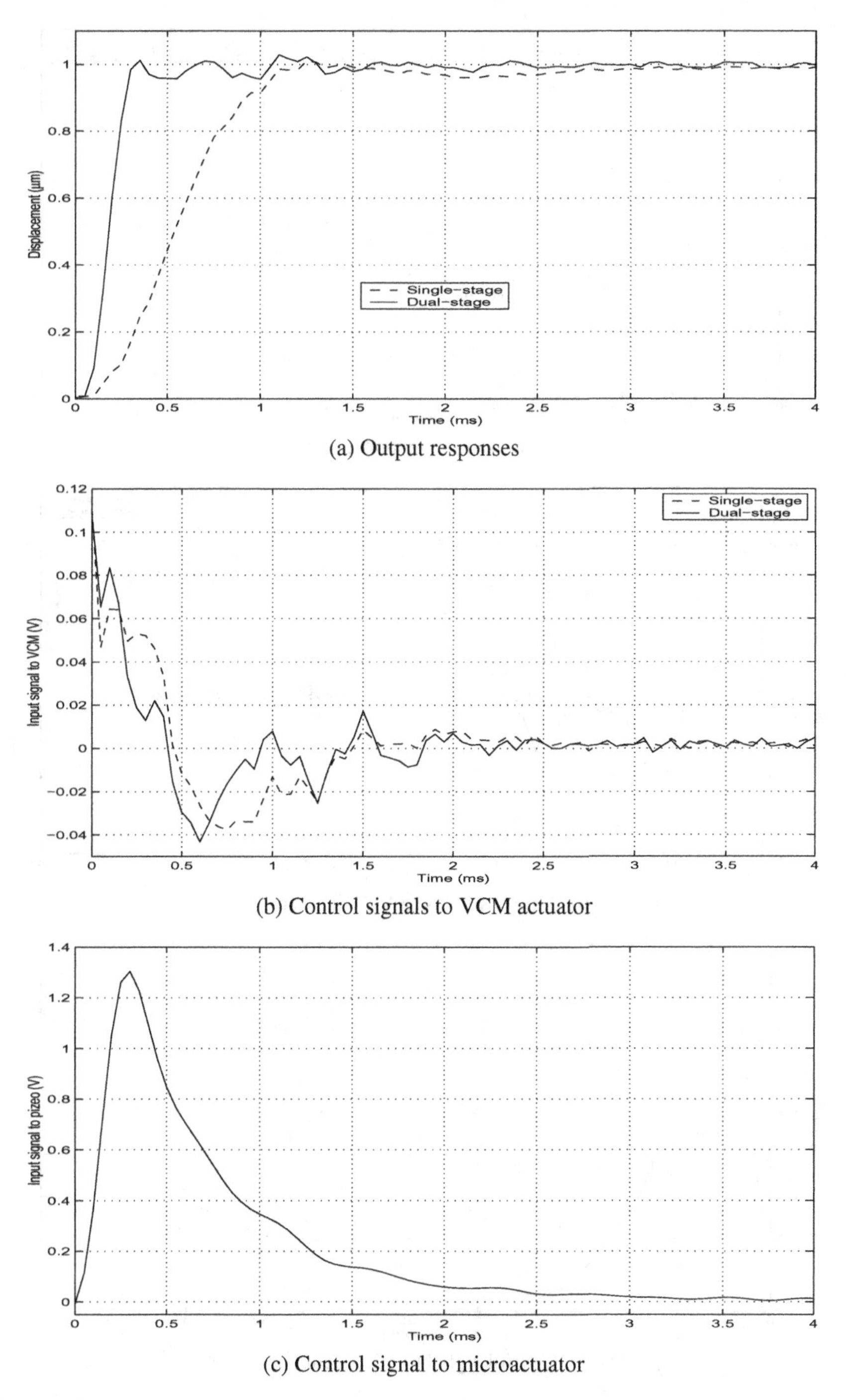

Figure 8.12. Implementation results of single- and dual-stage servo systems with RPT control

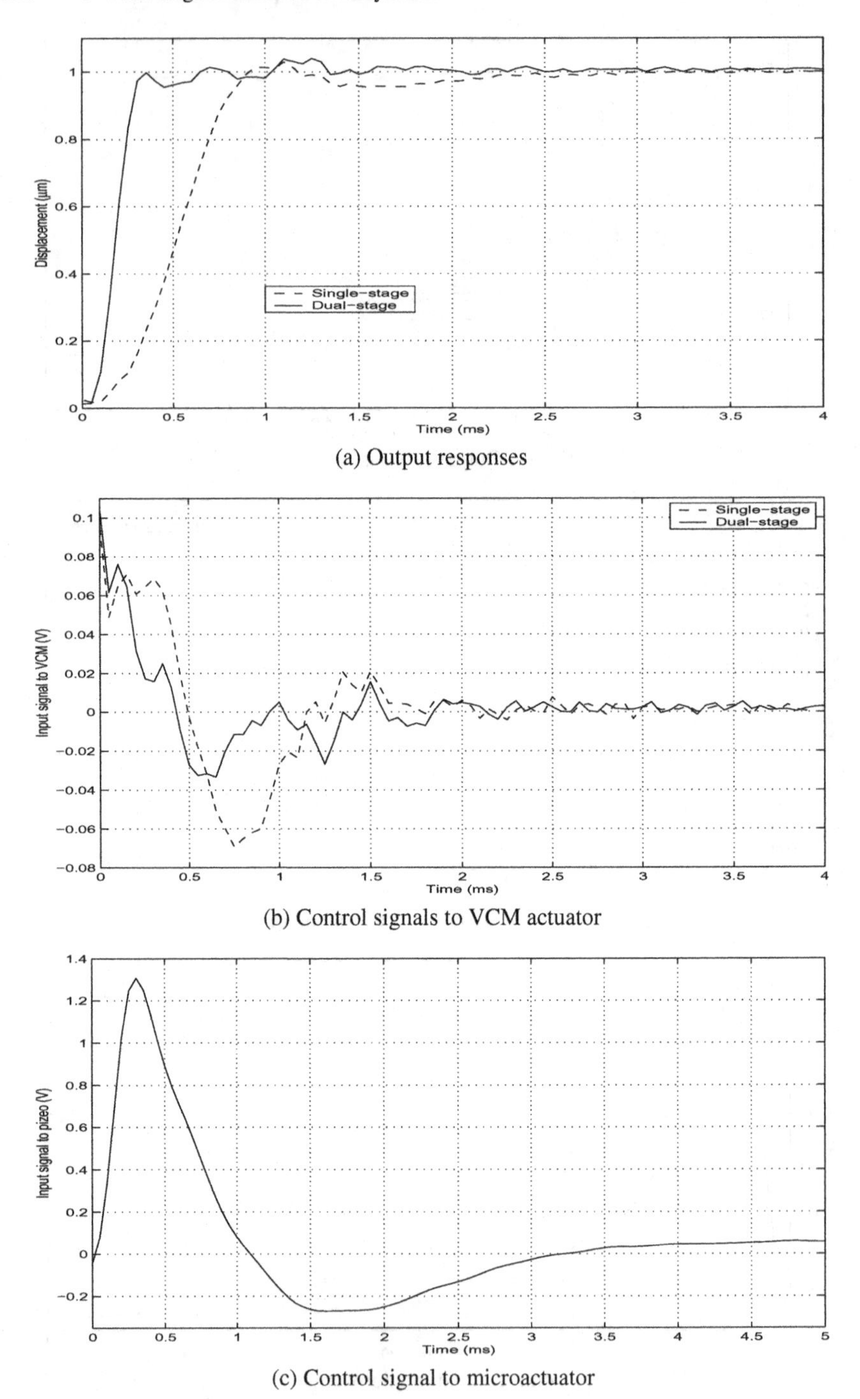

Figure 8.13. Implementation results of single- and dual-stage servo systems with CNF control

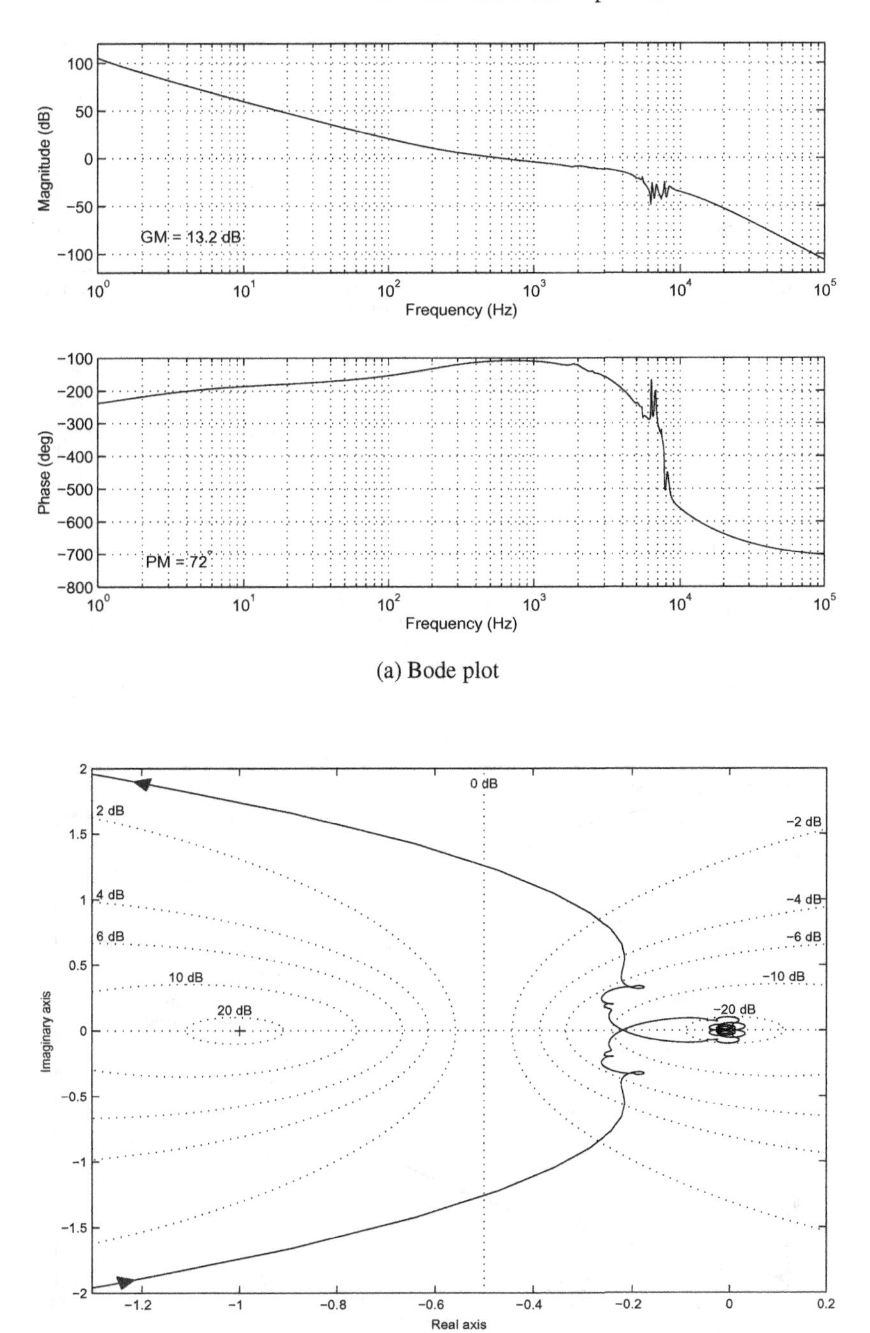

(a) Bode plot

(b) Nyquist plot

Figure 8.14. Bode and Nyquist plots of the PID control system

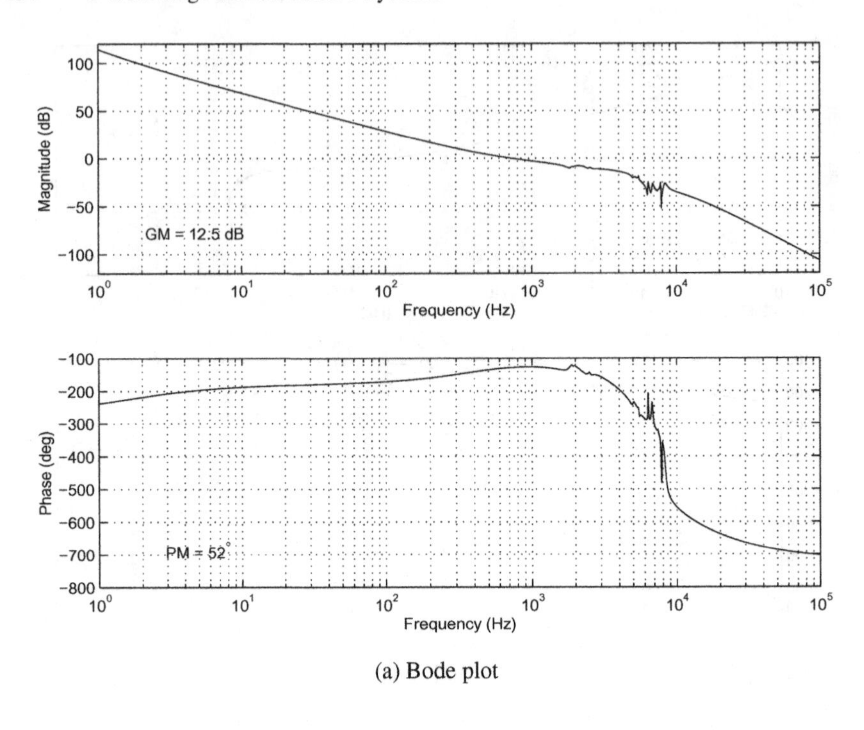

(a) Bode plot

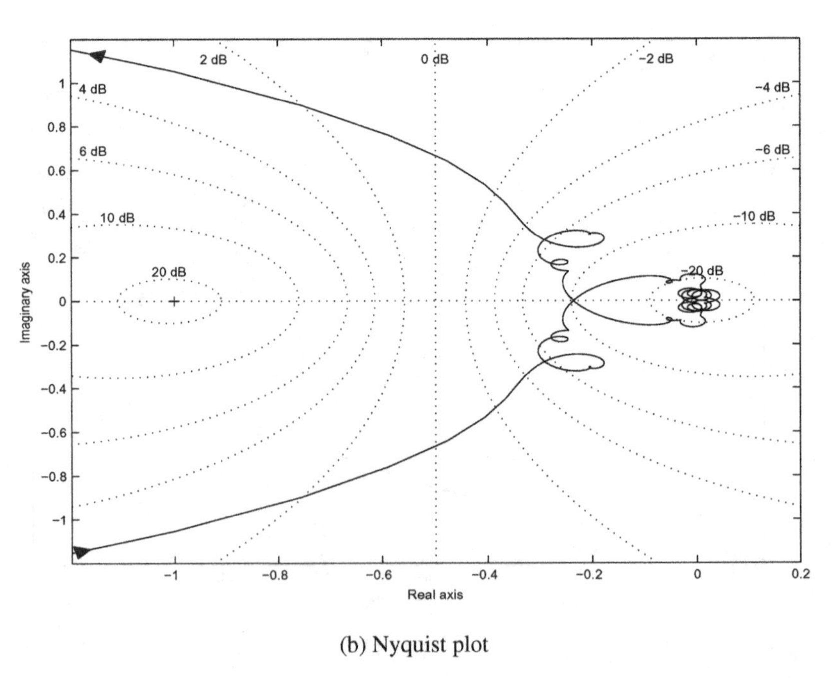

(b) Nyquist plot

Figure 8.15. Bode and Nyquist plots of the RPT control system

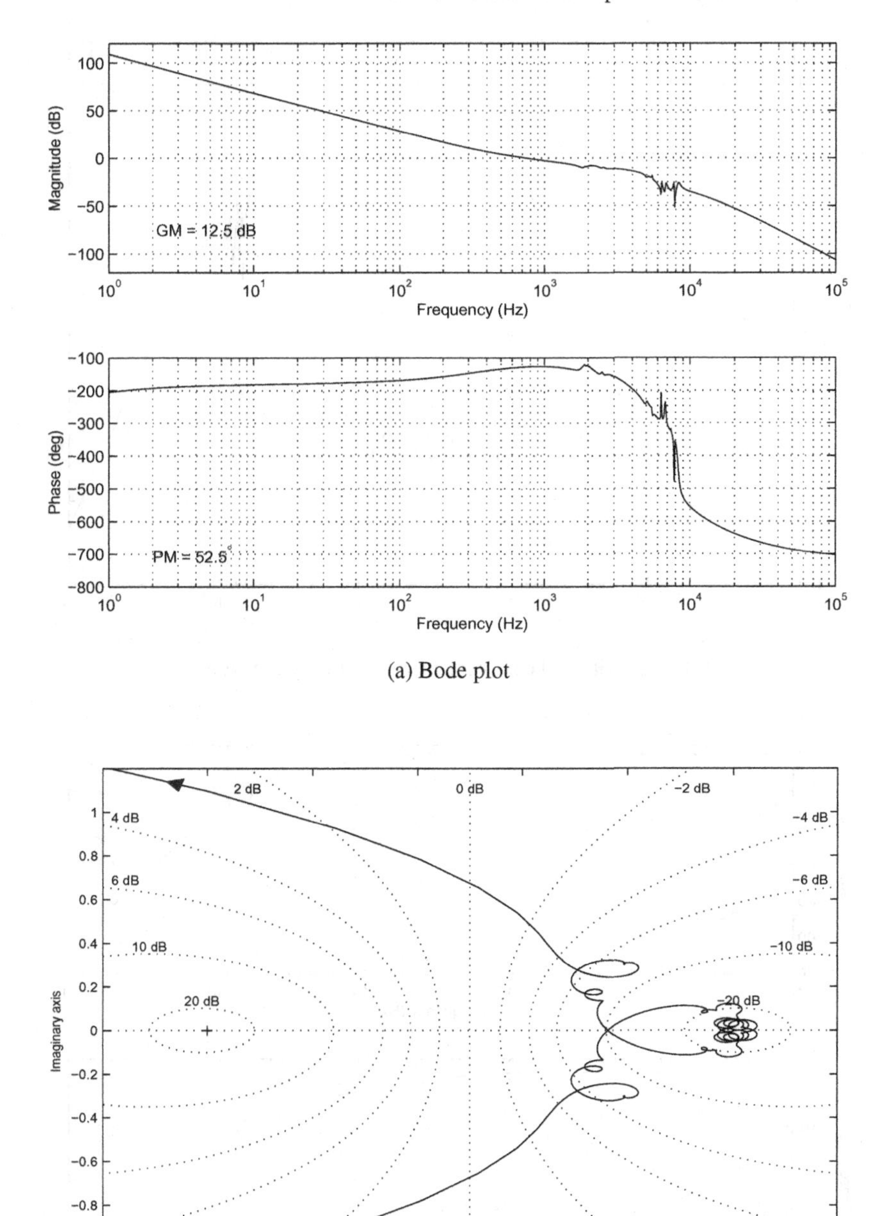

(a) Bode plot

(b) Nyquist plot

Figure 8.16. Bode and Nyquist plots of the CNF control system

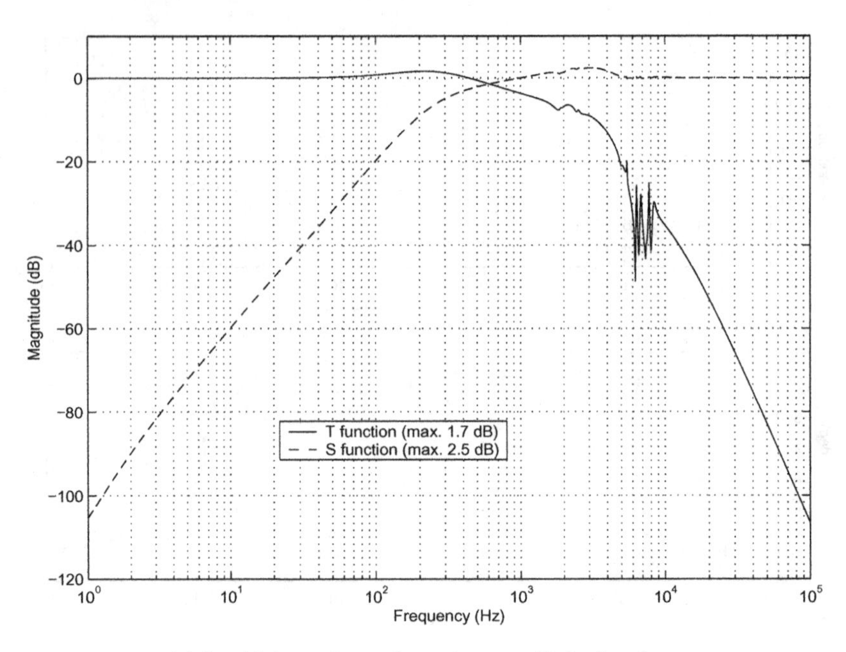

(a) Sensitivity and complementary sensitivity functions

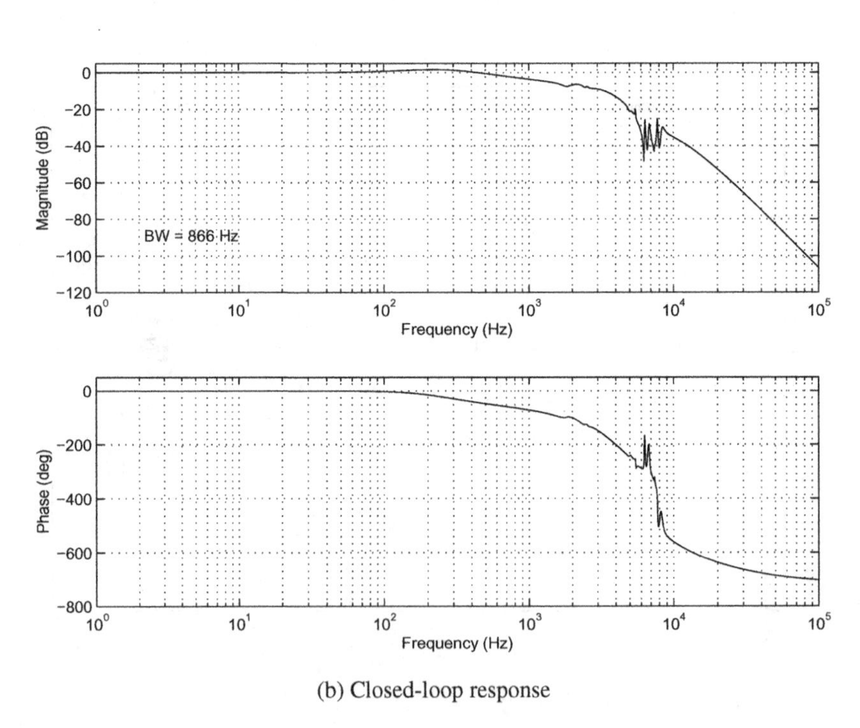

(b) Closed-loop response

Figure 8.17. Sensitivity functions and closed-loop transfer function of the PID control system

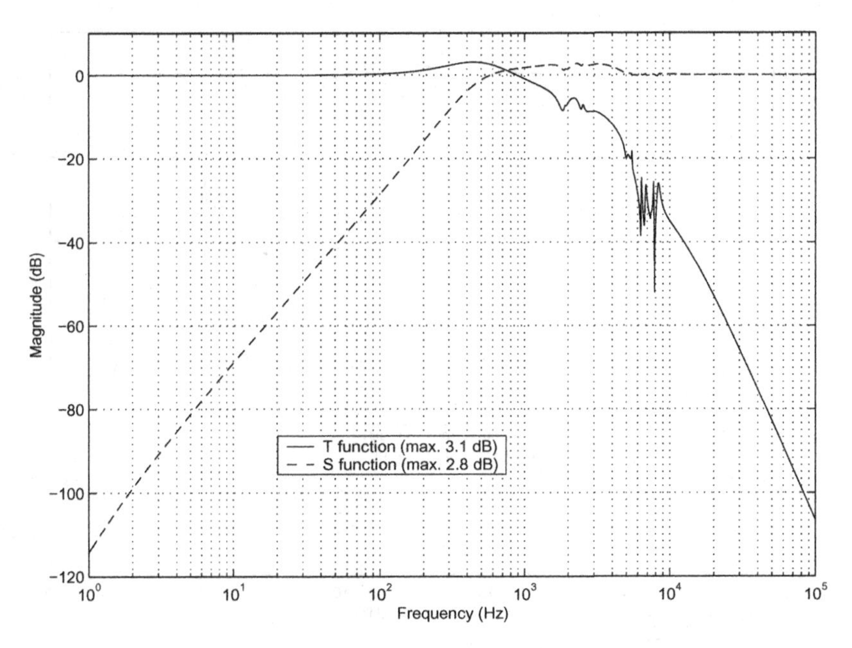

(a) Sensitivity and complementary sensitivity functions

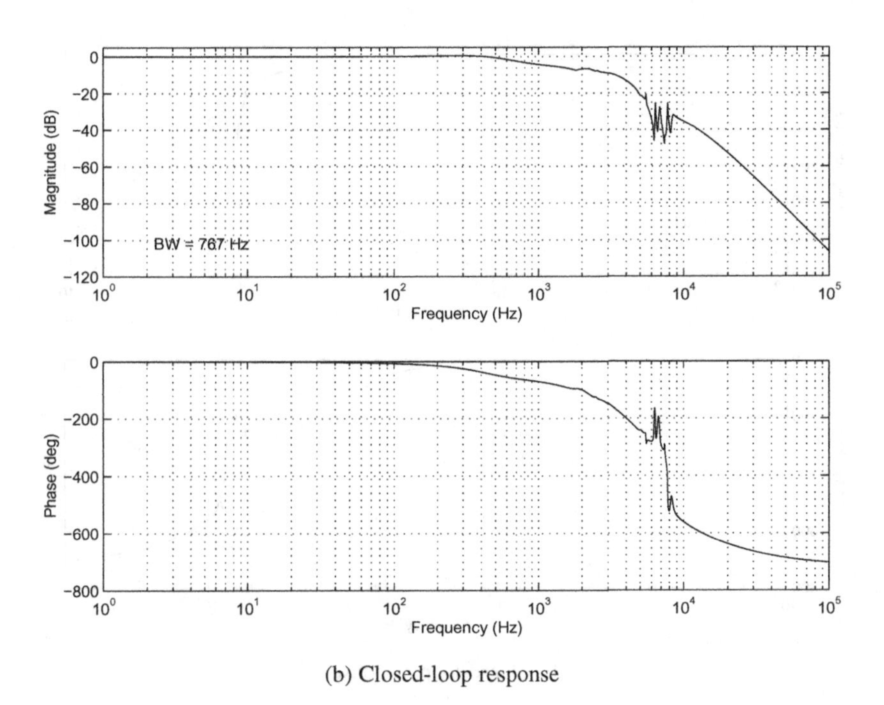

(b) Closed-loop response

Figure 8.18. Sensitivity functions and closed-loop transfer function of the RPT control system

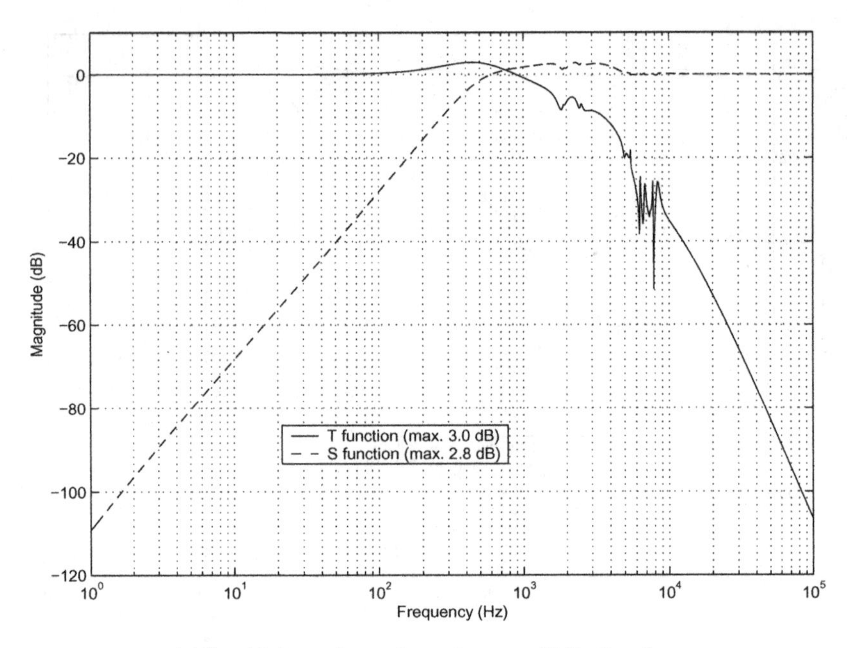

(a) Sensitivity and complementary sensitivity functions

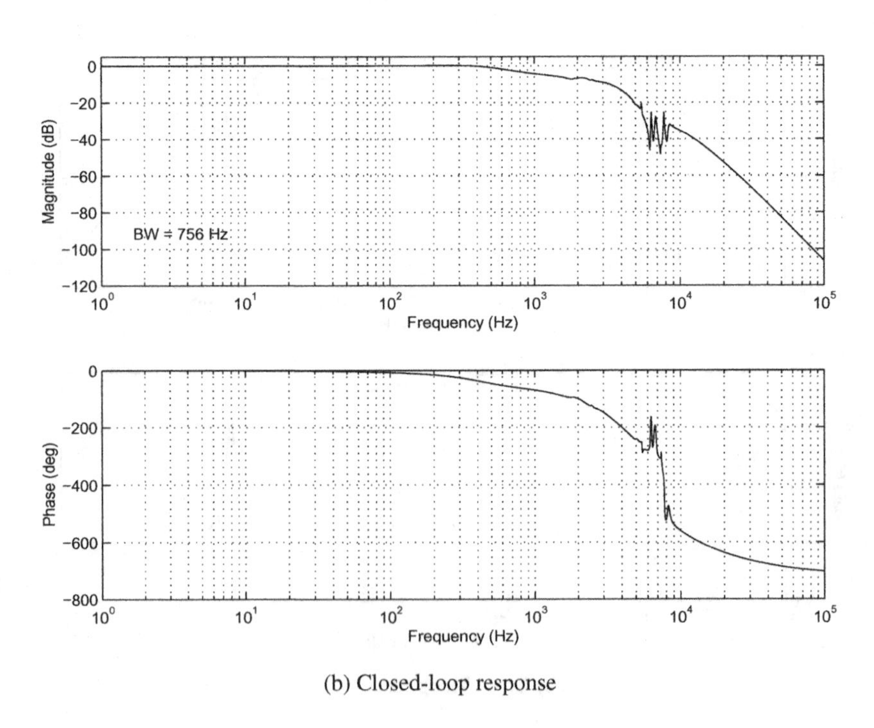

(b) Closed-loop response

Figure 8.19. Sensitivity functions and closed-loop transfer function of the CNF control system

which is the same as those in the previous chapters, to our servo systems. The implementation results of the corresponding responses are respectively shown in Figures 8.20 to 8.22.

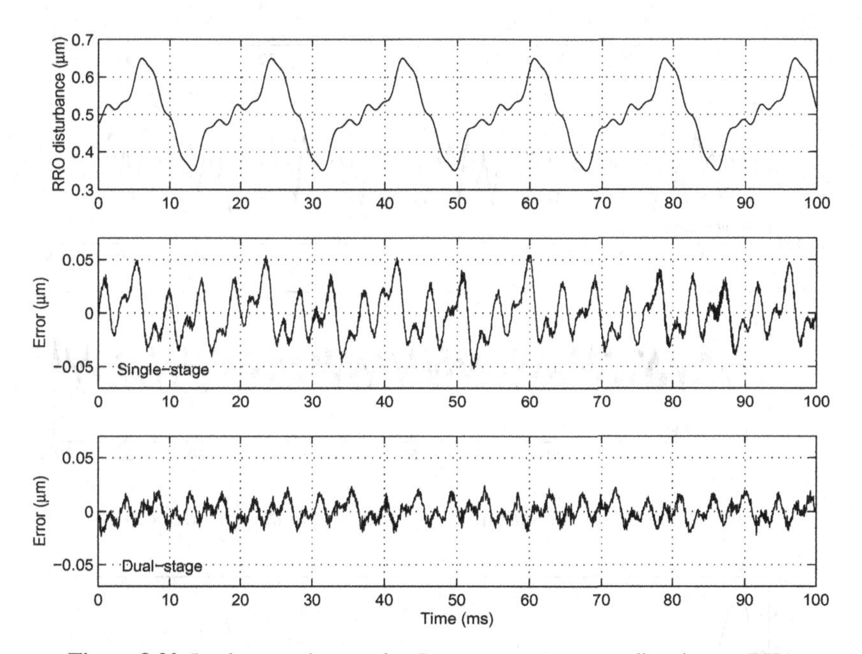

Figure 8.20. Implementation results: Responses to a runout disturbance (PID)

8.4.4 Position Error Signal Test

Lastly, as were done in Chapters 6 and 7, we conduct the PES tests for the complete single- and dual-stage actuated servo systems. The results, *i.e.* the histograms of the PES tests, are given in Figures 8.23 to 8.25. The $3\sigma_{\text{pes}}$ values of the PES tests, which are a measure of track misregistration (TMR) in HDDs and that are closely related to the maximum achievable track density, are summarized in Table 8.2.

Table 8.2. The $3\sigma_{\text{pes}}$ values of the PES tests

	$3\sigma_{\text{pes}}$ (μm)		
	PID control	RPT control	CNF control
Single-stage	0.0615	0.0375	0.0288
Dual-stage	0.0273	0.0204	0.0195

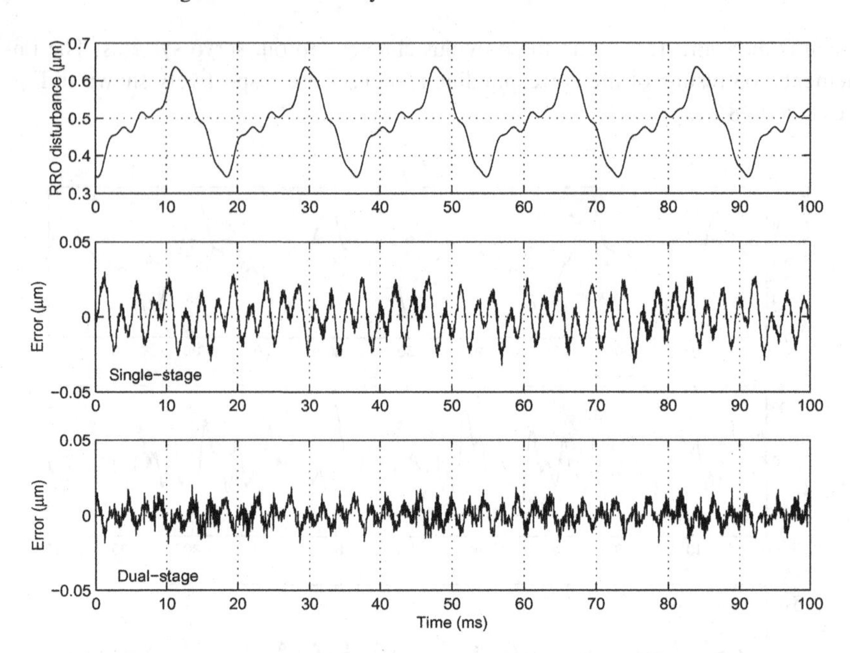

Figure 8.21. Implementation results: Responses to a runout disturbance (RPT)

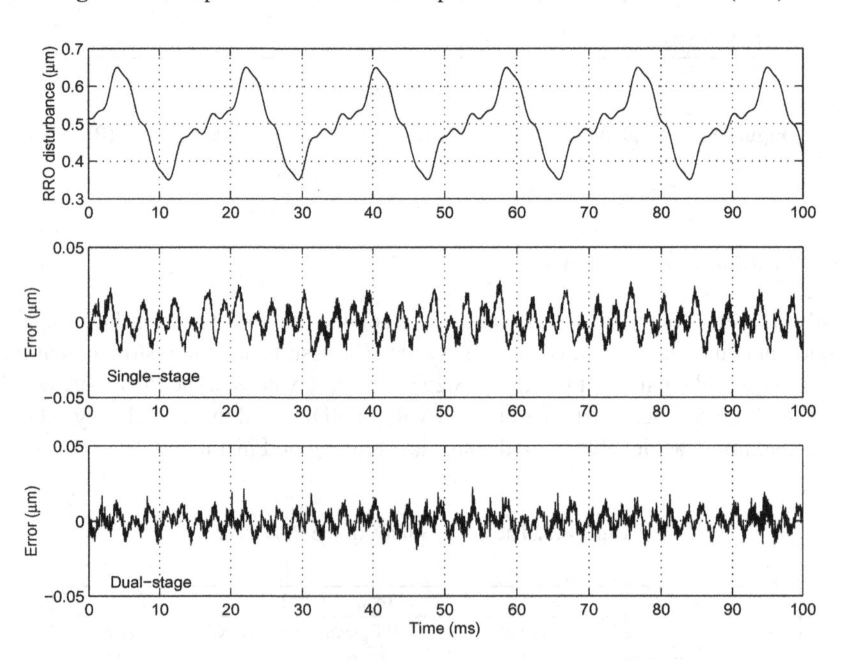

Figure 8.22. Implementation results: Responses to a runout disturbance (CNF)

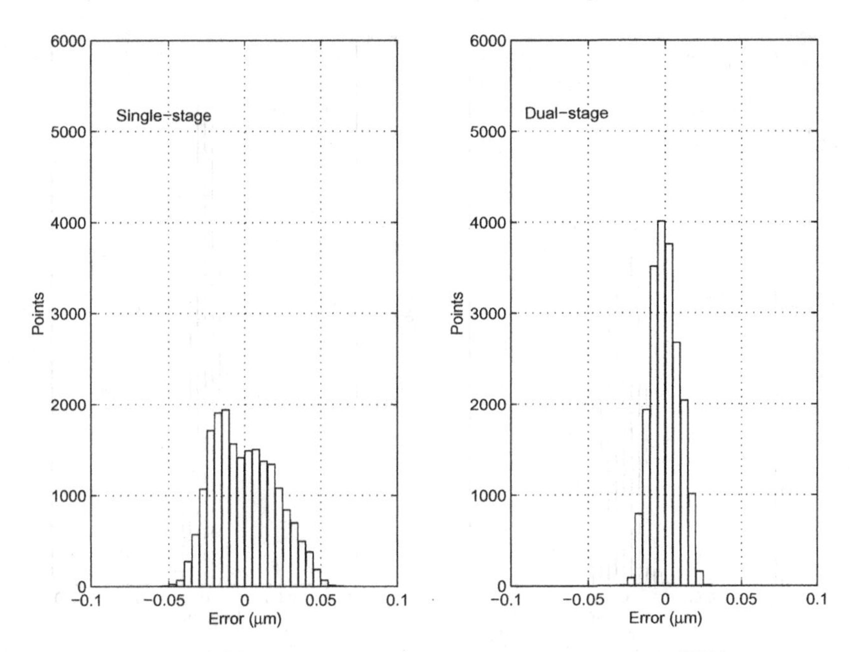

Figure 8.23. Implementation results: PES test histograms (PID)

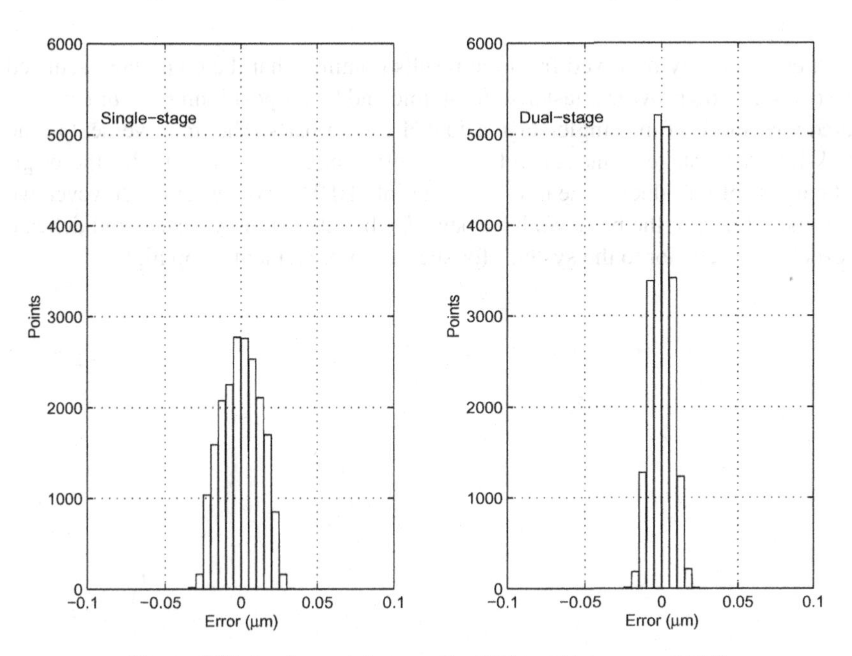

Figure 8.24. Implementation results: PES test histograms (RPT)

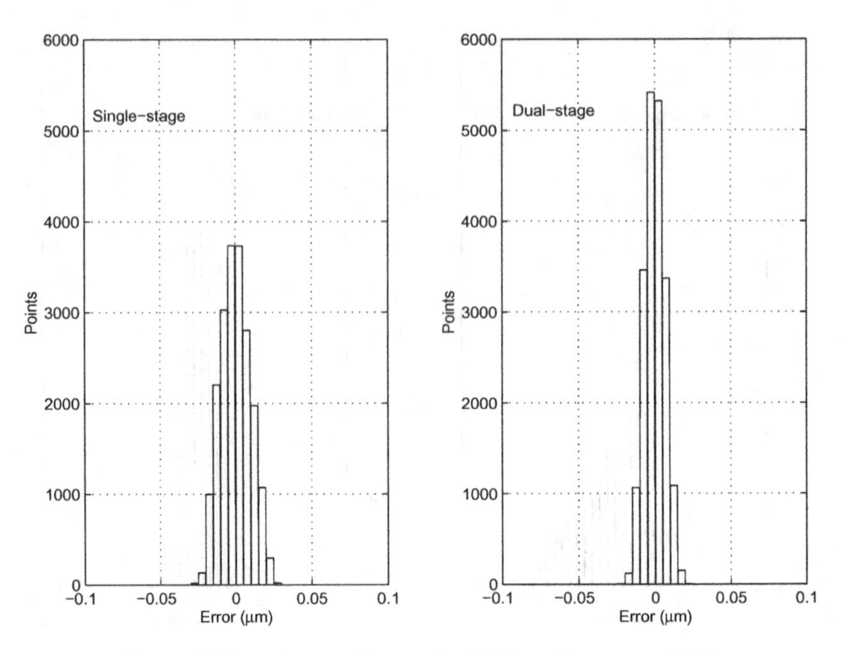

Figure 8.25. Implementation results: PES test histograms (CNF)

It can be easily observed from the results obtained that the dual-stage actuated servo systems do provide a faster settling time and better positioning accuracy compared with those of the single-stage actuated counterparts. The improvement in the track-following stage turns out to be very noticeable. This was actually the original purpose of introducing the microactuator into HDD servo systems. However, we personally feel that the price we have paid (*i.e.* by adding an expensive and delicate piezoelectric actuator to the system) for such an improvement is too high.

9

Modeling and Design of a Microdrive System

9.1 Introduction

Chapters 6 to 8 focus on the design of single- and dual-stage actuated hard drive servo systems. The hard drives considered are those used in normal desktop computers. As mentioned earlier in the introduction chapter, microdrives have become popular these days because of high demand from many new applications. Many factors such as frictional forces and nonlinearities, which are negligible for normal drives and thus ignored in servo systems given in Chapters 6 to 8, emerge as critical issues for microdrives. It can be observed that nonlinearities from friction in the actuator rotary pivot bearing and data flex cable in the VCM actuator (see Figure 9.1) generate large residual errors and deteriorate the performance of head positioning of HDD servo systems, which is much more severe in the track-following stage when the R/W head is moving from the current track to its neighborhood tracks. The desire to fully understand the behaviors of nonlinearities and friction in microdrives is obvious. Actually, this motivates us to carry out a complete study and modeling of friction and nonlinearities for the VCM-actuated HDD servo systems.

Friction is hard nonlinear and may result in residual errors, limit cycles and poor performance (see, *e.g.*, [168–171]. Friction exists in almost all servomechanisms, behaves in features of the Stribeck effect, hysteresis, stiction and varying break-away force, occurs in all mechanical systems and appears at the physical interface between two contact surfaces moving relative to each other. The features of friction have been extensively studied (see, *e.g.*, [168–174]), but there are significant differences among diverse systems. There has been a significantly increased interest in friction in the industry, which is driven by strong engineering needs in a wide range of industries and availability of both precise measurement and advanced control techniques.

The HDD industry persists in the need for companies to come up with devices that are cheaper and able to store more data and retrieve or write to them at faster speed. Decreasing the HDD track width is a feasible idea to achieve these objectives. But, the presence of friction in the rotary actuator pivot bearing results in large residual errors and high-frequency oscillations, which may produce a larger positioning error signal to hold back the further decreasing of the track width and to degrade

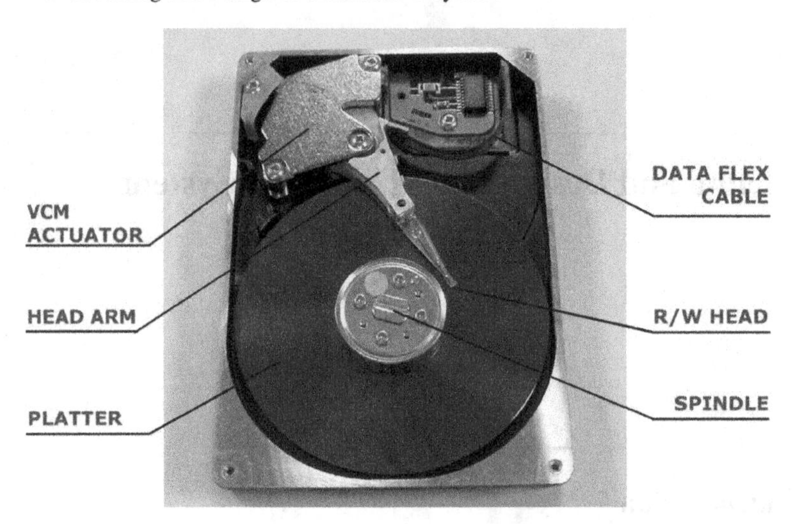

Figure 9.1. An HDD with a VCM actuator

the performance of the servo systems. This issue becomes more noticeable for small drives and is one of the challenges to design head positioning servo systems for small HDDs. Much effort has been put into the research on mitigation of the friction in the pivot bearing in the HDD industry in the last decade (see, *e.g.*, [8, 175–178]). It is still ongoing in the disk drive industry (see, *e.g.*, [69, 179, 180]).

Diverse modeling methods had been proposed (see, *e.g.*, [59, 69]) based on linear systems, where nonlinearities of plants are assumed to be tiny and can be neglected. As such, these methods cannot be directly applied to model plants with significant nonlinearities. Instead, in the first part of this chapter, we utilize the physical effect approach given in Chapter 2 to determine the structures of nonlinearities and friction associated with the VCM actuator in a typical HDD servo system. This is done by carefully examining and analyzing physical effects that occur in or between electromechanical parts. Then, we employ a Monte Carlo process (see Chapter 2) to identify the parameters in the structured model. We note that Monte Carlo methods are very effective in approximating solutions to a variety of mathematical problems, for which their analytical solutions are hard, if not impossible, to determine. Our simulation and experimental results show that the identified model of friction and nonlinearities using such approaches matches very well the behavior of the actual system.

The second part of this chapter focuses on the controller design for the HDD servo system. Our philosophy of designing servo systems is rather simple. Once the model of the friction and nonlinearities of the VCM actuator is obtained, we will try to cancel as much as possible all these unwanted elements in the servo system. As it is impossible to have perfect models for friction and nonlinearities, a perfect cancellation of these elements is unlikely to happen in the real world. We then formulate our design by treating the uncompensated portion as external disturbances. The PID and RPT control techniques of Chapter 3 and the CNF control technique of Chapter 5 are to be used to carry out our servo system design. We note that some of the results presented in this chapter have been reported earlier in [138].

9.2 Modeling of the Microdrive Actuator

The physical structure of a typical VCM actuator is shown in Figure 9.2. The motion of the coil is driven by a permanent magnet similar to typical DC motors. The stator of the VCM is built of a permanent magnet. The rotor consists of a coil, a pivot and a metal arm on which the R/W head is attached. A data flex cable is connected with the R/W head through the metal arm to transfer data read from or written to the HDD disc via the R/W head. Typically, the rotor has a deflected angle, α in rad, ranging up to 0.5 rad in commercial disk drives. We are particularly interested in the modeling of the friction and nonlinearities for the actuator in the track-following stage, in which the R/W head movement is within the neighborhood of its current track and thus $\alpha \ll 1$. An IBM microdrive (DMDM-10340) is used throughout for illustration.

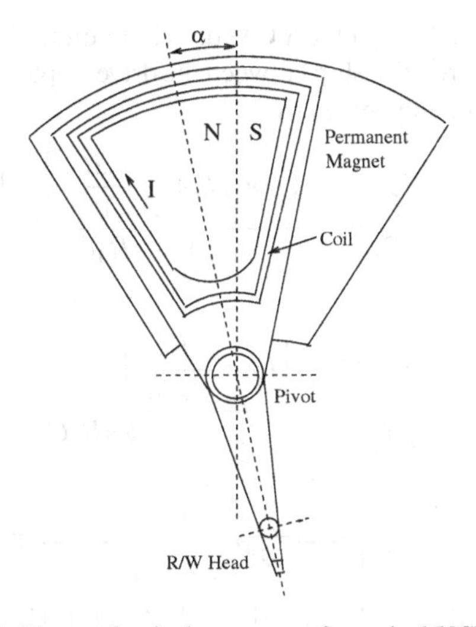

Figure 9.2. The mechanical structure of a typical VCM actuator

9.2.1 Structural Model of the VCM Actuator

We first adopt the physical effect analysis of Chapter 2 to determine the structures of nonlinearities in the VCM actuator. It is to analyze the effects between the components of the actuator, such as the stator, rotor and support plane as well as the VCM driver. The VCM actuator is designed to position the R/W head fast and precisely onto the target track, and is driven by a VCM driver, a full bridge power amplifier, which converts an input voltage into an electric current. The electrical circuit of a typical VCM driver is shown in Figure 9.3, where Z represents the coil of the VCM actuator and the external input voltage is exerted directly into the VCM driver to drive the coil.

In order to simplify our analysis, we assume that the physical system has the following properties: i) the permanent magnet is constant; and ii) the coil is assembled strictly along the radius and concentric circle of the pivot; Furthermore, we assume that the friction of a mechanical object consists of Coloumb friction and viscous damping, and is characterized by a typical friction function as follows:

$$f(F_{\rm N}, F_{\rm e}, v, \mu_{\rm d}, \mu_{\rm s}, \mu_{\rm v}) = \begin{cases} -\mu_{\rm d} F_{\rm N}\,\mathrm{sgn}(v) - \mu_{\rm v} v, & v \neq 0 \\ -F_{\rm e}, & v = 0,\ |F_{\rm e}| \leq \mu_{\rm s} F_{\rm N} \\ -\mu_{\rm s} F_{\rm N}\,\mathrm{sgn}(F_{\rm e}), & v = 0,\ |F_{\rm e}| > \mu_{\rm s} F_{\rm N} \end{cases} \tag{9.1}$$

where f is the friction force, $F_{\rm N}$ is the normal force, *i.e.* the force perpendicular to the contacted surfaces of the objects, $F_{\rm e}$ is the external force applied to the object, v is the relative moving speed between two contact surfaces, and $\mu_{\rm s} F_{\rm N}$ is the breakaway force. Furthermore, $\mu_{\rm d}$, $\mu_{\rm s}$ and $\mu_{\rm v}$ are, respectively, the dynamic, static and viscous coefficients of friction.

Through a detailed analysis of the VCM driver circuit in Figure 9.3, it is straightforward to verify that the relationship between the driver input voltage and the current and voltage of the VCM coil is given by

$$\frac{R_{\rm s}+(1+A_{\rm c})(1+A_{\rm b})R}{R_{\rm s}A_{\rm c}(1+A_{\rm c})(1+A_{\rm b})R}\cdot\frac{R_2}{R_1}u = \frac{1+s\left[R_3+\dfrac{R_2}{A_{\rm c}(1+A_{\rm b})}\right]C_3}{1+sR_3C_3}I+\mathcal{E}(s)u_{\rm c} \tag{9.2}$$

where

$$\mathcal{E}(s) = \frac{A_{\rm b}}{(1+A_{\rm c})(1+A_{\rm b})R}\cdot\frac{1+s\left[R_3+\dfrac{R_2}{A_{\rm c}A_{\rm b}}\left(1+\dfrac{R(1+A_{\rm c})}{R_{\rm s}}\right)\right]C_3}{1+sR_3C_3} \tag{9.3}$$

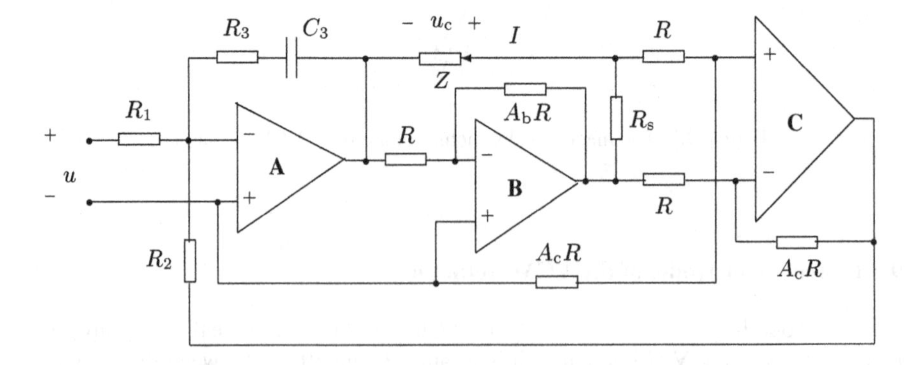

Figure 9.3. The electrical circuit of a typical VCM driver

u is the input voltage to the VCM driver, I and u_c are, respectively, the VCM coil current and voltage. For the IBM microdrive (DMDM-10340) used in our experiment, $R_1 = 8.2$ kΩ, $R_2 = 1$ kΩ, $R_3 = 270$ kΩ, $R_s = 1$ Ω, $R = 16$ kΩ, $C_3 = 270$ pF, and the amplifier gains $A_c = 2$ and $A_b = 4.7$. For such a drive, we have

$$\mathcal{E}(s) = \frac{1.7178 \times 10^{-5}(1 + 1.45 \times 10^{-3}s)}{1 + 7.29 \times 10^{-5}s} \tag{9.4}$$

which has a magnitude response ranging from −95 dB (for frequency less than 110 Hz) to −69 dB (for frequency greater than 2.2 kHz), and

$$\frac{1 + s\left[R_3 + \dfrac{R_2}{A_c(1 + A_b)}\right]C_3}{1 + sR_3C_3} \approx 1, \quad \text{for all } s \tag{9.5}$$

Such a property generally holds for all commercial disk drives. As such, it is safe to approximate the relationship of u and I of the VCM driver as

$$I = \frac{R_2}{A_c R_1 R_s} u = k_{vd} u \tag{9.6}$$

For the IBM microdrive used in this work, $k_{vd} = 0.061\ \Omega^{-1}$.

Next, it is straightforward to derive that the torque T_m, relative to the center of the pivot and that moving anticlockwise is positive and produced by the permanent magnet B_{coil} in the coil with the electric current, is given by

$$T_m = (r_1^2 - r_2^2) B_{coil} n_{coil} k_{vd} u \tag{9.7}$$

r_1 and r_2 are, respectively, the outside and inside radius of the coil to the center of the pivot, and n_{coil} is the number of windings of the coil. The total external torque T_e applied to the VCM actuator is given as follows:

$$T_e = T_m - T_c(\alpha) \tag{9.8}$$

where $T_c(\alpha)$ is the spring torque produced by the data flex cable and is a function of the deflection angle α or the displacement of the R/W head. The friction torque in the VCM actuator comes from two major sources: One is the friction in the pivot bearing and the other is between the pivot bearing and the support plane. The friction torque in the pivot bearing can be characterized as

$$T_{f1} = f(F_{N1}, F_{e1}, r_3\dot{\alpha}, \mu_{d1}, \mu_{s1}, \mu_{v1}) r_3 \tag{9.9}$$

where F_{e1} is the external force, μ_{d1}, μ_{s1} and μ_{v1} are the related friction coefficients as defined in Equation 9.1, r_3 is the radius of the pivot to its center, and

$$F_{N1} = \left|F_r + m_r(\dot{\alpha})^2\right| \tag{9.10}$$

is the normal force, which consists of the centrifugal force of the rotor and the diametrical force, F_r. Furthermore, m_r is a constant dependent on the mass distribution of the rotor, and

$$F_{\mathrm{r}} = 2(r_1 - r_2)B_{\mathrm{coil}}n_{\mathrm{coil}}k_{\mathrm{vd}}u\alpha \tag{9.11}$$

is the force along the radius of the pivot bearing produced in the coil by the permanent magnet.

The friction torque between the pivot bearing and the support plane can be characterized as:

$$T_{\mathrm{f2}} = f(F_{\mathrm{N2}}, F_{\mathrm{e2}}, r_3\dot{\alpha}, \mu_{\mathrm{d2}}, \mu_{\mathrm{s2}}, \mu_{\mathrm{v2}})r_3 \tag{9.12}$$

where F_{e2} is the external force, μ_{d2}, μ_{s2} and μ_{v2} are the related friction coefficients as defined in Equation 9.1, and

$$F_{\mathrm{N2}} = |T_0|r_3^{-1} \tag{9.13}$$

is the normal force resulted from a static balance torque of the rotor, T_0. Thus, the total friction torque T_{f} presented in the VCM actuator is given by

$$T_{\mathrm{f}} = T_{\mathrm{f1}} + T_{\mathrm{f2}} = \begin{cases} -T_{\mathrm{v}}, & \dot{\alpha} \neq 0 \\ -T_{\mathrm{e}}, & \dot{\alpha} = 0 \text{ and } |T_{\mathrm{e}}| \leq T_{\mathrm{s}} \\ -T_{\mathrm{s}}\,\mathrm{sgn}(T_{\mathrm{e}}), & \dot{\alpha} = 0 \text{ and } |T_{\mathrm{e}}| > T_{\mathrm{s}} \end{cases} \tag{9.14}$$

where

$$T_{\mathrm{v}} = \Big[\mu_{\mathrm{d1}}\left|2(r_1 - r_2)B_{\mathrm{coil}}n_{\mathrm{coil}}k_{\mathrm{vd}}u\alpha + m_{\mathrm{r}}(\dot{\alpha})^2\right|r_3 + \mu_{\mathrm{d2}}|T_0|\Big]\mathrm{sgn}(\dot{\alpha}) + (\mu_{\mathrm{v1}} + \mu_{\mathrm{v2}})r_3^2\dot{\alpha} \tag{9.15}$$

and

$$T_{\mathrm{s}} = 2\mu_{\mathrm{s1}}\left|(r_1 - r_2)B_{\mathrm{coil}}n_{\mathrm{coil}}k_{\mathrm{vd}}u_0\alpha_0\right|r_3 + \mu_{\mathrm{s2}}|T_0| \tag{9.16}$$

is the breakaway torque, and where u_0 and α_0 are, respectively, the corresponding input voltage and the deflection angle for the situation when $\dot{\alpha} = 0$.

Lastly, it is simple to verify that the relative displacement of the R/W head, y, is given by

$$y = 2r_4\sin\frac{\alpha}{2} \approx r_4\alpha, \quad |\alpha| \ll 1 \tag{9.17}$$

where r_4 is the length from the R/W head to the center of the pivot. Following Newton's law of motion, $J\ddot{\alpha} = T_{\mathrm{e}} + T_{\mathrm{f}}$, where J is the moment of inertia of the VCM rotor, we have

$$\ddot{y} = -\tilde{T}_{\mathrm{c}} + \tilde{T}_{\mathrm{f}} + bu \tag{9.18}$$

where

$$\tilde{T}_{\mathrm{f}} = \begin{cases} -\left[\left|d_1buy + d_2(\dot{y})^2\right| + d_3\right]\mathrm{sgn}(\dot{y}) - d_0\dot{y}, & \dot{y} \neq 0 \\ -\tilde{T}_{\mathrm{e}}, & \dot{y} = 0,\ |\tilde{T}_{\mathrm{e}}| \leq \tilde{T}_{\mathrm{s}} \\ -\tilde{T}_{\mathrm{s}}\,\mathrm{sgn}(\tilde{T}_{\mathrm{e}}), & \dot{y} = 0,\ |\tilde{T}_{\mathrm{e}}| > \tilde{T}_{\mathrm{s}} \end{cases} \tag{9.19}$$

where

$$\left.\begin{aligned}
\tilde{T}_{\rm e} &= -\tilde{T}_{\rm c}(y) + bu \\
\tilde{T}_{\rm s} &= d_4 b|u_0 y_0| + d_5 \\
\tilde{T}_{\rm c} &= J^{-1} r_4 T_{\rm c}(\alpha) \\
b &= J^{-1} r_4 (r_1^2 - r_2^2) B_{\rm coil} n_{\rm coil} k_{\rm vd} \\
d_0 &= J^{-1} (\mu_{\rm v1} + \mu_{\rm v2}) r_3^2 \\
d_1 &= 2 r_3 \mu_{\rm d1} [r_4 (r_1 + r_2)]^{-1} \\
d_2 &= r_3 m_{\rm r} \mu_{\rm d1} (J r_4)^{-1} \\
d_3 &= J^{-1} r_4 \mu_{\rm d2} |T_0| \\
d_4 &= 2 r_3 \mu_{\rm s1} [r_4 (r_1 + r_2)]^{-1} \\
d_5 &= J^{-1} r_4 \mu_{\rm s2} |T_0|
\end{aligned}\right\} \tag{9.20}$$

with u_0 and y_0 being, respectively, the corresponding input voltage and the displacement for the case when $\dot{y} = 0$. It is clear now that the expressions in Equations 9.18–9.20 give a complete structure of the VCM model including friction and nonlinearities from the data flex cable. Our next task is to identify all these parameters for the IBM microdrive (DMDM-10340).

9.2.2 Identification and Verification of Model Parameters

We proceed to identify the parameters of the VCM actuator model given in Equations 9.18–9.20. We note that there are results available in the literature (see, *e.g.*, [168]) to estimate friction parameters for typical DC motors for which both velocity and displacement are measurable and without constraint. Unfortunately, for the VCM actuator studied in this chapter, it is impossible to measure the time responses in constant-velocity motions and only the relative displacement of the R/W head is measurable. As such, the method of [168] cannot be adopted to solve our problems. Instead, we employ the popular Monte Carlo method of Chapter 2 (see also, [63–65]), which has been widely used in solving engineering problems and is capable of providing good numerical solutions.

First, it is simple to obtain from Equation 9.18 at a steady state when $\dot{y} = 0$ and $\ddot{y} = 0$,

$$|u_0 - b^{-1}\tilde{T}_{\rm c}(y_0)| = b^{-1}|\tilde{T}_{\rm f}| \le b^{-1}\tilde{T}_{\rm s} = d_4|u_0 y_0| + b^{-1} d_5 \tag{9.21}$$

Our experimental results show that the right hand side of Equation 9.21 is very insignificant for small input signal u_0 and small displacement y_0. This will be verified later when the model parameters are fully identified. Thus, we have

$$u_0 \approx b^{-1}\tilde{T}_{\rm c}(y_0) := \tilde{T}_{\rm bc}(y_0) \tag{9.22}$$

which is used to identify $\tilde{T}_{\rm c}(y)$ or equivalently $T_{\rm c}(\alpha)$, the spring torque produced by the data flex cable. Next, for the small neighborhood of (u_0, y_0), we can rewrite the dynamic expression of Equation 9.18 as

$$\ddot{y} = -\left.\partial\tilde{T}_{\rm c}/\partial y\right|_{y=y_0} (y - y_0) + b(u - u_0) + \tilde{T}_{\rm f} \tag{9.23}$$

For small signals, and by omitting the nonlinear terms in $\tilde{T}_{\rm f}$, the system dynamics in Equation 9.23 can be approximated by a second-order linear system with a transfer function from $u - u_0$ to $y - y_0$:

$$G_{y_0,u_0}(s) = \frac{b}{s^2 + d_0 s + \omega_{\rm p}^2} \tag{9.24}$$

The natural frequency of the above transfer function (or roughly its peak frequency), $\omega_{\rm p}$, is given by

$$\omega_{\rm p} = \left(\partial \tilde{T}_{\rm c} / \partial y \Big|_{y=y_0} \right)^{\frac{1}{2}} \tag{9.25}$$

and its static gain is given by $k_{\rm dc} = b\omega_{\rm p}^{-2}$, which implies that

$$b = k_{\rm dc}\omega_{\rm p}^2 = 4\pi^2 k_{\rm dc} f_{\rm p}^2 \tag{9.26}$$

where $f_{\rm p} = \omega_{\rm p}/(2\pi)$. The expression in Equation 9.26 will be used to estimate the parameter b. More specifically, the parameters of the dynamic models of the VCM actuator will be identified using the following procedure:

1. The nonlinear characteristics of the data flex cable or equivalently $\tilde{T}_{\rm bc}(y)$ will be initially determined using Equation 9.22 with a set of input signal, u_0, and its corresponding output displacement, y_0. It will be fine tuned later using the Monte Carlo method.
2. The parameter b will be initially computed using measured static gains and peak frequencies as in Equation 9.26, resulting from the dynamical responses of the actuator to a set of small input signals. Again, the identified parameter will be fine tuned later using the Monte Carlo method.
3. All system parameters will then be identified using the Monte Carlo method to match the frequency response to small input signal;
4. The high-frequency resonance modes of the actuator, which have not been included in either Equation 9.18 or 9.23, will be determined from frequency responses to input signals at high frequencies.

The above procedure will yield a complete and comprehensive model including nominal dynamics, high-frequency resonance modes, friction and nonlinearities of the VCM actuator. In our experiments, the relative displacement of the R/W head is the only measurable output and is measured using a laser Doppler vibrometer (LDV). A dynamic signal analyzer (DSA) (Model SRS 785) is used to measure the frequency responses of the VCM actuator. The DSA is also used to record both input and output signals of time-domain responses. Square waves are generated with a dSpace DSP board installed in a personal computer.

The time-domain response of the VCM actuator to a typical square input signal about 1 Hz is shown in Figure 9.4. With a group of time-domain responses to a set of square input signals, we obtain the corresponding measurement data for the nonlinear function, $\tilde{T}_{\rm bc}(y_0)$, which can be matched nicely by an arctan function (see Figure 9.5) as follows:

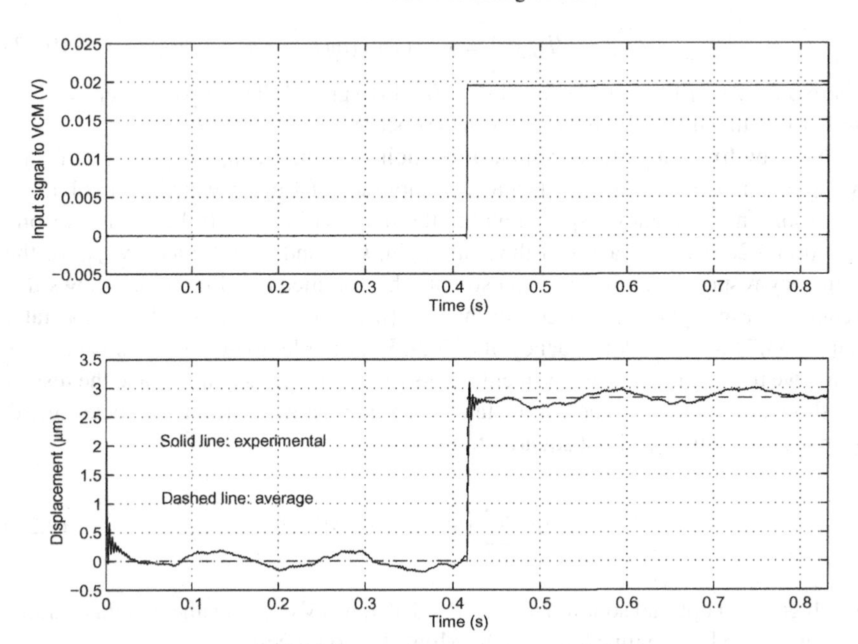

Figure 9.4. Time-domain response of the VCM actuator to a square wave input

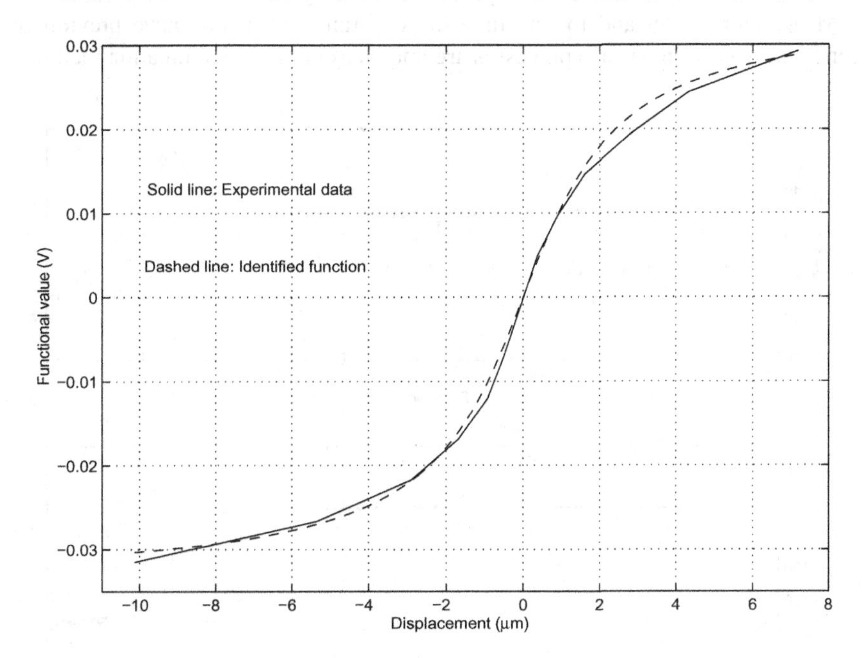

Figure 9.5. Nonlinear characteristics of the data flex cable

$$\tilde{T}_{\mathrm{bc}}(y) = a \arctan(cy) \tag{9.27}$$

where $a = 21.8748 \times 10^{-3}$ V and $c = 0.5351$ (μm)$^{-1}$. These parameters will be further fine tuned later in the Monte Carlo process.

Next, by fixing a particular input offset point u_0 and by injecting on top of u_0 a sweep of small sinusoidal signals with an amplitude of 1 mV, we are able to obtain a corresponding frequency response within the range of interest. It then follows from Equation 9.26 that the values of the static gain, k_{dc}, and peak frequency, f_{p}, of the frequency response can be used to estimate the parameter, b. Figure 9.6 shows the frequency response of the system for the pair $(u_0, y_0) = (0, 0)$, which gives a static gain of 63.71 and a peak frequency of 305.24 Hz. In order to obtain a more accurate result, we repeat the above experimental tests for several pairs (u_0, y_0) and the results are shown in Table 9.1. The parameter, b, can then be more accurately determined from these data using a least square fitting,

$$\min_{b} \sum_{i=1}^{7} \left[b - k_{\mathrm{dc},i} \omega_{\mathrm{p},i}^2 \right]^2 \tag{9.28}$$

which gives an optimal solution $b = 2.28 \times 10^8$ μm / (Vs^2). Nonetheless, this parameter will again be fine tuned later in the Monte Carlo process.

Lastly, we apply a Monte Carlo process to identify all other parameters of our VCM actuator model and to fine tune those parameters, which have previously been identified. Monte Carlo processes are known as numerical simulation methods

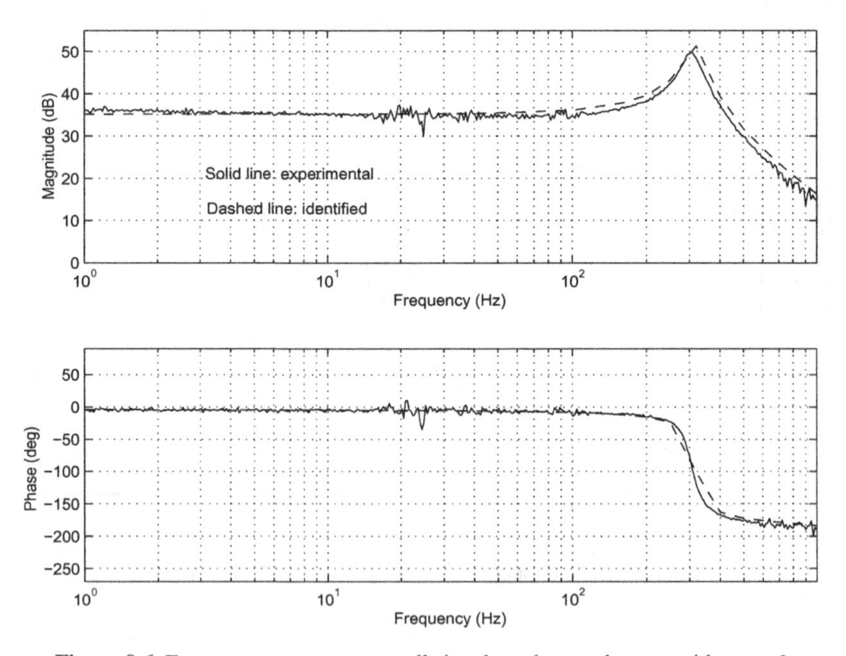

Figure 9.6. Frequency response to small signals at the steady state with $u_0 = 0$

Table 9.1. Static gains and peak frequencies of the actuator for small inputs

u_0 (mV)	-10	-5	0	5	10	15	20
k_{dc}	60.73	59.06	63.71	62.43	63.72	65.12	65.50
f_p (Hz)	310.30	310.63	305.24	303.88	299.56	296.99	295.43

that make use of random numbers and probability statistics to solve some complicated mathematical problems. The detailed treatments of Monte Carlo methods vary widely from field to field. Originally, a Monte Carlo experiment means to use random numbers to examine some stochastic problems. The idea can be extended to deterministic problems by presetting some parameters and conditions of the problems. The use of Monte Carlo methods for modeling physical systems allows us to solve more complicated problems, and provides approximate solutions to a variety of mathematical problems, whose analytical solutions are hard, if not impossible, to derive. In what follows, a Monte Carlo process is utilized to obtain time-domain responses of the VCM actuator model in Equation 9.18 with a set of preset parameters $(b, a, c, d_0, d_1, d_2, d_3, d_4, d_5)$ and input signals. The corresponding frequency responses are obtained through Fourier transformation of the obtained time-domain responses. Our idea of using the Monte Carlo process is to minimize the differences between simulated frequency responses and the experimental ones by iteratively adjusting the parameters of the physical model in Equation 9.18. The input signals in our simulations are again a combination of an offset u_0 and sinusoidal signals with a small amplitude 1 mV and several frequencies ranging from 1 Hz to 1 kHz.

Although Monte Carlo methods can only give locally minimal solutions, in our problem, however, the predetermined nonlinear characteristics of the data flex cable and the parameter, b, have given us a rough idea on what the true solution should be. The solution within the neighborhood of the previously identified parameters are given by

$$\left.\begin{aligned} b &= 2.35 \times 10^8 \ \mu\text{m} / (\text{V}\cdot\text{s}^2) \\ a &= 0.02887 \ \text{V} \\ c &= 0.5886 \ (\mu\text{m})^{-1} \\ d_0 &= 282.6 \ \text{s}^{-1} \\ d_1 &= 0.005 \ (\mu\text{m})^{-1} \\ d_2 &= 0.01 \ (\mu\text{m})^{-1} \\ d_3 &= 1.5 \times 10^4 \ \mu\text{m} / \text{s}^2 \\ d_4 &= 0.0055 \ (\mu\text{m})^{-1} \\ d_5 &= 1.65 \times 10^4 \ \mu\text{m} / \text{s}^2 \end{aligned}\right\} \tag{9.29}$$

These parameters will be used for further verifications using the experimental setup of the actual system.

So far, we have only focused on the low-frequency components of the VCM actuator model. In fact, there are many high-frequency resonance modes, which are crucial to the overall performance of HDD servo systems. The high-frequency resonance modes of the VCM actuator can be obtained from frequency responses of the system in the high-frequency region (see Figure 9.7). The transfer function that matches the frequency responses given in Figure 9.7 is identified using the standard least square estimation method of Chapter 2 and is characterized by

$$G(s) = \frac{2.35 \times 10^8}{s^2} G_{\mathrm{r.m.1}}(s) G_{\mathrm{r.m.2}}(s) G_{\mathrm{r.m.3}}(s) G_{\mathrm{r.m.4}}(s) G_{\mathrm{r.m.5}}(s) \tag{9.30}$$

with the resonance modes being given as

$$G_{\mathrm{r.m.1}}(s) = \frac{0.8709s^2 + 1726s + 1.369 \times 10^9}{s^2 + 1480s + 1.369 \times 10^9} \tag{9.31}$$

$$G_{\mathrm{r.m.2}}(s) = \frac{0.9332s^2 - 805.8s + 1.739 \times 10^9}{s^2 + 125.1s + 1.739 \times 10^9} \tag{9.32}$$

$$G_{\mathrm{r.m.3}}(s) = \frac{1.072s^2 + 925.1s + 1.997 \times 10^9}{s^2 + 536.2s + 1.997 \times 10^9} \tag{9.33}$$

$$G_{\mathrm{r.m.4}}(s) = \frac{0.9594s^2 + 98.22s + 2.514 \times 10^9}{s^2 + 1805s + 2.514 \times 10^9} \tag{9.34}$$

and

$$G_{\mathrm{r.m.5}}(s) = \frac{7.877 \times 10^9}{s^2 + 6212s + 7.877 \times 10^9} \tag{9.35}$$

Finally, for easy reference, we conclude this section by explicitly expressing the identified rigid model of VCM actuator:

$$\ddot{y} = 2.35 \times 10^8 u - 6.78445 \times 10^6 \arctan(0.5886y) + \tilde{T}_{\mathrm{f}} \tag{9.36}$$

where

$$\tilde{T}_{\mathrm{f}} = \begin{cases} -\tilde{T}_{\mathrm{e}}, & \dot{y}=0,\ |\tilde{T}_{\mathrm{e}}| \le \tilde{T}_{\mathrm{s}} \\ -\tilde{T}_{\mathrm{s}}\,\mathrm{sgn}(\tilde{T}_{\mathrm{e}}), & \dot{y}=0,\ |\tilde{T}_{\mathrm{e}}| > \tilde{T}_{\mathrm{s}} \\ -\left[\left|1.175 \times 10^6 uy + (0.1\dot{y})^2\right| + 15000\right]\mathrm{sgn}(\dot{y}) - 282.6\dot{y}, & \dot{y} \ne 0 \end{cases} \tag{9.37}$$

and where

$$\tilde{T}_{\mathrm{e}} = 2.35 \times 10^8 \left[-0.02887 \arctan(0.5886y) + u\right] \tag{9.38}$$

$$\tilde{T}_{\mathrm{s}} = 1.293 \times 10^6 |u_0 y_0| + 1.65 \times 10^4 \tag{9.39}$$

with u_0 and y_0 being, respectively, the corresponding input voltage and the displacement for the case when $\dot{y} = 0$. Note that in the above model, the input signal u is in voltage and the output displacement y is in micrometers. Together with the high-frequency resonance modes of Equations 9.31–9.35, the above model presents a comprehensive characterization of the VCM actuator studied. This model will be further verified using experimental tests on the actual system.

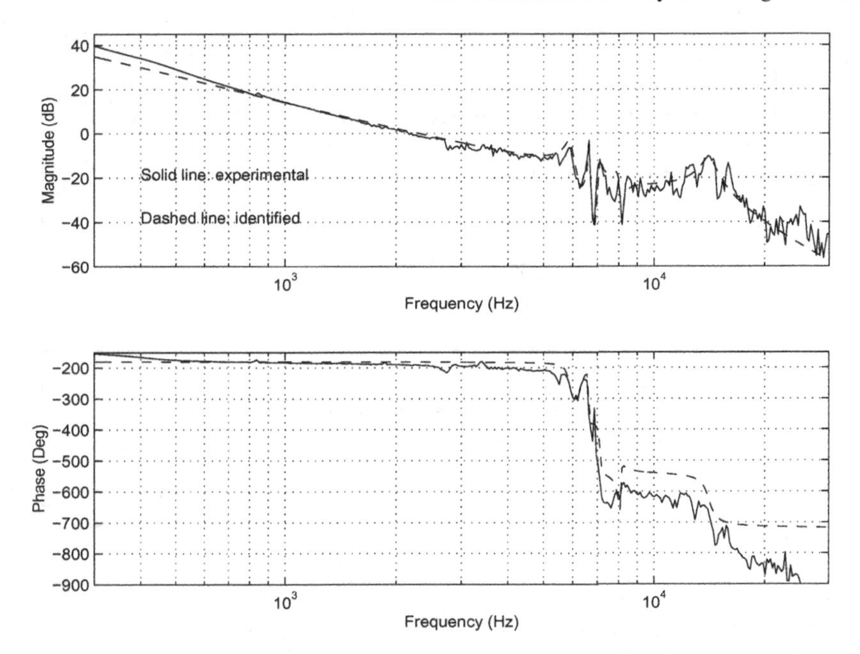

Figure 9.7. Frequency responses of the VCM actuator in the high-frequency region

In order to verify the validity of the established model of the VCM actuator, we carry out a series of comparisons between the experimental results and computed results of the time-domain responses and frequency-domain responses of the actuator. The comparison of the frequency responses between the experimental result and the identified result for inputs consisting of $u_0 = 5$ mV and sine waves with amplitude of 1 mV is shown in Figure 9.8. It clearly shows that the result of the identified model matches well with the experimental result. The comparison of the time-domain responses for an input signal consisting of $u_0 = -5$ mV and a sine wave with an amplitude of 5 mV is given in Figure 9.9. It shows that the simulation results match the trends and values of those obtained from experiments. The noises associated with experimental results in Figures 9.8 and 9.9 are drift noises caused by the LDV and/or DSA. The comparisons of both frequency-domain and time-domain responses demonstrate that the identified model of the VCM actuator indeed describes the features of the actuator.

9.3 Microdrive Servo System Design

We proceed to design a servo system for the microdrive identified in Section 9.2. As mentioned earlier, our design philosophy is rather simple. We make full use of the obtained model of the friction and nonlinearities of the VCM actuator to design a precompensator, which would cancel as much as possible all the unwanted elements in the servo system. As it is impossible to have perfect models for friction

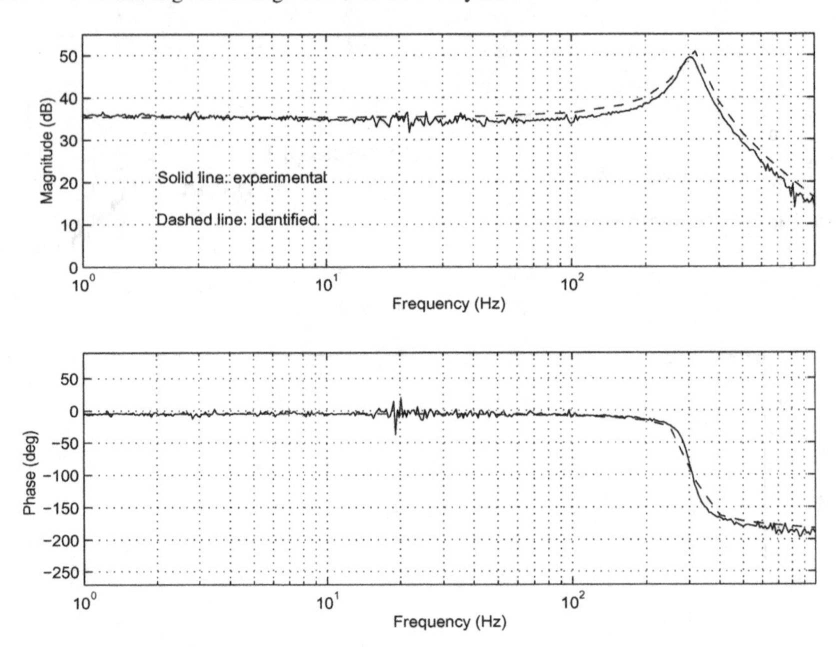

Figure 9.8. Comparison of frequency responses to small signals of actuator with $u_0 = 5$ mV

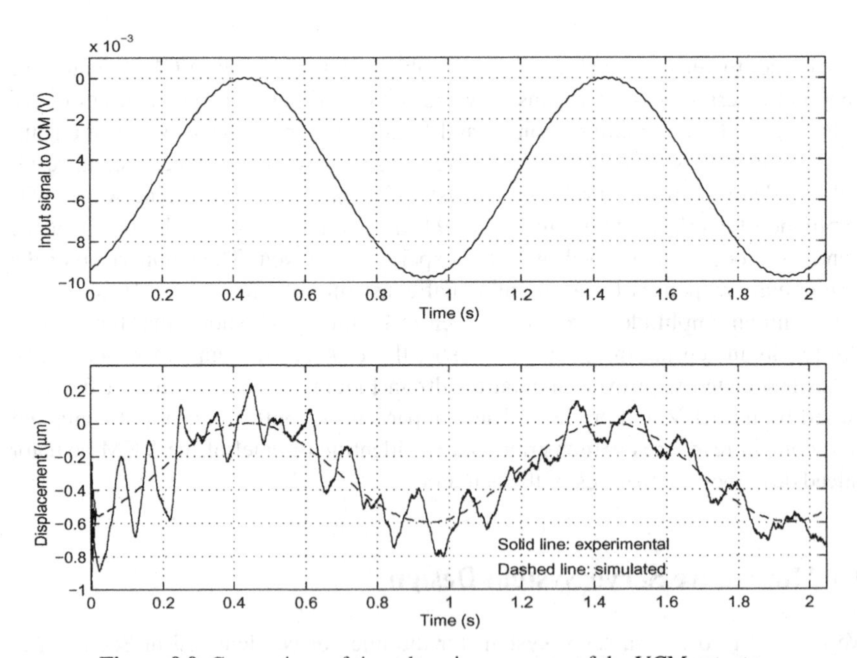

Figure 9.9. Comparison of time-domain responses of the VCM actuator

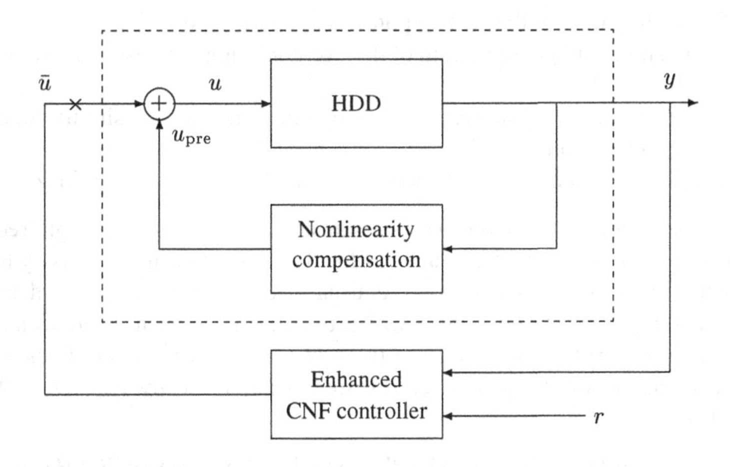

Figure 9.10. Control scheme for the HDD servo system

and nonlinearities, a perfect cancellation of these elements is unlikely to happen in the real world. We then formulate our design by treating the uncompensated portion as external disturbances. The enhanced CNF control technique of Chapter 5 is then employed to design an effective tracking controller. The overall control scheme for the servo system is depicted in Figure 9.10. Although we focus our attention here on HDD, it is our belief that such an approach can be adopted to solve other servo problems.

Examining the model of Equation 9.36, it is easy to obtain a precompensation,

$$u_{\rm pre} = u - \bar{u} = 0.0288737 \arctan(0.5886\, y) \tag{9.40}$$

which would eliminate the majority of nonlinearities in the data flex cable. The HDD model of Equation 9.18 can then be simplified as follows:

$$\begin{cases} \dot{x} = \begin{bmatrix} 0 & 1 \\ 0 & 0 \end{bmatrix} x + \begin{bmatrix} 0 \\ 2.35 \times 10^8 \end{bmatrix} \text{sat}(\bar{u}) + \begin{bmatrix} 0 \\ 2.35 \times 10^8 \end{bmatrix} w \\ y = h = [1 \quad 0]\, x \end{cases} \tag{9.41}$$

where the disturbance, w, represents uncompensated nonlinearities, and $y = h$ is the relative displacement of the R/W head (in micrometers). The control input, $\bar{u}$, is to be limited within $\pm\bar{u}_{\rm max}$ with $\bar{u}_{\rm max} = 3$ V.

We design a microdrive servo system that meets the following design constraints and specifications:

1. the control input does not exceed ± 3 V owing to physical constraints on the actual VCM actuator;
2. the overshoot and undershoot of the step response are kept to less than 5% as the R/W head can start to read or write within $\pm 5\%$ of the target;

3. the 5% settling time in the step response is as short as possible;
4. the gain margin and phase margin of the overall design are, respectively, greater than 6 dB and 30°;
5. the maximum peaks of the sensitivity and complementary sensitivity functions are less than 6 dB; and
6. the sampling frequency in implementing the actual controller is 20 kHz.

It turns out that for the microdrive its resonance modes are at very high frequencies that are far above the working range of the drive. It is thus not necessary to add a notch filter to minimize their effects. As usual, we consider a second-order nominal model of Equation 9.41 for the VCM actuator. The resonance modes and the notch filter will be put back to evaluate the performance of the overall design. As in Chapters 6 and 8, we design our servo system using, respectively, PID, RPT and CNF control.

1. The PID control law (discretized with a sampling frequency of 20 kHz) is given by

$$\bar{u}(k) = \frac{0.3145z^2 - 0.6095z + 0.2951}{z^2 - 1.091z + 0.09091} \cdot \frac{0.4665}{z - 0.5335}\Big[r(k) - y(k)\Big] \tag{9.42}$$

2. The RPT controller is given by

$$\dot{x}_{\rm v} = \begin{bmatrix} 0 & 0 \\ -2220.7 & -12911 \end{bmatrix} x_{\rm v} + \begin{bmatrix} -1 & 1 \\ 222.56 & -1255.4 \end{bmatrix} \begin{pmatrix} r \\ y \end{pmatrix} \tag{9.43}$$

and

$$\bar{u} = [\,-0.94496 \quad -2.0897\,]\, x_{\rm v} + [\,0.094705 \quad -0.26188\,] \begin{pmatrix} r \\ y \end{pmatrix} \tag{9.44}$$

3. Finally, the CNF control law is given as follows:

$$\begin{pmatrix} \dot{x}_{\rm i} \\ \dot{x}_{\rm v} \end{pmatrix} = \begin{bmatrix} 0 & 0 \\ 0 & -8000 \end{bmatrix} \begin{pmatrix} x_{\rm i} \\ x_{\rm v} \end{pmatrix} - \begin{bmatrix} -1 \\ 6.4\times 10^7 \end{bmatrix} y + \begin{bmatrix} 0 \\ 2.35\times 10^8 \end{bmatrix} \mathrm{sat}(\bar{u}) - \begin{pmatrix} r \\ 0 \end{pmatrix} \tag{9.45}$$

and

$$\begin{aligned} \bar{u} = \rho(e)\,[\,1.3607 \quad 0.13607 \quad 2.4063\times 10^{-5}\,] \begin{pmatrix} z \\ y - r \\ x_{\rm v} + 8000y \end{pmatrix} \\ - [\,1.3607\times 10^{-4} \quad 0.13607 \quad 2.6470\times 10^{-5}\,] \begin{pmatrix} z \\ y - r \\ x_{\rm v} + 8000y \end{pmatrix} \end{aligned} \tag{9.46}$$

where

$$\rho(e) = -0.3164\left|\mathrm{e}^{-|e|} - 0.3679\right| - 0.3 \tag{9.47}$$

9.4 Simulation and Implementation Results

As in the early chapters, the following tests are performed on the obtained microdrive servo systems: i) the track-following test of the closed-loop systems, ii) the frequency-domain test including the Bode and Nyquist plots as well as the plots of the resulting sensitivity and complementary sensitivity functions, iii) the runout disturbance test, and lastly iv) the PES test. Our controller was implemented on an open microdrive with a sampling rate of 20 kHz with the actual implementation setup as depicted in Figure 1.7.

9.4.1 Track-following Test

The simulation result and actual implementation result of the closed-loop responses for the control systems are, respectively, shown in Figures 9.11 and 9.12. As expected, the PID control generates large overshoots in both simulation and implementation, while the systems with the RPT and CNF control have very little overshoot. We summarize the resulting 5% settling time in Table 9.2. Clearly, the CNF control once again gives the best performance in the time domain compared to those of the other two systems.

Table 9.2. Performances of the track-following controllers

	Settling time (ms)		
	PID control	RPT control	CNF control
Simulation	1.25	0.60	0.50
Implementation	1.50	0.65	0.60

9.4.2 Frequency-domain Test

To verify the frequency-domain properties of our design, we present in Figures 9.13 to 9.18, respectively, the Bode plot, the Nyquist plot, and the sensitivity and complementary sensitivity functions, as well as the closed-loop transfer functions of the resulting control systems. The results show that all these designs meet the frequency-domain specifications.

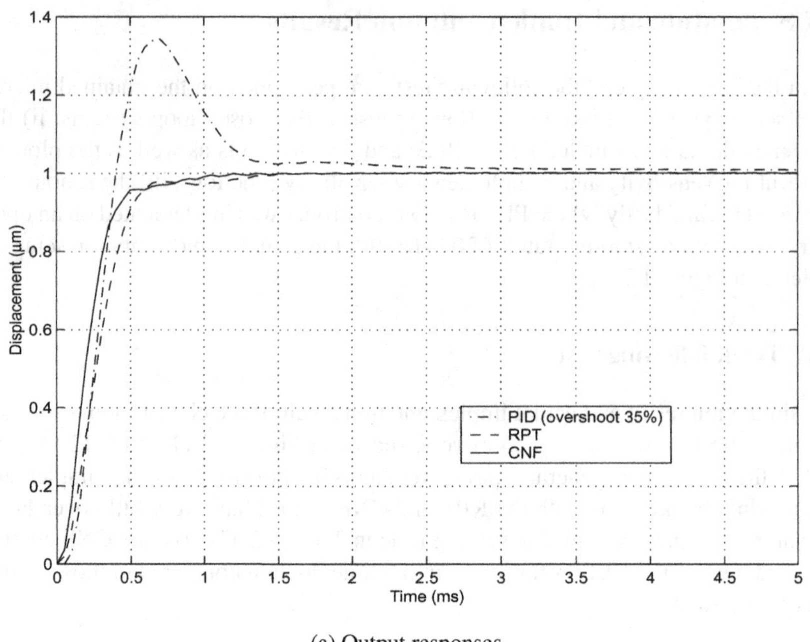

(a) Output responses

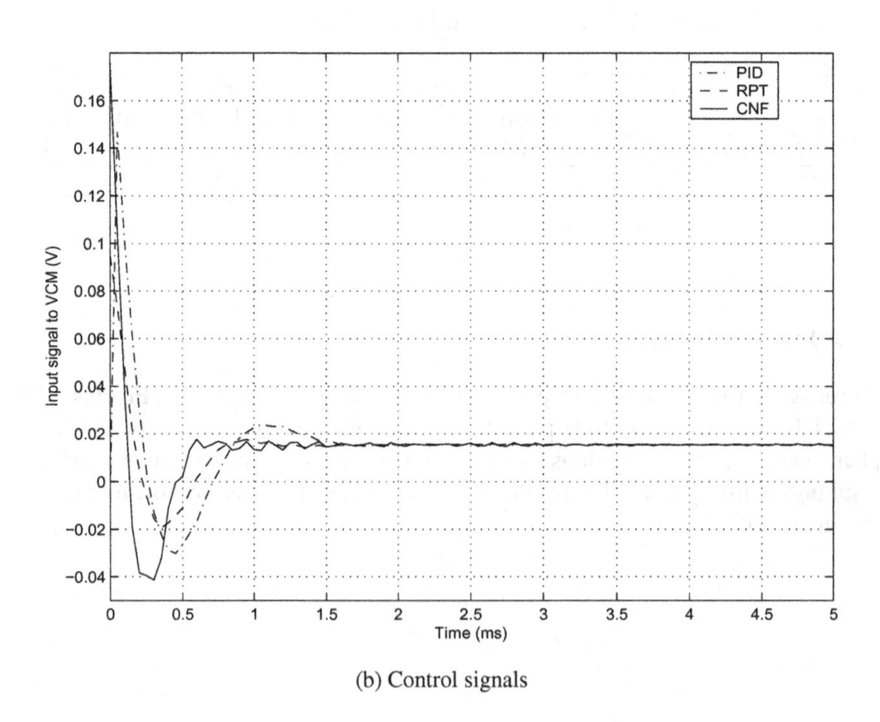

(b) Control signals

Figure 9.11. Simulation result: step responses with PID, RPT and CNF control

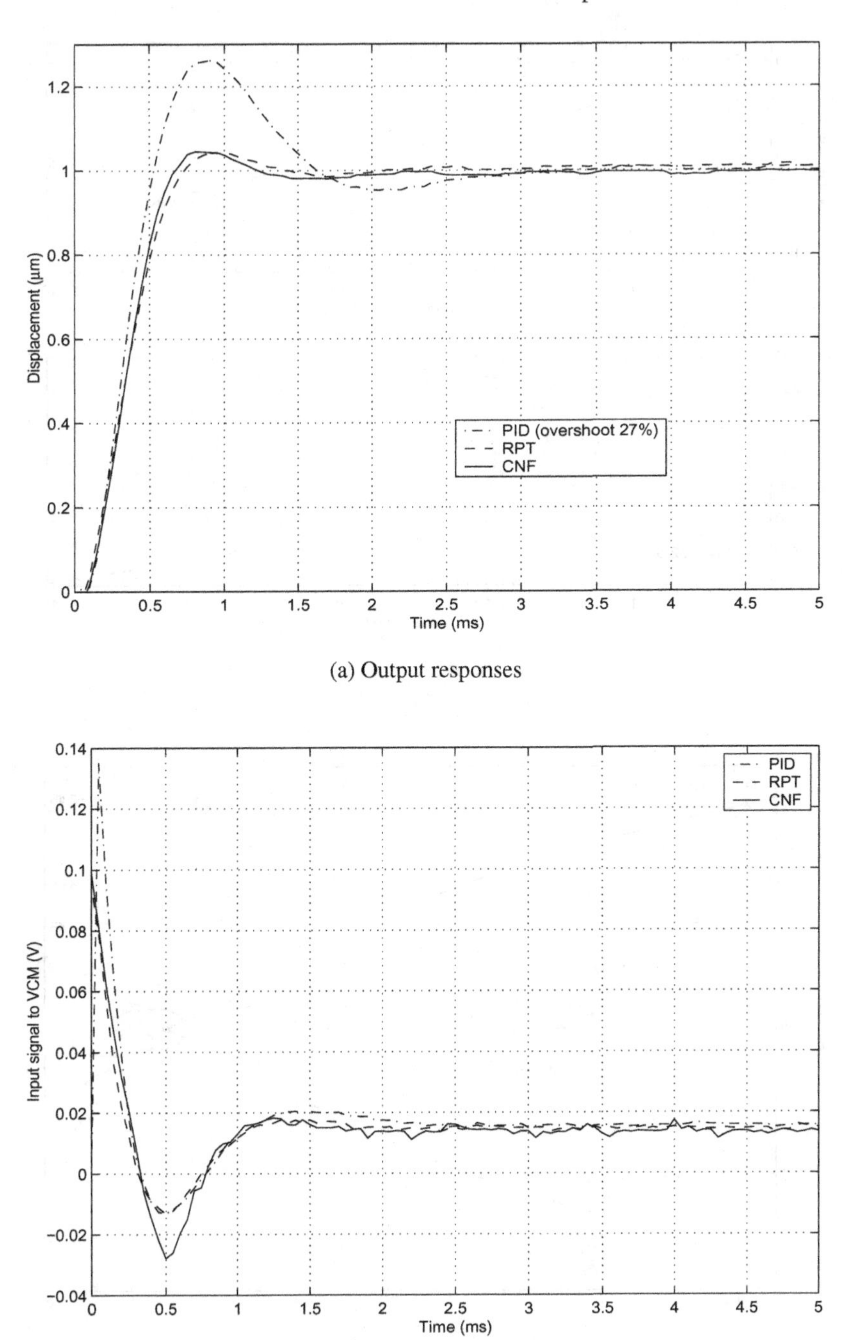

(a) Output responses

(b) Control signals

Figure 9.12. Implementation result: step responses with PID, RPT and CNF control

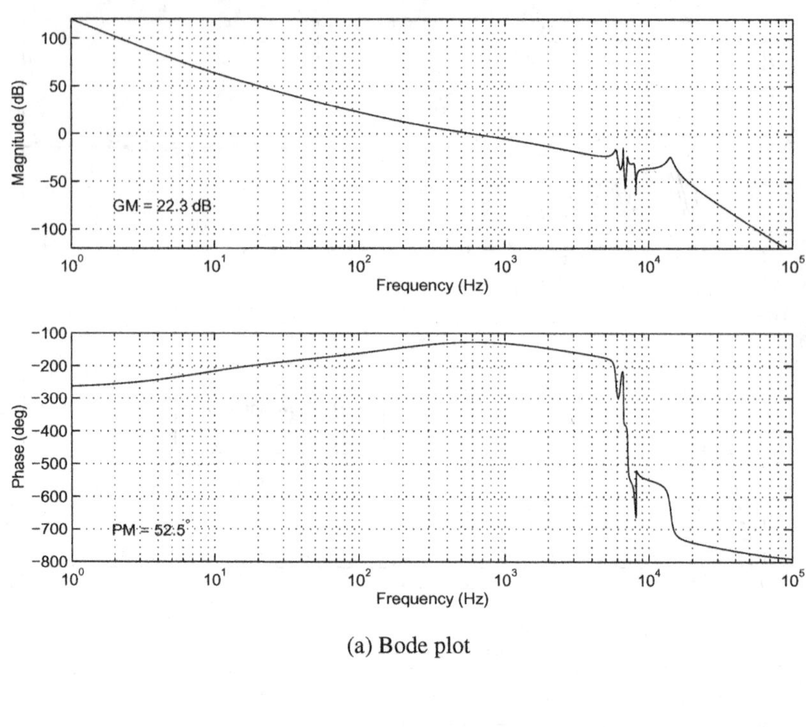

(a) Bode plot

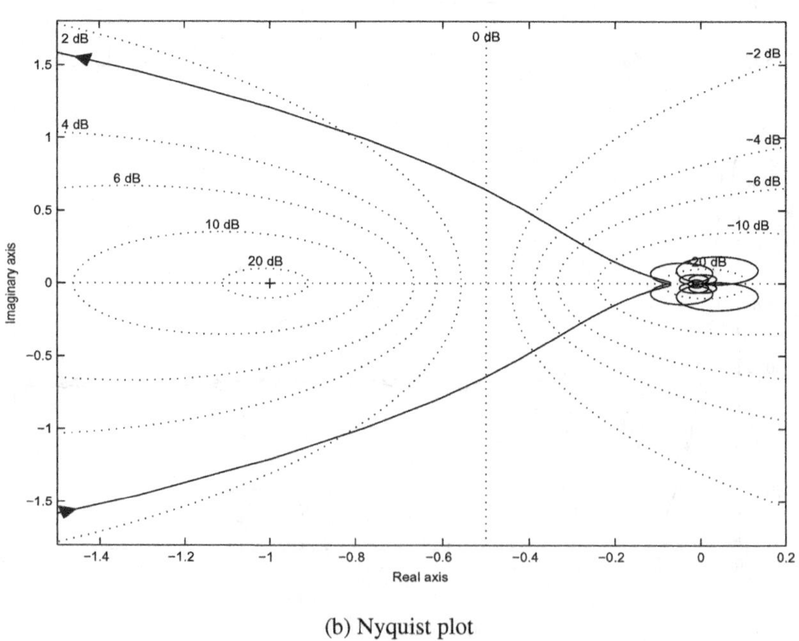

(b) Nyquist plot

Figure 9.13. Bode and Nyquist plots of the PID control system

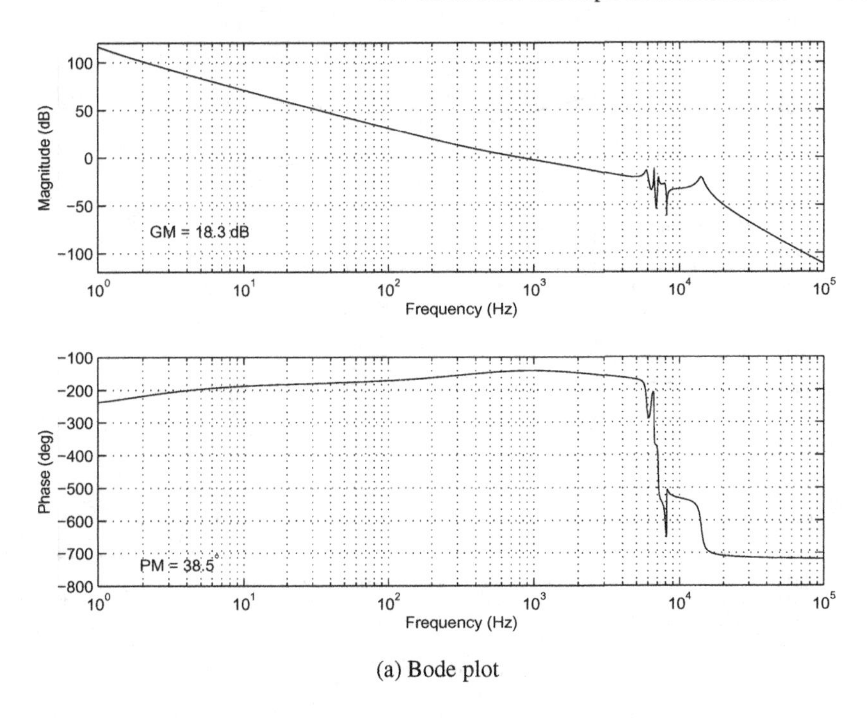

(a) Bode plot

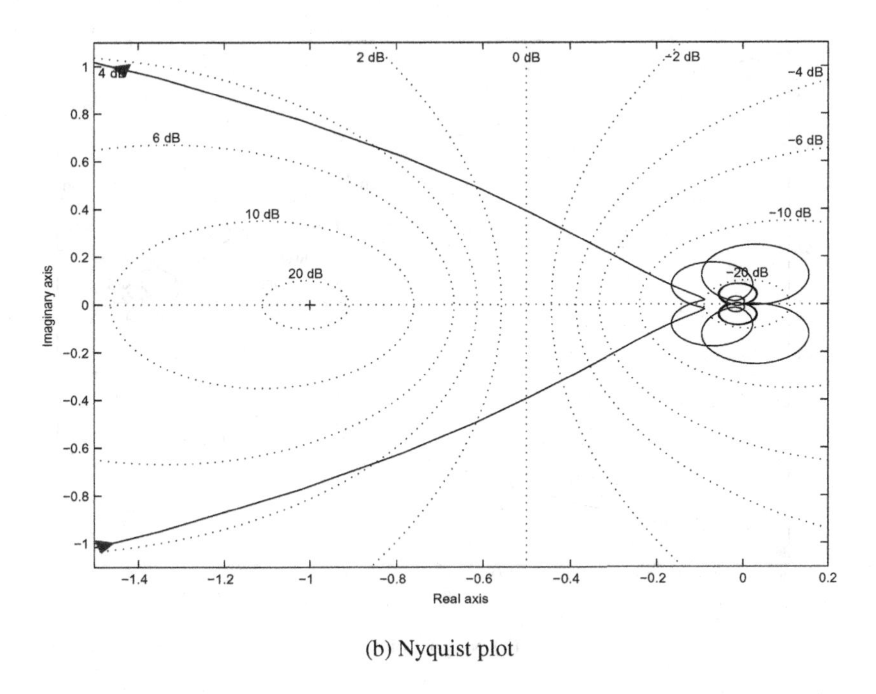

(b) Nyquist plot

Figure 9.14. Bode and Nyquist plots of the RPT control system

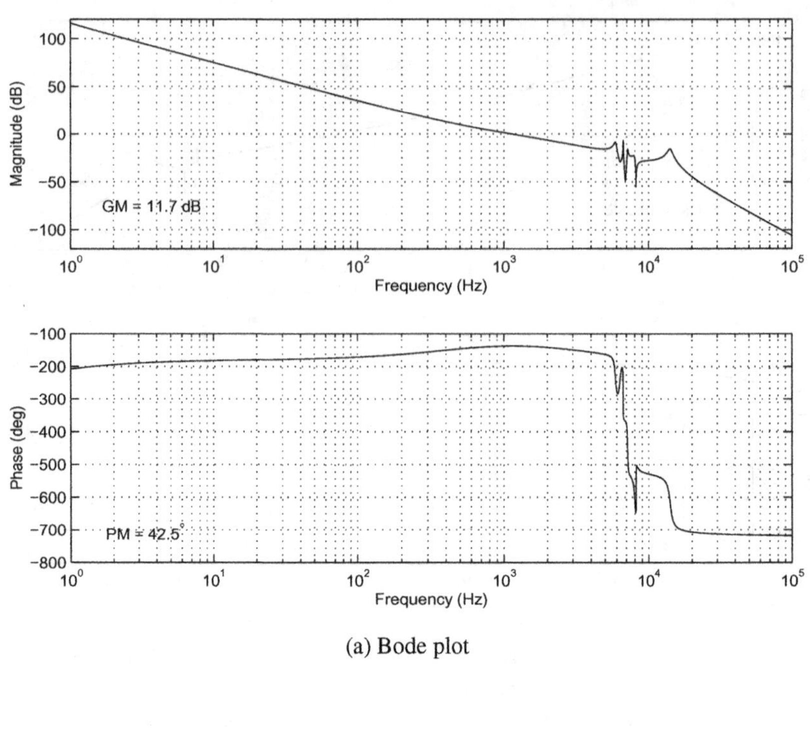

(a) Bode plot

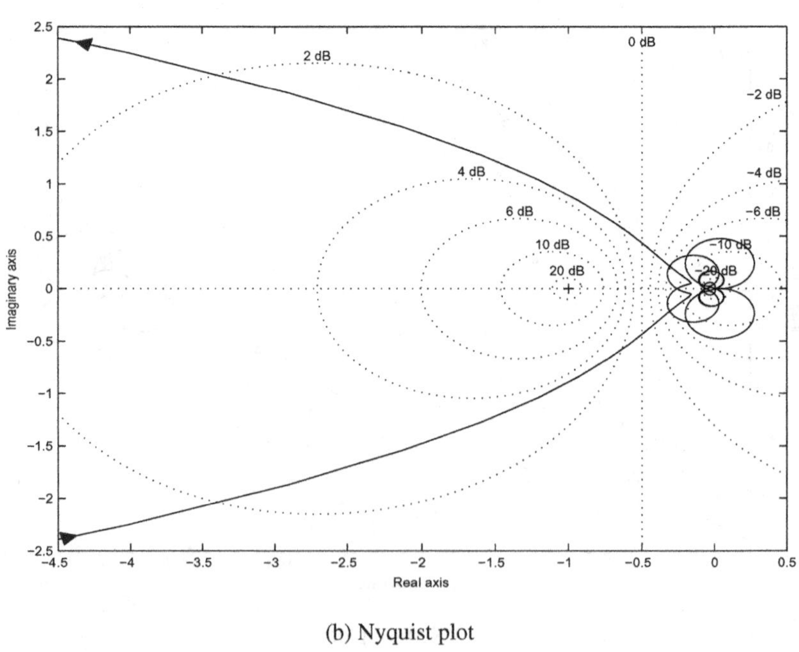

(b) Nyquist plot

Figure 9.15. Bode and Nyquist plots of the CNF control system

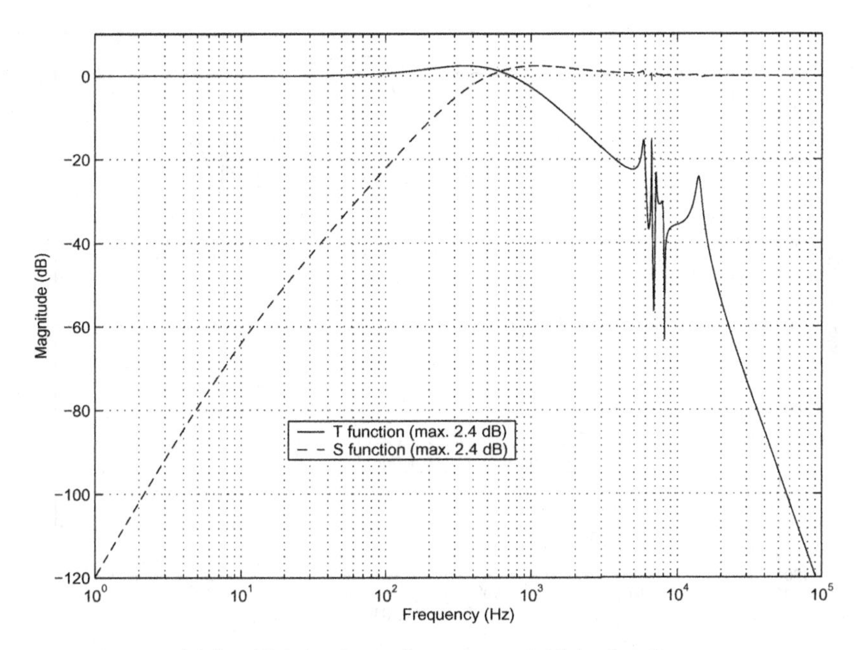

(a) Sensitivity and complementary sensitivity functions

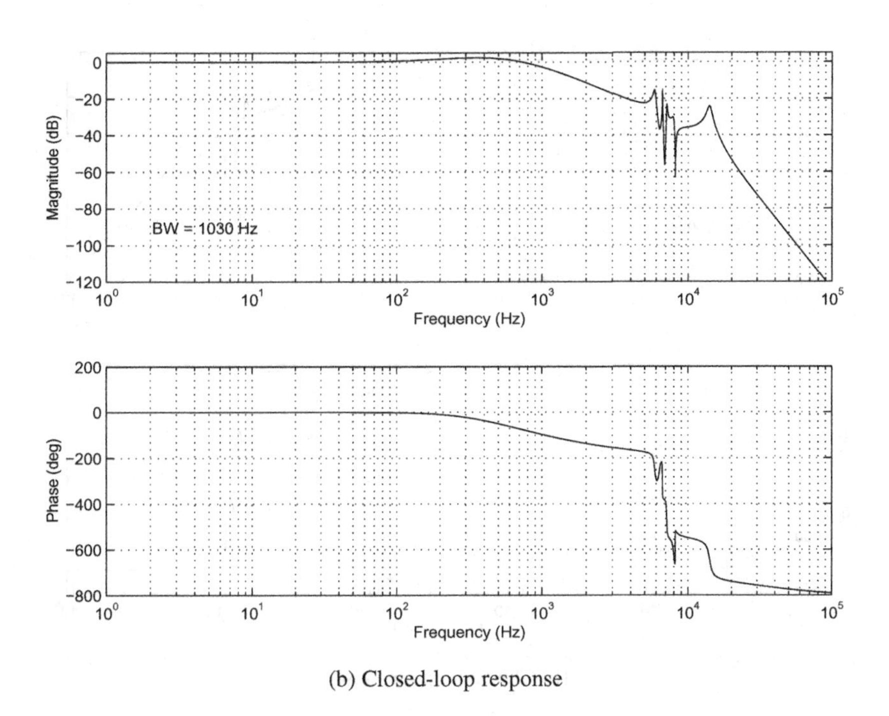

(b) Closed-loop response

Figure 9.16. Sensitivity functions and closed-loop transfer function of the PID control system

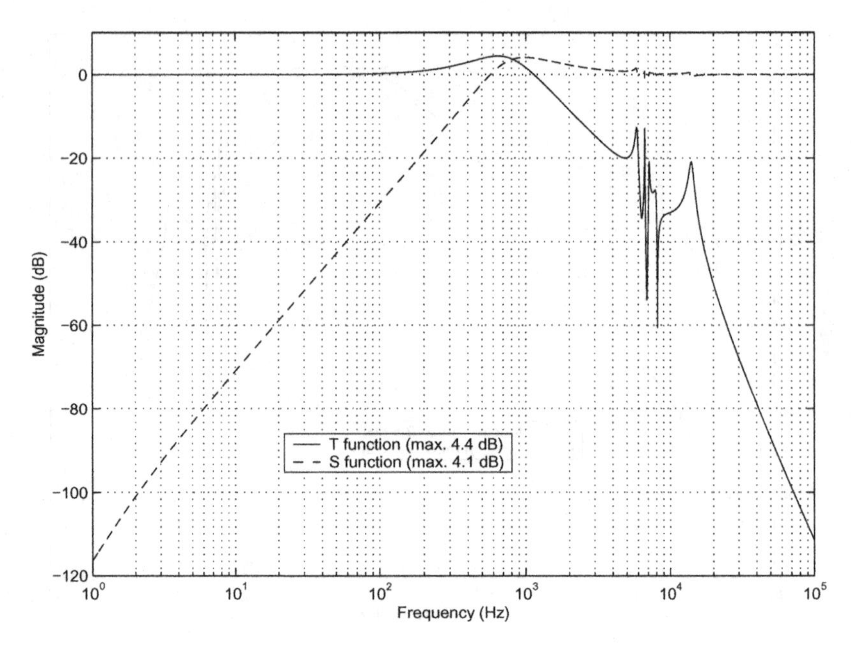

(a) Sensitivity and complementary sensitivity functions

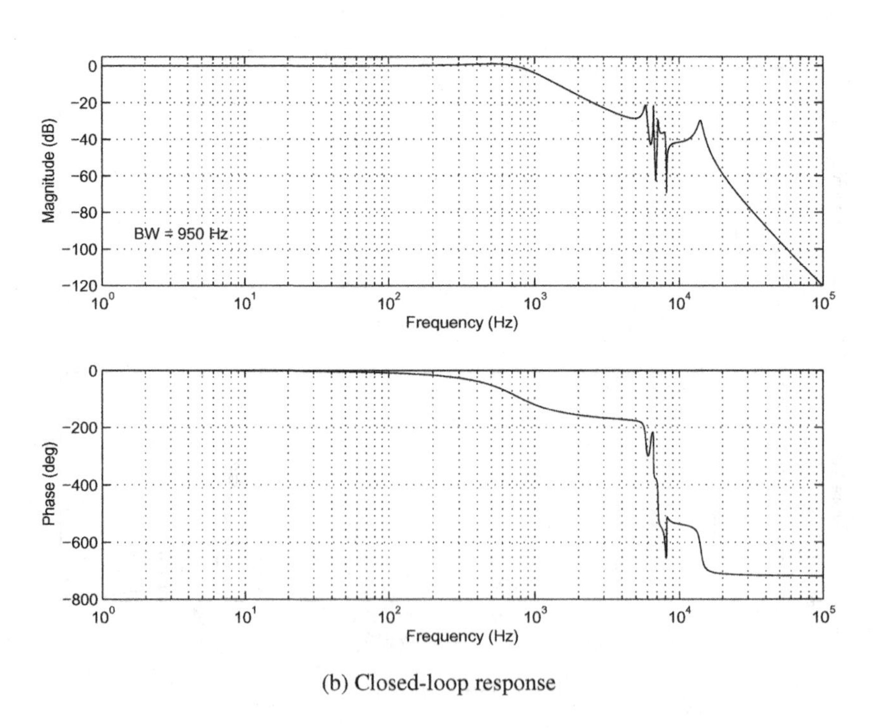

(b) Closed-loop response

Figure 9.17. Sensitivity functions and closed-loop transfer function of the RPT control system

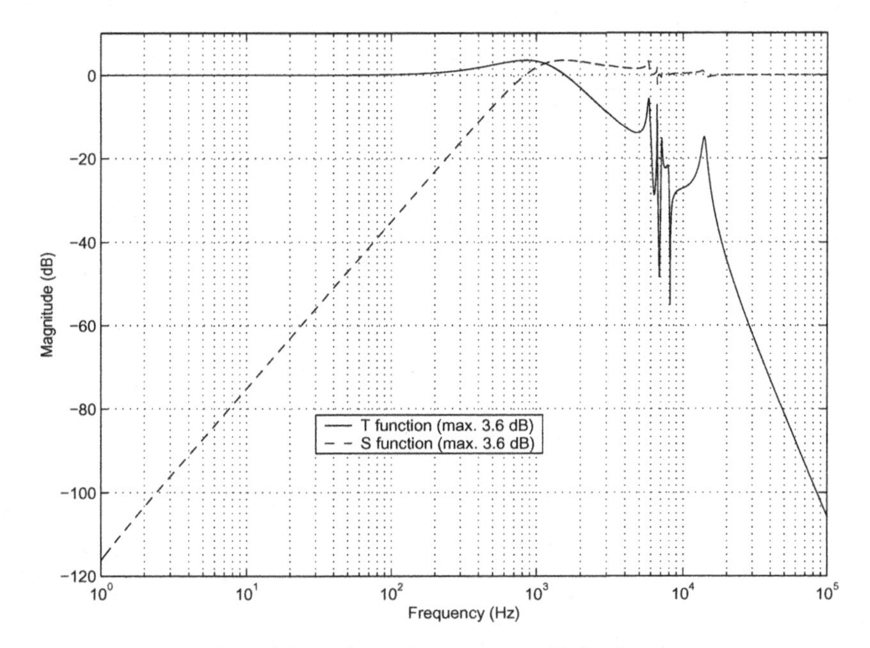

(a) Sensitivity and complementary sensitivity functions

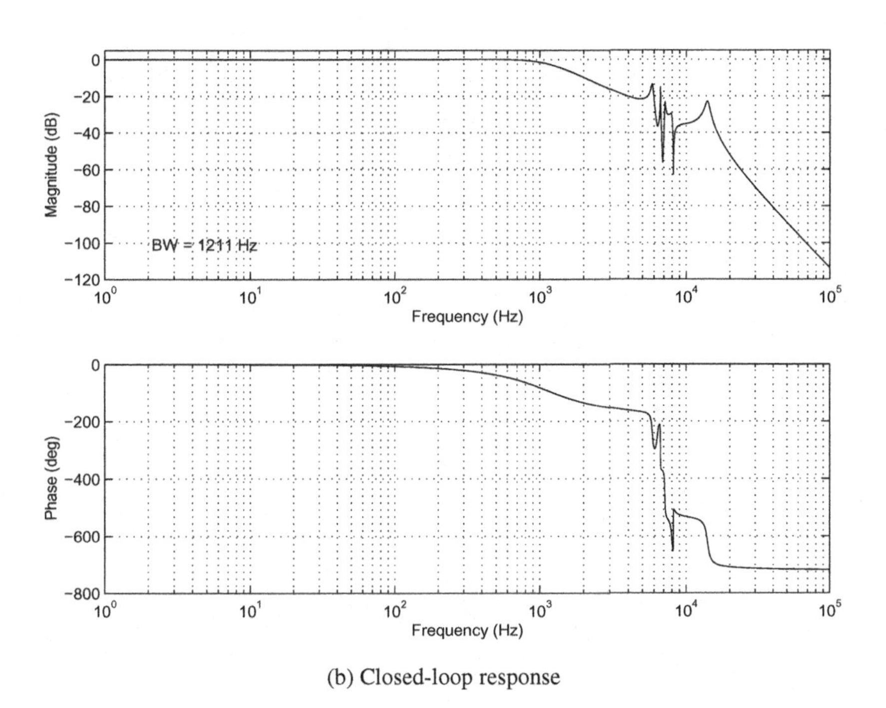

(b) Closed-loop response

Figure 9.18. Sensitivity functions and closed-loop transfer function of the CNF control system

10

Design of a Piezoelectric Actuator System

10.1 Introduction

We present in this chapter a case study on a piezoelectric bimorph actuator control system design using an H_∞ optimization approach, which was originally reported by Chen *et al.* [20]. Piezoelectricity is a fundamental process in electromechanical energy conversion. It relates electric polarization to mechanical stress/strain in piezoelectric materials. Under the direct piezoelectric effect, an electric charge can be observed when the materials are deformed. The converse, or the reciprocal piezoelectric effect, is that the application of an electric field can cause mechanical stress/strain in the piezo materials. There are numerous piezoelectric materials available today, including PZT (lead zirconate titanate), PLZT (lanthanum-modified lead zirconate titanate), and PVDF (piezoelectric polymeric polyvinylidene fluoride), to name a few (see Low and Guo [181]).

Piezoelectric structures are widely used in applications that require electrical to mechanical energy conversion coupled with size limitations, precision, and speed of operation. Typical examples are microsensors, micropositioners, speakers, medical diagnostics, shutters and impact print hammers. In most applications, bimorph or stack piezoelectric structures are used because of the relatively high stress/strain to input electric field ratio (see Low and Guo [181]).

The work was motivated by the possibility of applying piezoelectric microactuators in magnetic recording. The focus of this chapter is on the control issues involved in dealing with the nonlinear hysteresis behavior displayed by most piezoelectric actuators. More specifically, we consider a robust controller design for a piezoelectric bimorph actuator as depicted in Figure 10.1. A scaled-up model of this piezoelectric actuator, which is targeted for use in the secondary stage of a dual-stage actuator for magnetic recording, was actually built and modeled by Low and Guo [181]. It has two pairs of bimorph beams that are subjected to bipolar excitation. The dynamics of the actuator were identified in [181] as a second-order linear model coupled with a hysteresis. The linear model is given by

$$m\ddot{x}_1 + b\dot{x}_1 + kx_1 = k(du - z) \tag{10.1}$$

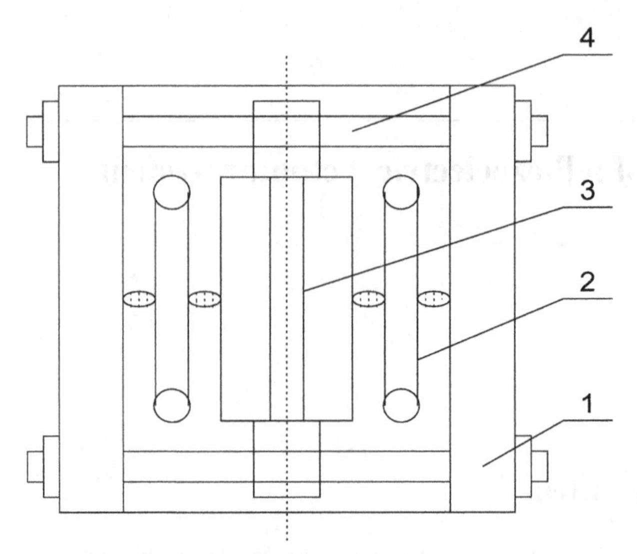

1-base; 2-piezoelectric bimorph beams; 3-moving plate; and 4-guides

Figure 10.1. Structure of the piezoelectric bimorph actuator

where m, b, k and d are the tangent mass, damping, stiffness and effective piezoelectric coefficients; u is the input voltage that generates excitation forces in the actuator system. The variable x_1 is the displacement of the actuator and is also the only measurement we can have in this system. It is noted that the working range of the displacement of this actuator is within ±1 μm. The variable z arises from hysteretic nonlinear dynamics [181] and is governed by

$$\dot{z} = \alpha d\dot{u} - \beta|\dot{u}|z - \gamma\dot{u}|z| \tag{10.2}$$

where α, β and γ are some constants that control the shape of the hysteresis. For the actuator system that we are considering here, the above coefficients are identified as follows:

$$\left.\begin{array}{rcc} m &=& 0.01595 \text{ kg} \\ b &=& 1.169 \text{ N s/m} \\ k &=& 4385 \text{ N/m} \\ d &=& 8.209 \times 10^{-7} \text{ m/V} \\ \alpha &=& 0.4297 \\ \beta &=& 0.03438 \\ \gamma &=& -0.002865 \end{array}\right\} \tag{10.3}$$

For a more detailed description of this piezoelectric actuator system and the identifications of the above parameters, we refer interested readers to the work of Low and Guo [181]. Our goal in this chapter is to design a robust controller, as in Figure 10.2, that meets the following design specifications.

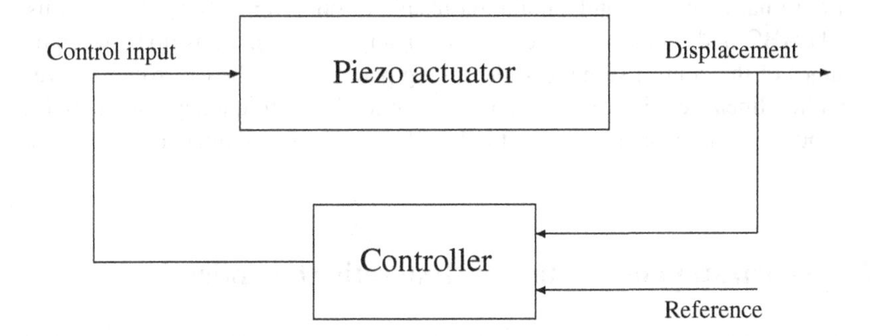

Figure 10.2. Piezoelectric bimorph actuator plant with controller

1. The steady-state tracking error of the displacement is less than 1% for any input reference signals that have frequencies ranging from 0 to 30 Hz, as the actuator is to be used to track certain colored noise types of signal in disk drive systems.
2. The 1% settling time is as short as possible (we are able to achieve a 1% settling time of less than 0.003 s in our design).
3. The control input signal $u(t)$ does not exceed 112.5 V because of the physical limitations of the piezoelectric materials.

Our approach is as follows: we first use the stochastic equivalent linearization method proposed by Chang [182] to obtain a linearized model for the nonlinear hysteretic dynamics. Then we reformulate our design into an H_∞ almost disturbance decoupling problem in which the disturbance inputs are the reference input and the error between the hysteretic dynamics and that of its linearized model, and where the controlled output is simply the double integration of the tracking error. Thus, our task boils down to designing a controller such that, when it is applied to the piezoelectric actuator, the overall system is asymptotically stable, and the controlled output, which corresponds to the tacking error, is as small as possible and decays as fast as possible. Such an approach is quite innovative and can be adopted to tackle similar problems for other types of piezo materials and applications. Ever since its original publication in [20], the work has been frequently cited in the literature (see, for example, [3, 132, 133, 183–201]).

The outline of this chapter is as follows. In Section 10.2 a first-order linearized model is obtained for the nonlinear hysteresis using the stochastic equivalent linearization method. A simulation result is also given to show the match between the nonlinear and linearized models. In Section 10.3 we formulate our controller design into a standard almost disturbance decoupling problem by properly defining the disturbance input and the controlled output. Two integrators are augmented into the original plant to enhance the performance of the overall system. Then a robust controller that is explicitly parameterized by a certain tuning parameter, and that solves the proposed almost disturbance decoupling problem, is carried out using a so-called asymptotic time scale and eigenstructure assignment technique. In Section 10.4 we present the final controller and simulation results of our overall control system using MATLAB® and Simulink®. We also obtain an explicit relationship between the peak values of the control signal and the tuning parameter of the controller, as well as an explicit linear relationship of the maximum trackable frequency, for which the corresponding tracking error can be settled to 1%, versus the tuning parameter of the controller.

10.2 Linearization of Nonlinear Hysteretic Dynamics

We proceed to linearize the nonlinear hysteretic dynamics of Equation 10.2 in this section. As pointed out by Chang [182], there are basically three methods available in the literature to linearize the hysteretic types of nonlinear system. These are (i) the Fokker–Planck equation approach (see, *e.g.*, Caughey [202]), (ii) the perturbation techniques (see, *e.g.*, Crandall [203] and Lyon [204]) and (iii) the stochastic linearization approach. All of them have certain advantages and limitations. However, the stochastic linearization technique has the widest range of applications compared with the other methods. This method is based on the concept of replacing the nonlinear system with an "equivalent" linear system in such a way that the "difference" between these two systems is minimized in a certain sense. The technique was initiated by Booton [205]. In this chapter, we just follow the stochastic linearization method given by Chang [182] to obtain a linear model of the following form:

$$\dot{z} = k_1 \dot{u} + k_2 z \tag{10.4}$$

for the hysteretic dynamics of Equation 10.2, where k_1 and k_2 are the linearization coefficients and are to be determined. The procedure is quite straightforward and proceeds as follows. First, we introduce a so-called "difference" function e between $\dot{z}$ of Equation 10.2 and $\dot{z}$ of Equation 10.4:

$$e(k_1, k_2) = \alpha d\dot{u} - \beta|\dot{u}|z - \gamma\dot{u}|z| - (k_1\dot{u} + k_2 z) \tag{10.5}$$

Then minimizing $\boldsymbol{E}[e^2]$, where $\boldsymbol{E}$ is the expectation operator, with respect to k_1 and k_2, we obtain

$$\frac{\partial \boldsymbol{E}[e^2]}{\partial k_1} = \frac{\partial \boldsymbol{E}[e^2]}{\partial k_2} = 0 \tag{10.6}$$

from which the stochastic linearization coefficients k_1 and k_2 are determined. It turns out that if z and $\dot{u}$ are of zero means and jointly Gaussian, then k_1 and k_2 can be easily obtained. Let us assume that z and $\dot{u}$ have a joint probability density function

$$f_{\dot{u}z}(\dot{u}, z) = \frac{1}{2\pi\sigma_{\dot{u}}\sigma_z\sqrt{1-\rho_{\dot{u}z}^2}} \exp\left\{-\frac{\sigma_{\dot{u}}^2 z^2 - 2\sigma_{\dot{u}}\sigma_z\rho_{\dot{u}z}\dot{u}z + \sigma_z^2\dot{u}^2}{2\sigma_{\dot{u}}^2\sigma_z^2(1-\rho_{\dot{u}z}^2)}\right\}$$

where $\rho_{\dot{u}z}$ is the normalized covariance of $\dot{u}$ and z, and $\sigma_{\dot{u}}$ and σ_z are the standard deviations of $\dot{u}$ and z, respectively. Then the linearization coefficients k_1 and k_2 can be expressed as follows:

$$k_1 = \alpha d - \beta c_1 - \gamma c_2 \tag{10.7}$$

and

$$k_2 = -\beta c_3 - \gamma c_4 \tag{10.8}$$

where c_1, c_2, c_3 and c_4 are given by

$$c_1 = 0.79788456\sigma_z \cos\left[\tan^{-1}\left(\frac{\sqrt{1-\rho_{\dot{u}z}^2}}{\rho_{\dot{u}z}}\right)\right] \tag{10.9}$$

$$c_2 = 0.79788456\sigma_z, \quad c_4 = 0.79788456\rho_{\dot{u}z}\sigma_{\dot{u}} \tag{10.10}$$

and

$$c_3 = 0.79788456\sigma_{\dot{u}}\left\{1-\rho_{\dot{u}z}^2 + \rho_{\dot{u}z}\cos\left[\tan^{-1}\left(\frac{\sqrt{1-\rho_{\dot{u}z}^2}}{\rho_{\dot{u}z}}\right)\right]\right\} \tag{10.11}$$

After a few iterations, we found that a sinusoidal excitation $\dot{u}$ with frequencies ranging from 0 to 100 Hz (the expected working frequency range) and peak magnitude of 50 V, which has a standard deviation of $\sigma_{\dot{u}} = 35$, would yield a suitable linearized model for Equation 10.2. For this excitation, we obtain $\sigma_z = 5 \times 10^{-7}$, $\rho_{\dot{u}z} = 5 \times 10^{-3}$

$$c_1 = 1.9947 \times 10^{-9}, \quad c_2 = 3.9894 \times 10^{-7}, \quad c_3 = 27.9260 \tag{10.12}$$

and

$$c_4 = 0.1396, \quad k_1 = 3.5382 \times 10^{-7}, \quad k_2 = -0.9597 \tag{10.13}$$

The stochastic linearization model of the given nonlinear hysteretic dynamics of Equation 10.2 is then given by

$$\dot{\hat{z}} = k_1\dot{u} + k_2\hat{z} = 3.5382 \times 10^{-7}\dot{u} - 0.9597\hat{z} \tag{10.14}$$

For future use, let us define the linearization error as

$$e_z = z - \hat{z} \tag{10.15}$$

Figure 10.3 shows the open-loop simulation results of the nonlinear hysteresis and its linearized model, as well as the error for a typical sine wave input signal u. The results are quite satisfactory. Here we note that, because of the nature of our approach in controller design later in the next section, the variation of the linearized model within a certain range, which might result in larger linearization error e_z, does not much affect the overall performance of the closed-loop system. We formulate e_z as a disturbance input and our controller automatically rejects it from the output response.

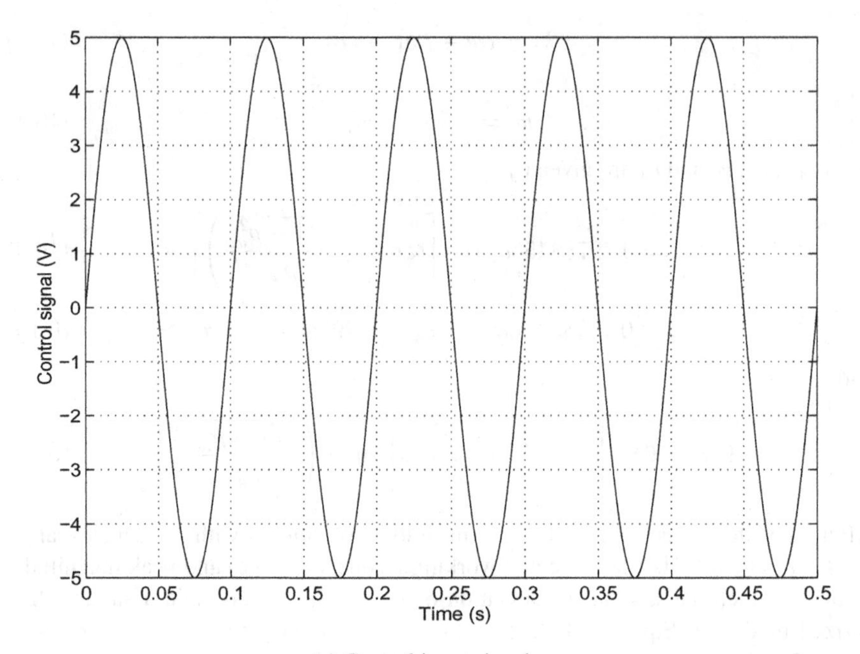

(a) Control input signal, u

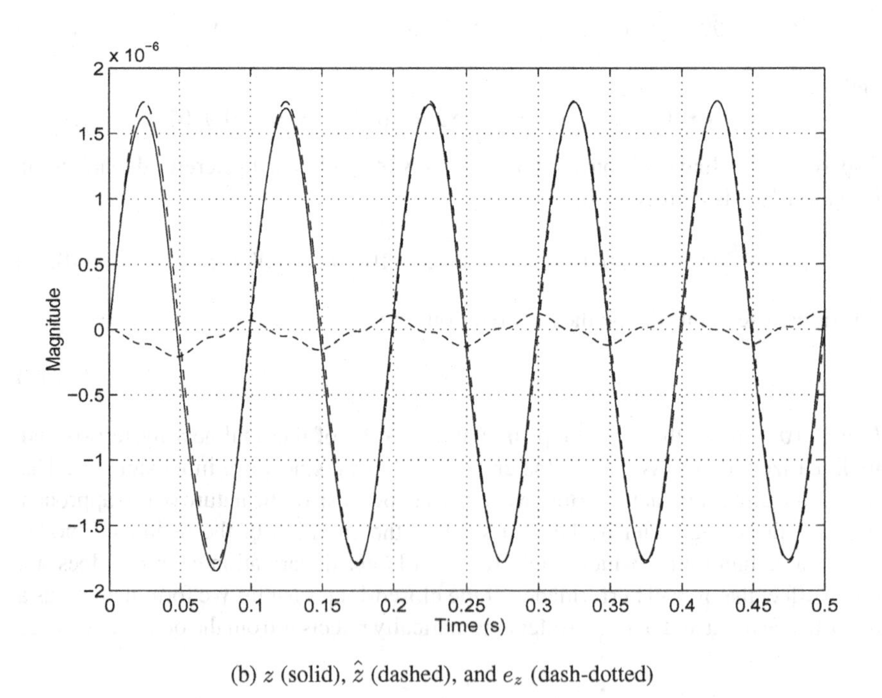

(b) z (solid), $\hat{z}$ (dashed), and e_z (dash-dotted)

Figure 10.3. Responses of hysteresis and its linearized model to a sine input

10.3 Almost Disturbance Decoupling Controller Design

This section is the heart of this chapter. We first formulate our control system design for the piezoelectric bimorph actuator into a standard H_∞ almost disturbance decoupling problem, and then apply the results of Chapter 3 to check the solvability of the proposed problem. Finally, we utilize the results in Chapter 3 to find an internally stabilizing controller that solves the proposed almost disturbance decoupling problem. Of course, most importantly, the resulting closed-loop system and its responses have to meet all the design specifications as stated at the beginning of this chapter. To do this, we have to convert the dynamic model of Equation 10.1 with the linearized model of the hysteresis into a state-space form. Let us first define a new state variable

$$v = \hat{z} - k_1 u \tag{10.16}$$

Then, from Equation 10.14, we have

$$\dot{v} = \dot{\hat{z}} - k_1 \dot{u} = k_2 \hat{z} = k_2 v + k_1 k_2 u \tag{10.17}$$

Substituting Equations 10.15 and 10.16 into Equation 10.1, we obtain

$$\ddot{x}_1 + \frac{b}{m}\dot{x}_1 + \frac{k}{m}x_1 + \frac{k}{m}v = \frac{k(d-k_1)}{m}u - \frac{k}{m}e_z \tag{10.18}$$

The overall controller structure of our approach is then depicted in Figure 10.4. Note that in Figure 10.4 we have augmented two integrators after e, the tracking error between the displacement x_1 and the reference input signal r. We have observed a very interesting property of this problem, *i.e.* the more integrators that we augment after the tracking error e, the smaller the tracking error we can achieve for the same level of control input u. Because our control input u is limited to the range from -112.5 to 112.5 V, it turns out that two integrators are needed in order to meet all the design specifications. It is clear to see that the augmented system has an order of five. Next, let us define the state of the augmented system as

$$x = (\, x_1 \quad \dot{x}_1 \quad v \quad x_4 \quad x_5 \,)' \tag{10.19}$$

and the measurement output

$$y = \begin{pmatrix} y_1 \\ y_2 \\ y_3 \end{pmatrix} = \begin{pmatrix} x_1 \\ x_4 \\ x_5 \end{pmatrix} \tag{10.20}$$

i.e. the original measurement of displacement x_1 plus two augmented states. The auxiliary disturbance input is

$$w = \begin{pmatrix} e_z \\ r \end{pmatrix} \tag{10.21}$$

and the output to be controlled, h, is simply the double integration of the tracking error. The state-space model of the overall augmented system is then given by

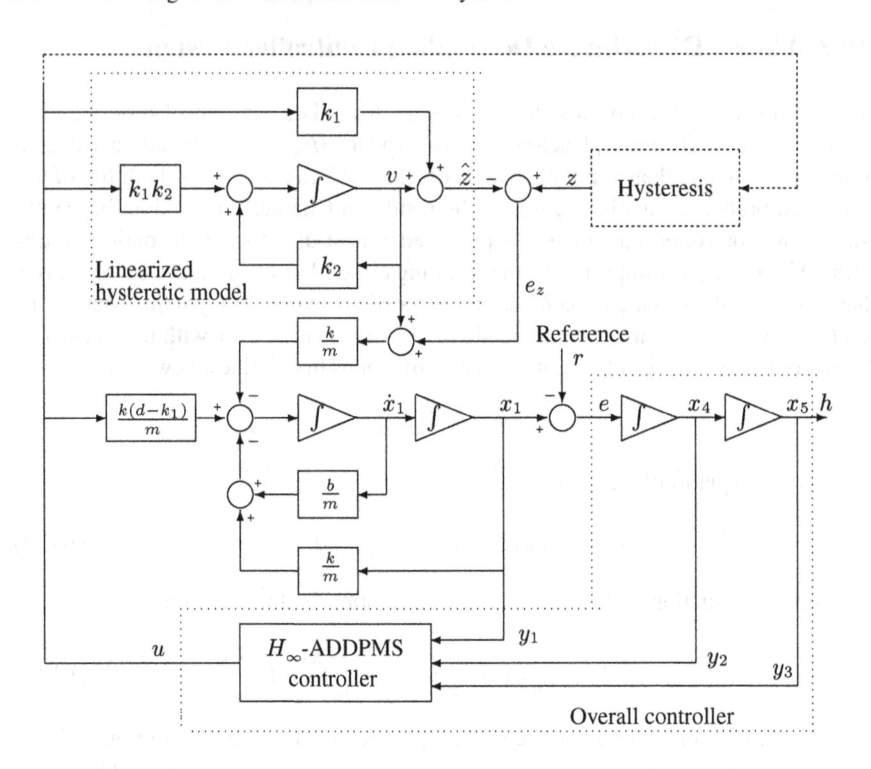

Figure 10.4. Augmented linearized model with controller

$$\Sigma : \begin{cases} \dot{x} = A\ x + B\ u + E\ w \\ y = C_1\ x \qquad\qquad + D_1\ w \\ h = C_2\ x + D_2\ u \end{cases} \tag{10.22}$$

with

$$C_1 = \begin{bmatrix} 1 & 0 & 0 & 0 & 0 \\ 0 & 0 & 0 & 1 & 0 \\ 0 & 0 & 0 & 0 & 1 \end{bmatrix}, \quad D_1 = \begin{bmatrix} 0 & 0 \\ 0 & 0 \\ 0 & 0 \end{bmatrix} \tag{10.23}$$

$$C_2 = [0 \ \ 0 \ \ 0 \ \ 0 \ \ 1], \quad D_2 = 0 \tag{10.24}$$

$$A = \begin{bmatrix} 0 & 1 & 0 & 0 & 0 \\ -k/m & -b/m & -k/m & 0 & 0 \\ 0 & 0 & k_2 & 0 & 0 \\ 1 & 0 & 0 & 0 & 0 \\ 0 & 0 & 0 & 1 & 0 \end{bmatrix}$$

$$= \begin{bmatrix} 0 & 1 & 0 & 0 & 0 \\ -274921.63 & -73.2915 & -274921.63 & 0 & 0 \\ 0 & 0 & -0.9597 & 0 & 0 \\ 1 & 0 & 0 & 0 & 0 \\ 0 & 0 & 0 & 1 & 0 \end{bmatrix} \tag{10.25}$$

$$B = \begin{bmatrix} 0 \\ k(d-k_1)/m \\ k_1 k_2 \\ 0 \\ 0 \end{bmatrix} = \begin{bmatrix} 0 \\ 0.12841 \\ -3.39561 \times 10^{-7} \\ 0 \\ 0 \end{bmatrix} \tag{10.26}$$

$$E = \begin{bmatrix} 0 & 0 \\ -k/m & 0 \\ 0 & 0 \\ 0 & -1 \\ 0 & 0 \end{bmatrix} = \begin{bmatrix} 0 & 0 \\ -274921.63 & 0 \\ 0 & 0 \\ 0 & -1 \\ 0 & 0 \end{bmatrix} \tag{10.27}$$

For the problem that we are considering here, it is simple to verify that the system Σ of Equation 10.22 has the following properties.

1. The subsystem (A, B, C_2, D_2) is invertible and of minimum phase with one invariant zero at -1.6867. It also has one infinite zero of order four.
2. The subsystem (A, E, C_1, D_1) is left invertible and of minimum phase with one invariant zero at -0.9597 and two infinite zeros of orders one and two.

Then, it follows from the results of Section 3.5 that the H_∞-ADDPMS for the system in Equation 10.22 is solvable. In fact, one can design either a full-order observer-based controller or a reduced-order observer-based controller to solve this problem. For the full-order observer-based controller, the order of the disturbance decoupling controller (see Figure 10.4) is five and the order of the final overall controller (again see Figure 10.4) is seven (the disturbance decoupling controller plus two integrators). On the other hand, if we use a reduced-order observer in the disturbance decoupling controller, the total order of the resulting final overall controller will be reduced to four. From the practical point of view, the latter is much more desirable than the former. Thus, in what follows we only focus on the controller design based on a reduced-order observer. We can separate our controller design into two steps:

1. In the first step, we assume that all five states of Σ in Equation 10.22 are available and then design a static and parameterized state feedback control law,

$$u = F(\varepsilon)x \tag{10.28}$$

   such that it solves the almost disturbance decoupling problem for the state feedback case, *i.e.* $y = x$, by adjusting the tuning parameter ε to an appropriate value;
2. In the second step, we design a reduced-order observer-based controller. It has a parameterized reduced-order observer gain matrix $K_2(\varepsilon)$ that can be tuned to recover the performance achieved by the state feedback control law in the first step.

We use the structural decomposition approach of Chapter 3 to construct both the state feedback law and the reduced-order observer gain. We would like to note that, in principle, one can also apply the ARE-based H_∞ optimization technique (see, *e.g.*, Zhou and Khargonekar [93]) to solve this problem. However, because the numerical conditions of our system Σ are very poor, we are unable to obtain any satisfactory solution from the ARE approach. We cannot get any meaningful solution for the associated H_∞ continuous-time ARE in MATLAB®. The following is a closed-form solution of the static state feedback parameterized gain matrix obtained using the method given in Chapter 3.

$$F(\varepsilon) = \Big[(2.1410\times 10^6 - 62.3004/\varepsilon^2) \quad (570.7619 - 31.1502/\varepsilon) \quad 2.1410\times 10^6 \quad -62.3004/\varepsilon^3 \quad -31.1502/\varepsilon^4 \Big] \tag{10.29}$$

where ε is the tuning parameter that can be adjusted to achieve almost disturbance decoupling. It can be verified that the closed-loop system matrix, $A + BF(\varepsilon)$ is asymptotically stable for all $0 < \varepsilon < \infty$ and the closed-loop transfer function from the disturbance w to the controlled output h, $T_{hw}(\varepsilon, s)$, satisfying

$$\|T_{hw}(\varepsilon, s)\|_\infty = \|[C_2 + D_2F(\varepsilon)][sI - A - BF(\varepsilon)]^{-1}E\|_\infty \to 0 \tag{10.30}$$

as $\varepsilon \to 0$.

The next step is to design a reduced-order observer-based controller that recovers the performance of the above state feedback control law. First, let us perform the following nonsingular (permutation) state transformation to the system Σ of Equation 10.22,

$$x = T\tilde{x} \tag{10.31}$$

where

$$T = \begin{bmatrix} 1 & 0 & 0 & 0 & 0 \\ 0 & 0 & 0 & 1 & 0 \\ 0 & 0 & 0 & 0 & 1 \\ 0 & 1 & 0 & 0 & 0 \\ 0 & 0 & 1 & 0 & 0 \end{bmatrix} \tag{10.32}$$

such that the transformed measurement matrix has the form of

$$C_1T = \begin{bmatrix} 1 & 0 & 0 & 0 & 0 \\ 0 & 1 & 0 & 0 & 0 \\ 0 & 0 & 1 & 0 & 0 \end{bmatrix} = [\, I_3 \quad 0 \,] \tag{10.33}$$

Clearly, the first three states of the transformed system, or x_1, x_4 and x_5 of the original system Σ in Equation 10.22, need not be estimated as they are already available from the measurement output. Let us now partition the transformed system as follows:

$$T^{-1}AT = \left[\begin{array}{c|c} A_{11} & A_{12} \\ \hline A_{21} & A_{22} \end{array}\right] = \left[\begin{array}{ccc|cc} 0 & 0 & 0 & 1 & 0 \\ 1 & 0 & 0 & 0 & 0 \\ 0 & 1 & 0 & 0 & 0 \\ \hline -274921.63 & 0 & 0 & -73.2915 & -274921.63 \\ 0 & 0 & 0 & 0 & -0.9597 \end{array}\right] \tag{10.34}$$

$$T^{-1}B = \left[\frac{B_1}{B_2}\right] = \left[\begin{array}{c} 0 \\ 0 \\ 0 \\ \hline 0.12841 \\ -3.39561\times 10^{-7} \end{array}\right] \tag{10.35}$$

$$T^{-1}E = \left[\frac{E_1}{E_2}\right] = \left[\begin{array}{cc} 0 & 0 \\ 0 & -1 \\ 0 & 0 \\ \hline -274921.63 & 0 \\ 0 & 0 \end{array}\right] \tag{10.36}$$

Also, we partition

$$F(\varepsilon)T = \left[F_1(\varepsilon)\middle|F_2(\varepsilon)\right] \tag{10.37}$$

$$= \Big[(2.14\times 10^6 - 62.3/\varepsilon^2) \quad -62.3/\varepsilon^3 \quad -31.15/\varepsilon^4 \quad \Big| \quad (570.76 - 31.15/\varepsilon) \quad 2.14\times 10^6\Big] \tag{10.38}$$

Then the reduced-order observer-based controller (see Chapter 3) is given in the form of

$$\Sigma_{\rm cmp} : \begin{cases} \dot{v} = A_{\rm cmp}(\varepsilon)\, v + B_{\rm cmp}(\varepsilon)\, y \\ u = C_{\rm cmp}(\varepsilon)\, v + D_{\rm cmp}(\varepsilon)\, y \end{cases} \tag{10.39}$$

with

$$A_{\rm cmp}(\varepsilon) = A_{22} + K_2(\varepsilon)A_{12} + B_2F_2(\varepsilon) + K_2(\varepsilon)B_1F_2(\varepsilon) \tag{10.40}$$

$$B_{\rm cmp}(\varepsilon) = A_{21} + K_2(\varepsilon)A_{11} - \big[A_{22} + K_2(\varepsilon)A_{12}\big]K_2(\varepsilon) + \big[B_2 + K_2(\varepsilon)B_1\big]\big[F_1(\varepsilon) - F_2(\varepsilon)K_2(\varepsilon)\big] \tag{10.41}$$

$$C_{\rm cmp}(\varepsilon) = F_2(\varepsilon) \tag{10.42}$$

$$D_{\rm cmp}(\varepsilon) = F_1(\varepsilon) - F_2(\varepsilon)K_2(\varepsilon) \tag{10.43}$$

where $K_2(\varepsilon)$ is the parameterized reduced-order observer gain matrix and is to be designed such that $A_{22} + K_2(\varepsilon)A_{12}$ is asymptotically stable for sufficiently small ε and also

$$\|[sI - A_{22} - K_2(\varepsilon)A_{12}]^{-1}[E_2 + K_2(\varepsilon)E_1]\|_\infty \to 0 \tag{10.44}$$

as $\varepsilon \to 0$. Again, using the software package of [53], we obtained the following parameterized reduced-order observer gain matrix

$$K_2(\varepsilon) = \begin{bmatrix} 73.2915 - 1/\varepsilon & 0 & 0 \\ 0 & 0 & 0 \end{bmatrix} \tag{10.45}$$

Then the explicitly parameterized matrices of the state-space model of the reduced-order observer-based controller are given by

$$A_{\rm cmp}(\varepsilon) = \begin{bmatrix} 73.2915 - 4/\varepsilon - 1/\varepsilon & 0 \\ -1.9381 \times 10^{-4} + 1.0577 \times 10^{-5}/\varepsilon & -1.6867 \end{bmatrix}$$

$$C_{\rm cmp}(\varepsilon) = [\,570.7619 - 31.1502/\varepsilon \quad 2140967\,]$$

$$D_{\rm cmp}(\varepsilon) = [\,2099135.4 + 2853.81/\varepsilon - 93.45/\varepsilon^2 \quad -62.3/\varepsilon^3 \quad -31.15/\varepsilon^4\,]$$

$$B_{\rm cmp}(\varepsilon) = \begin{bmatrix} \psi_1 & -8/\varepsilon^3 & -4/\varepsilon^4 \\ \psi_2 & 2.1155 \times 10^{-5}/\varepsilon^3 & 1.0577 \times 10^{-5}/\varepsilon^4 \end{bmatrix}$$

where

$$\psi_1 = -5731.6533 - 13/\varepsilon^2 + 439.7492/\varepsilon \tag{10.46}$$

and

$$\psi_2 = -0.7128 + 3.1732 \times 10^{-5}/\varepsilon^2 - 9.6904 \times 10^{-4}/\varepsilon \tag{10.47}$$

The overall closed-loop system comprising the system Σ of Equation 10.22 and the above controller would be asymptotically stable as long as $\varepsilon \in (0, \infty)$. In fact, the closed-loop poles are exactly located at -1.6867, two pairs at $-1/\varepsilon \pm j1/\varepsilon$, -0.9597 and $-1/\varepsilon$. The plots of the maximum singular values of the closed-loop transfer function matrix from the disturbance w to the controlled output h, namely $T_{hw}(\varepsilon, s)$, for several values of ε, *i.e.* $\varepsilon = 1/100$, $\varepsilon = 1/400$ and $\varepsilon = 1/3000$, in Figure 10.5 show that as ε tends to smaller values, the H_∞-norm of $T_{hw}(\varepsilon, s)$ also becomes smaller. Hence, almost disturbance decoupling is indeed achieved. These are the properties of our control system in the frequency domain. In the next section we will address its time-domain properties, which are, of course, much more important, as all the design specifications are in the time domain.

10.4 Final Controller and Simulation Results

In this section we put our design of the previous section into a final controller as depicted in Figure 10.2. It is simple to derive the state-space model of the final overall controller by observing its interconnection with the disturbance decoupling controller $\Sigma_{\rm cmp}$ of Equation 10.39 (see Figure 10.4). We also present simulation results of the responses of the overall design to several different types of reference input signal. They clearly show that all the design specifications are successfully achieved. Furthermore, because our controller is explicitly parameterized by a tuning parameter, it is very easy to adjust it to meet other design specifications without going through it all over again from the beginning. This is also discussed next.

As mentioned earlier, the final overall controller of our design is of the order of four, of which two are from the disturbance decoupling controller and two from the augmented integrators. It has two inputs: one is the displacement x_1 and the other is the reference signal r. It is straightforward to verify that the state-space model of the final overall controller is given by

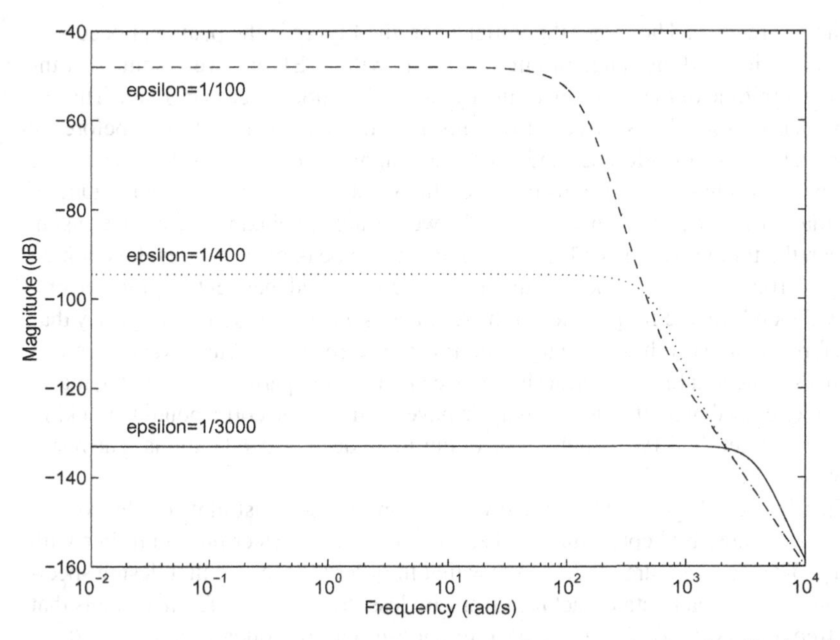

Figure 10.5. Maximum singular values of closed-loop transfer function $T_{hw}(\varepsilon, s)$

$$\Sigma_{\rm oc}(\varepsilon): \quad \begin{cases} \dot{v} = A_{\rm oc}(\varepsilon)\ v + B_{\rm oc}(\varepsilon)\ x_1 + G_{\rm oc}\ r \\ u = C_{\rm oc}(\varepsilon)\ v + D_{\rm oc}(\varepsilon)\ x_1 \end{cases} \tag{10.48}$$

where

$$A_{\rm oc}(\varepsilon) = \begin{bmatrix} 73.29 - 5/\varepsilon & 0 & -8/\varepsilon^3 & -4/\varepsilon^4 \\ 1.06\times10^{-5}/\varepsilon & -1.69 & 2.12\times10^{-5}/\varepsilon^3 & 1.08\times10^{-5}/\varepsilon^4 \\ 0 & 0 & 0 & 0 \\ 0 & 0 & 1 & 0 \end{bmatrix} \tag{10.49}$$

$$G_{\rm oc} = \begin{bmatrix} 0 \\ 0 \\ -1 \\ 0 \end{bmatrix}, \quad B_{\rm oc}(\varepsilon) = \begin{bmatrix} \psi_1 \\ \psi_2 \\ 1 \\ 0 \end{bmatrix} \tag{10.50}$$

with ψ_1 and ψ_2 given by Equations 10.46 and 10.47, respectively,

$$C_{\rm oc}(\varepsilon) = [\,570.76 - 31.15/\varepsilon \quad 2140967 \quad -62.3/\varepsilon^3 \quad -31.15/\varepsilon^4\,] \tag{10.51}$$

and

$$D_{\rm oc}(\varepsilon) = 2099135.4 - 93.45/\varepsilon^2 + 2853.81/\varepsilon \tag{10.52}$$

There are some very interesting and useful properties of this parameterized controller. After repeatedly simulating the overall design, we found that the maximum peak values of the control signal u are independent of the frequencies of the reference signals. They are only dependent on the initial error between displacement x_1

and the reference r. The larger the initial error, the bigger is the peak that occurs in u. Because the working range of our actuator is within ± 1 μm, we assume that the largest magnitude of the initial error in any situation is not larger than 1 μm. This assumption is reasonable, as we can always reset our displacement x_1 to zero before the system is to track any reference and hence the magnitude of initial tracking error can never be larger than 1 μm. Let us consider the worst case, *i.e.* that the magnitude of the initial error is 1 μm. Then, interestingly, we are able to obtain a clear relationship between the tuning parameter $1/\varepsilon$ and the maximum peak of u. The result is plotted in Figure 10.6. We also found that the tracking error is independent of initial errors. It only depends on the frequencies of the references, *i.e.* the larger the frequency that the reference signal r has, the larger the tracking error that occurs. Again, we can obtain a simple and linear relationship between the tuning parameter ε and the maximum frequency that a reference signal can have such that the corresponding tracking error is no larger than 1%, which is one of our main design specifications. The result is plotted in Figure 10.7.

Clearly, from Figure 10.6, we know that owing to the constraints on the control input, *i.e.* it must be kept within ± 112.5 V, we have to select our controller with $\varepsilon > 1/3370$. From Figure 10.7, we know that in order to meet the first design specification, *i.e.* the steady-state tracking errors are less than 1% for reference inputs that have frequencies up to 30 Hz, we have to choose our controller with $\varepsilon < 1/2680$. Hence, the final controller as given in Equations 10.48 to 10.52 meets all the design goals for our piezoelectric actuator system of Equations 10.1 and 10.2, for all $\varepsilon \in (1/3370, 1/2680)$. Let us choose $\varepsilon = 1/3000$. We obtain the overall controller as in the form of Equation 10.48 with

$$A_{\rm oc} = \begin{bmatrix} -14926.71 & 0 & -2.16 \times 10^{11} & -3.24 \times 10^{14} \\ 0.0315 & -1.69 & 5.71 \times 10^{5} & 8.57 \times 10^{8} \\ 0 & 0 & 0 & 0 \\ 0 & 0 & 1 & 0 \end{bmatrix} \tag{10.53}$$

$$B_{\rm oc} = \begin{bmatrix} -1.16 \times 10^{8} \\ 281.97 \\ 1 \\ 0 \end{bmatrix}, \quad G_{\rm oc} = \begin{bmatrix} 0 \\ 0 \\ -1 \\ 0 \end{bmatrix}, \quad D_{\rm oc} = -8.3 \times 10^{8} \tag{10.54}$$

$$C_{\rm oc} = [\,-92879.9 \quad 2140967 \quad -1.68 \times 10^{12} \quad -2.52 \times 10^{15}\,] \tag{10.55}$$

The simulation results presented in the following are done using the Simulink® package in MATLAB®, which is widely available these days. Two different reference inputs are simulated using the Runge–Kutta 5 method in Simulink® with a minimum step size of 10 μs and a maximum step size of 100 μs as well as a tolerance of 10^{-5}. These references are: 1) a cosine signal with a frequency of 30 Hz and peak magnitude of 1 μm, and 2) a sine signal with a frequency of 34 Hz and peak magnitude of 1 μm. The results for the cosine signal are given in Figures 10.8 to 10.10. In Figure 10.8, the solid-line curve is x_1 and the dash-dotted curve is the reference. The tracking error and the control signal corresponding to this reference are given in Figures 10.9 and 10.10, respectively. Similarly, Figures 10.11 to 10.13

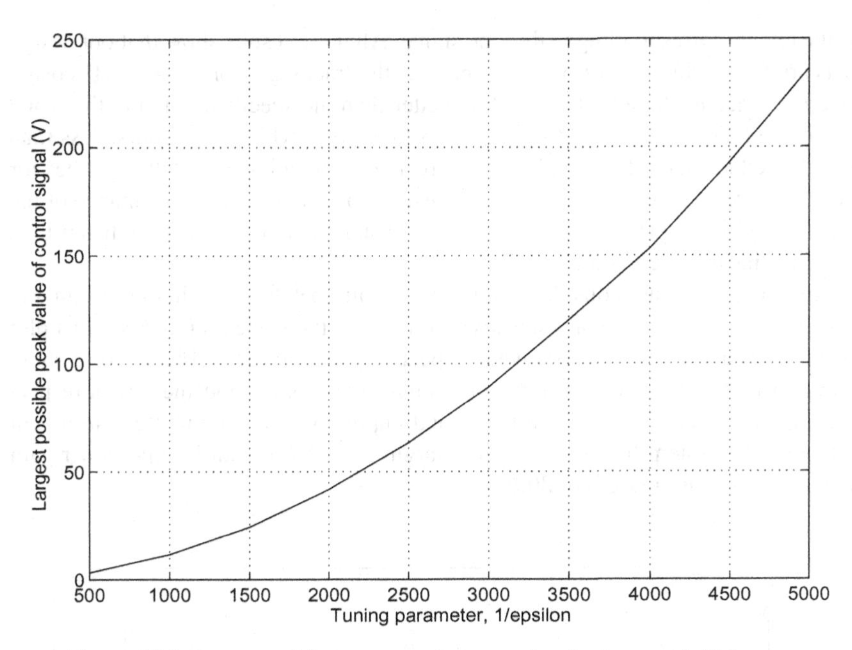

Figure 10.6. Parameter $1/\varepsilon$ versus maximum peaks of u in worst initial errors

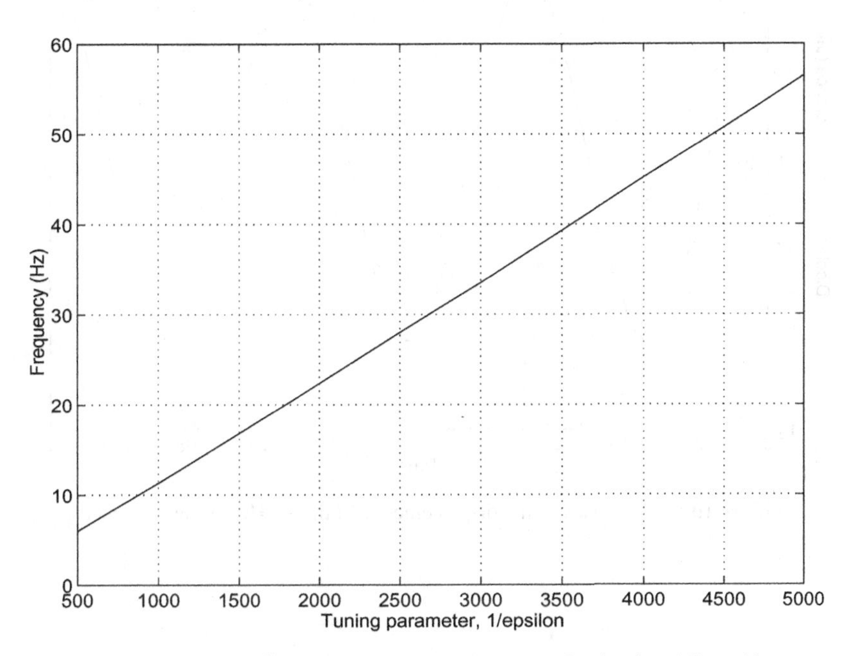

Figure 10.7. Parameter $1/\varepsilon$ versus maximum frequency of r that has 1% tracking error

are the results corresponding to the sine signal. All these results show that our design goals are fully achieved. To be more specific, the tracking error for a 30-Hz cosine wave reference is about 0.8%, which is better than the specification, and the worst peak magnitude of the control signal is less than 90 V, which is of course less than the saturated level, *i.e.* 112.5 V. Furthermore, the 1% tracking error settling times for both cases are less than 0.003 s. It is interesting to note that the performance of the actual closed-loop system with the nonlinear hysteretic dynamics is even better than that of its linearized counterpart.

Although we do not consider frequency-domain specifications in our design, the sensitivity and complementary sensitivity functions, the Bode plot and Nyquist plot (near the origin for verification) given, respectively, in Figures 10.14 and 10.15 show that the overall control system turns out to have impressive robustness with respect to disturbances and measurement noise, and impressive gain and phase margins. In particular, the system has a lower gain margin of 0.487 and an infinite upper gain margin, and a phase margin of 30.58°.

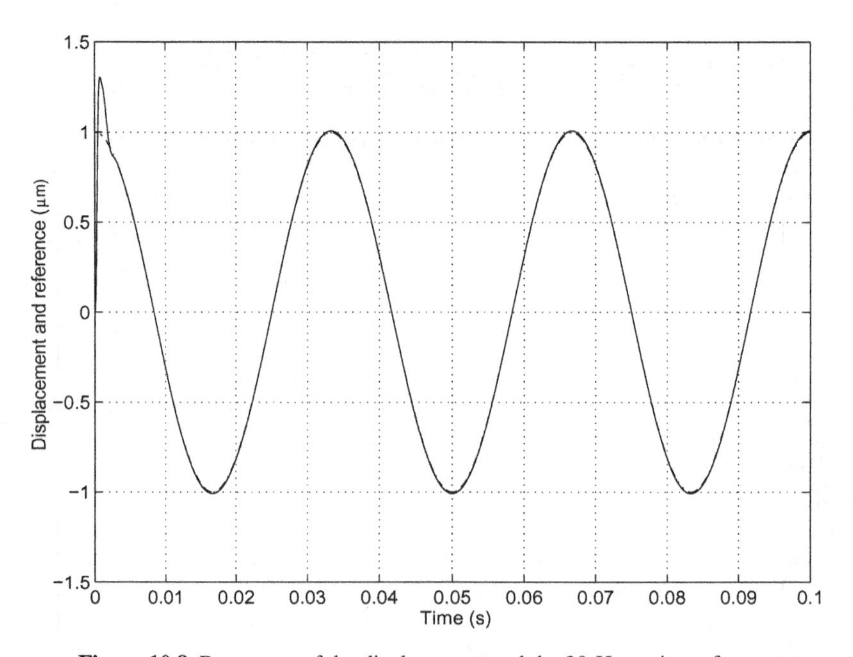

Figure 10.8. Responses of the displacement and the 30-Hz cosine reference

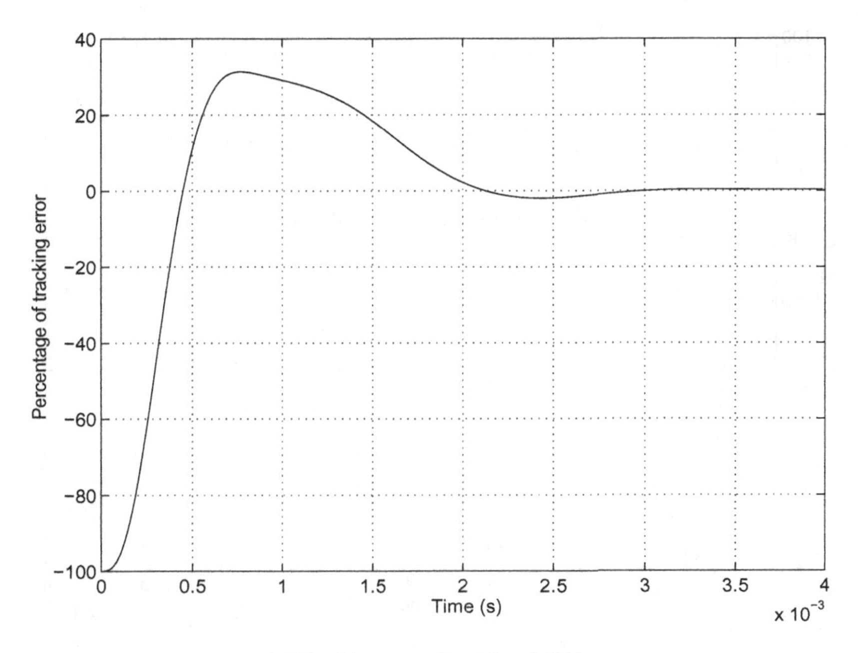

(a) Tracking error from 0 to 0.004 s

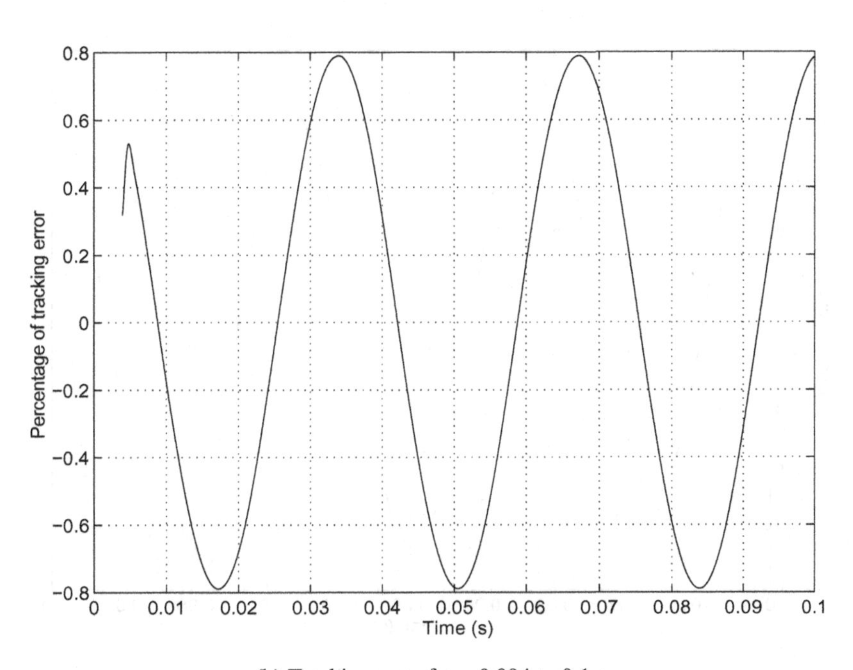

(b) Tracking error from 0.004 to 0.1 s

Figure 10.9. Tracking error for the 30-Hz cosine reference

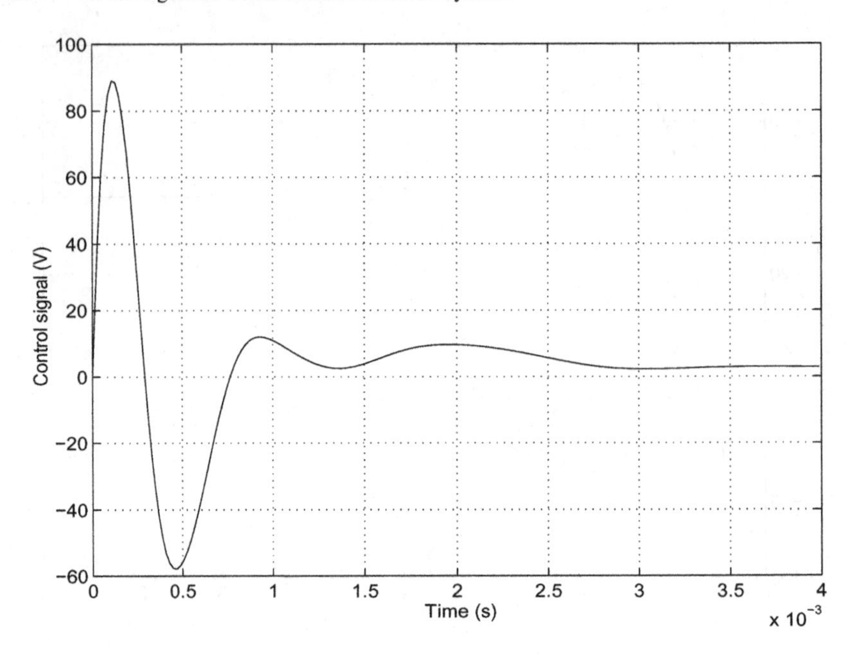

(a) Control signal from 0 to 0.004 s

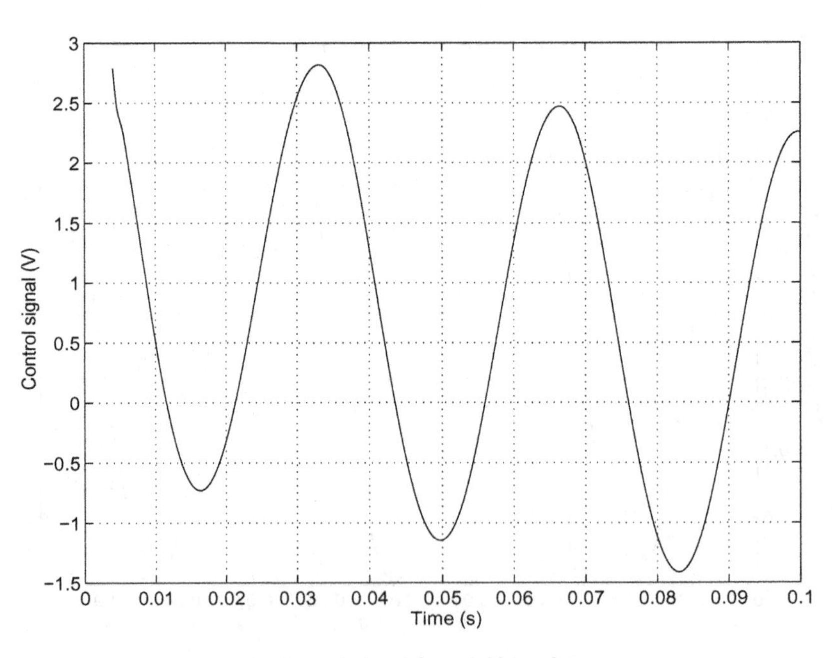

(b) Control signal from 0.004 to 0.1 s

Figure 10.10. Control signal for the 30-Hz cosine reference

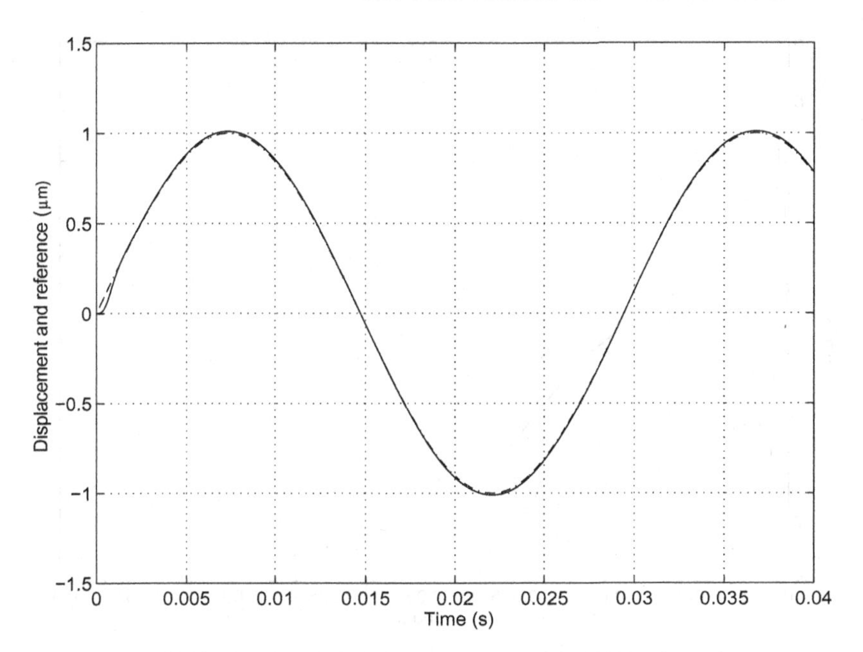

Figure 10.11. Responses of the displacement and the 34-Hz sine reference

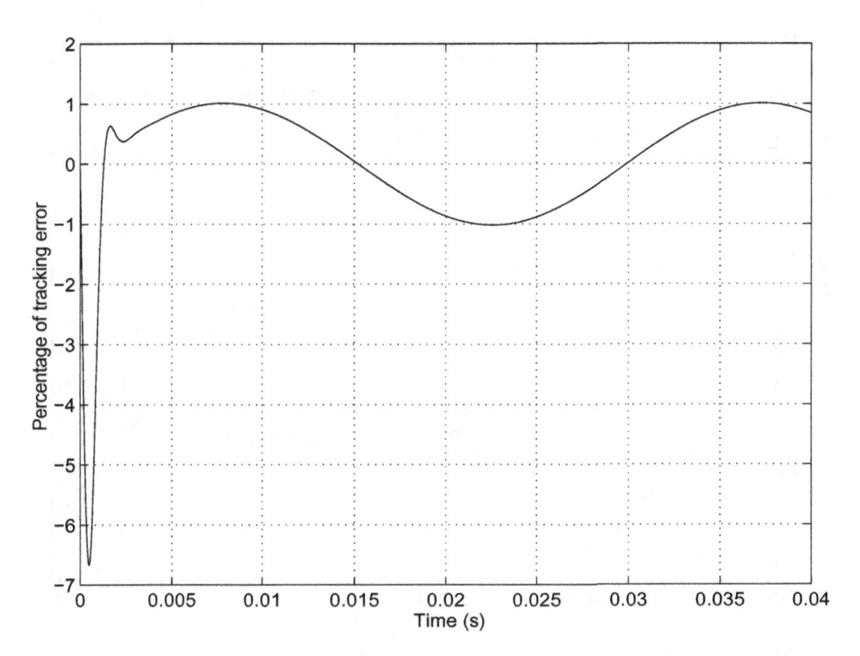

Figure 10.12. Tracking error for the 34-Hz sine reference

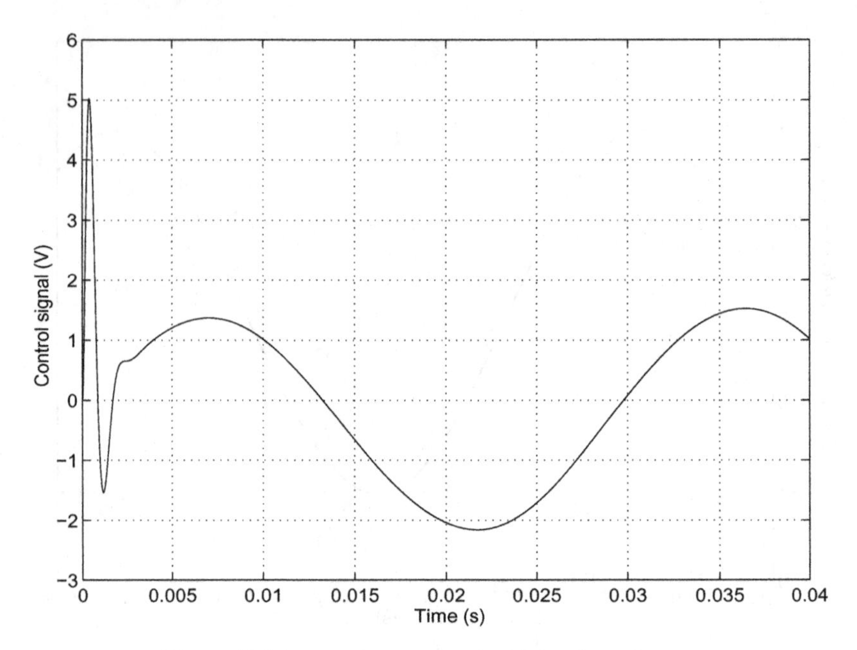

Figure 10.13. Control signal for the 34-Hz sine reference

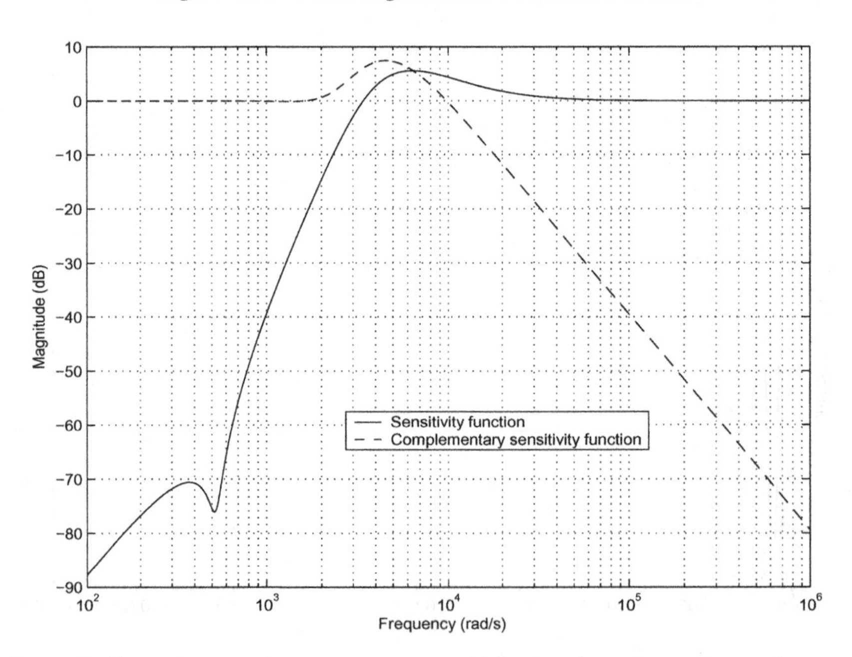

Figure 10.14. Sensitivity and complementary sensitivity functions of overall control system

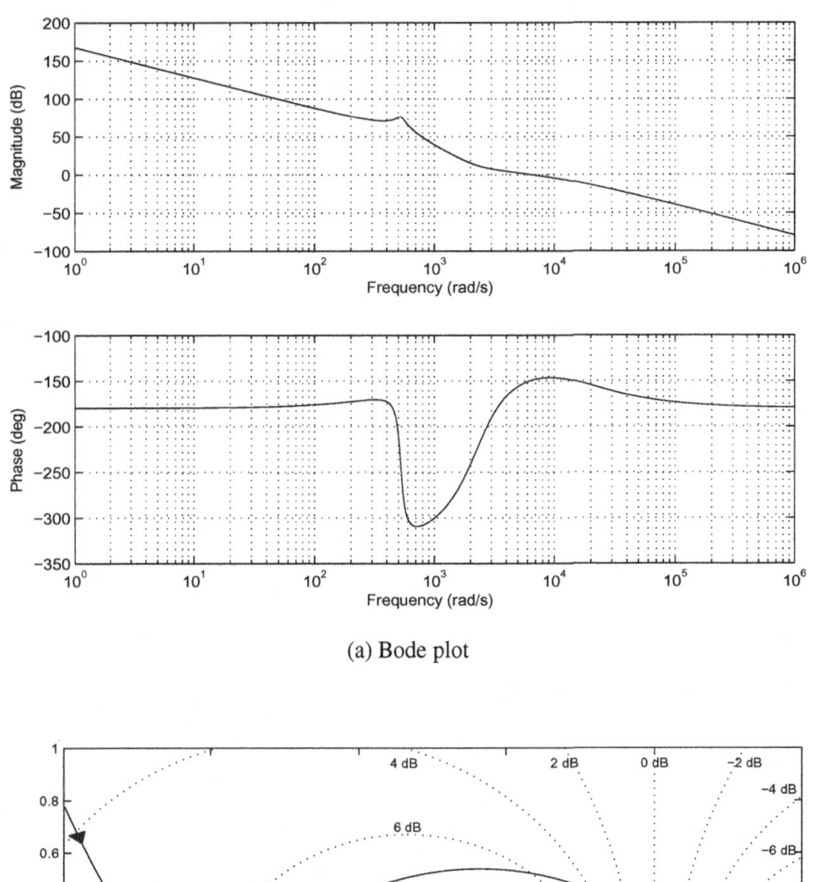

(a) Bode plot

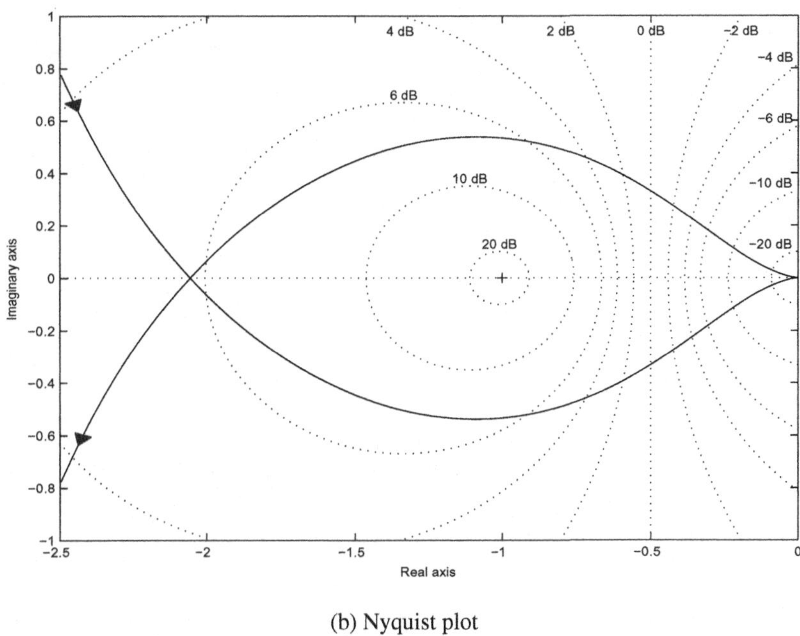

(b) Nyquist plot

Figure 10.15. Bode and Nyquist plots of the overall control system

11

A Benchmark Problem

Before ending this book, we post in this chapter a typical HDD servo control design problem. The problem has been tackled in the previous chapters using several design methods, such as PID, RPT, CNF, PTOS and MSC control. We feel that it can serve as an interesting and excellent benchmark example for testing other linear and nonlinear control techniques.

We recall that the complete dynamics model of a Maxtor (Model 51536U3) hard drive VCM actuator can be depicted as in Figure 11.1:

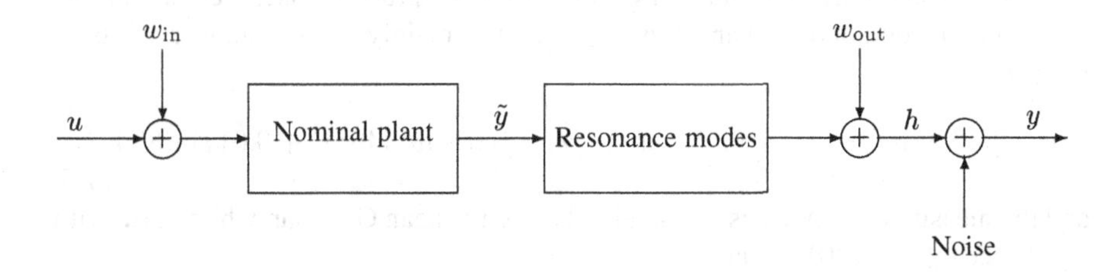

Figure 11.1. Block diagram of the dynamical model of the hard drive VCM actuator

The nominal plant of the HDD VCM actuator is characterized by the following second-order system:

$$\dot{x} = \begin{bmatrix} 0 & 1 \\ 0 & 0 \end{bmatrix} x + \begin{bmatrix} 0 \\ 6.4013 \times 10^7 \end{bmatrix} \left(\text{sat}(u) + w_{\text{in}}\right) \tag{11.1}$$

and

$$\tilde{y} = [\,1 \quad 0\,]\, x \tag{11.2}$$

where the control input u is limited within ± 3 V and w_{in} is an unknown input disturbance with $|w_{\text{in}}| \le 3$ mV. For simplicity and for simulation purpose, we assume that the unknown disturbance $w_{\text{in}} = -3$ mV. The measurement output available for control, *i.e.* y (in μm), is the measured displacement of the VCM R/W head and is given by

$$y = \left[\prod_{i=1}^{4} G_{\mathrm{r},i}(s)\right] \tilde{y} + w_{\mathrm{out}} + \text{Noise} \tag{11.3}$$

where the transfer functions of the resonance modes are given by

$$\left.\begin{aligned}
G_{\mathrm{r},1}(s) &= \frac{0.912s^2 + 457.4s + 1.433(1+\delta)\times 10^8}{s^2 + 359.2s + 1.433(1+\delta)\times 10^8} \\
G_{\mathrm{r},2}(s) &= \frac{0.7586s^2 + 962.2s + 2.491(1+\delta)\times 10^8}{s^2 + 789.1s + 2.491(1+\delta)\times 10^8} \\
G_{\mathrm{r},3}(s) &= \frac{9.917(1+\delta)\times 10^8}{s^2 + 1575s + 9.917(1+\delta)\times 10^8} \\
G_{\mathrm{r},4}(s) &= \frac{2.731(1+\delta)\times 10^9}{s^2 + 2613s + 2.731(1+\delta)\times 10^9}
\end{aligned}\right\} \tag{11.4}$$

with $-20\% \le \delta \le 20\%$ represents the variation of the resonance modes of the actual actuators whose resonant dynamics change from time to time and also from disk to disk in a batch of million drives. Note that many new hard drives in the market nowadays might have resonance modes at much higher frequencies (such as those for the IBM microdrives studied in Chapter 9). But, structurewise, they are almost the same. The output disturbance (in μm), which is mainly the repeatable runouts, is given by

$$w_{\mathrm{out}} = 0.1\sin(110\pi t) + 0.05\sin(220\pi t) + 0.02\sin(440\pi t) + 0.01\sin(880\pi t) \tag{11.5}$$

and the measurement noise is assumed to be a zero-mean Gaussian white noise with a variance $\sigma_{\mathrm{n}}^2 = 9\times 10^{-6}$ (μm)^2.

The problem is to design a controller such that when it is applied to the VCM actuator system, the resulting closed-loop system is asymptotically stable and the actual displacement of the actuator, *i.e.* h, tracks a reference $r = 1$ μm. The overall design has to meet the following specifications:

1. the overshoot of the actual actuator output is less than 5%;
2. the mean of the steady-state error is zero;
3. the gain margin and phase margin of the overall design are, respectively ,greater than 6 dB and 30°; and
4. the maximum peaks of the sensitivity and complementary sensitivity functions are less than 6 dB.

The results of Chapter 6 show that the 5% settling times of our design using the CNF control technique are, respectively, 0.80 ms in simulation and 0.85 ms in actual hardware implementation. We note that the simulation result can be further improved if we do not consider actual hardware constraints in our design. For example, the

CNF control law given below meets all design specifications and achieves a 5% settling time of 0.68 ms. It is obtained by using the toolkit of [55] under the option of the pole-placement method with a damping ratio of 0.1 and a natural frequency of 2800 rad/sec together with a diagonal matrix $W = \text{diag}\{1.5,\ 0.01,\ 2 \times 10^{-10}\}$. The dynamic equation of the control law is given by

$$\begin{pmatrix} \dot{x}_{\rm i} \\ \dot{x}_{\rm v} \end{pmatrix} = \begin{bmatrix} 0 & 0 \\ 0 & -4000 \end{bmatrix} \begin{pmatrix} x_{\rm i} \\ x_{\rm v} \end{pmatrix} - \begin{bmatrix} -10 \\ 1.6 \times 10^7 \end{bmatrix} y + \begin{bmatrix} 0 \\ 6.4013 \times 10^7 \end{bmatrix} \text{sat}(\tilde{u}) - \begin{pmatrix} 10r \\ 0 \end{pmatrix} \tag{11.6}$$

$$\begin{aligned} \tilde{u} = \rho(e)\,[0.61237 \quad 0.049449 \quad 8.4622 \times 10^{-5}] \begin{pmatrix} x_{\rm i} \\ y - r \\ x_{\rm v} + 4000y \end{pmatrix} \\ - [1.2248 \quad 0.12335 \quad 1.0310 \times 10^{-5}] \begin{pmatrix} x_{\rm i} \\ y - r \\ x_{\rm v} + 4000y \end{pmatrix} \end{aligned} \tag{11.7}$$

where

$$\rho(e) = -2.65 \left| \mathrm{e}^{-0.7|e|} - \mathrm{e}^{-0.7} \right| \tag{11.8}$$

and

$$u = G_{\text{notch}}(s) \cdot \tilde{u} \tag{11.9}$$

with $G_{\text{notch}}(s)$ being given as in Equation 6.9.

The simulation results obtained with $\delta = 0$ given in Figures 11.2 to 11.4 show that all the design specifications have been achieved. In particular, the resulting 5% settling time is 0.68 ms, the gain margin is 7.85 dB and the phase margin is 44.7°, and finally, the maximum values of the sensitivity and complementary sensitivity functions are less than 5 dB. The overall control system can still produce a satisfactory result and satisfy all the design specifications by varying the resonance modes with the value of δ changing from −20% to 20%.

Nonetheless, we invite interested readers to challenge our design. Noting that for the track-following case, *i.e.* when $r = 1$ µm, the control signal is far below its saturation level. Because of the bandwidth constraint of the overall system, it is not possible (and not necessary) to utilize the full scale of the control input to the actuator in the track-following stage. However, in the track-seeking case or equivalently by setting a larger target reference, say $r = 500$ µm, the very problem can serve as a good testbed for control techniques developed for systems with actuator saturation. Interested readers are referred to Chapter 7 for more information on track seeking of HDD servo systems.

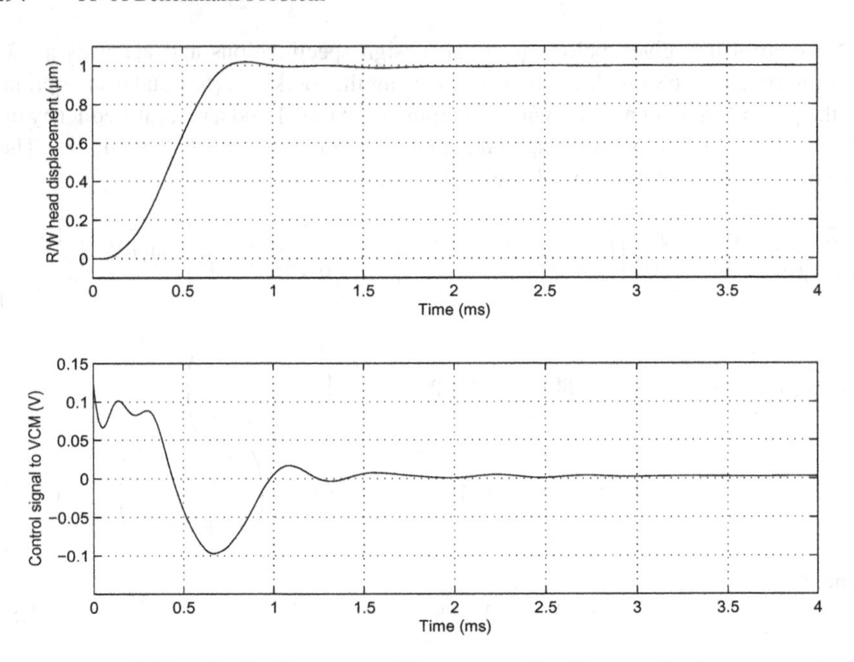

(a) h and u for the system without output disturbance and noise

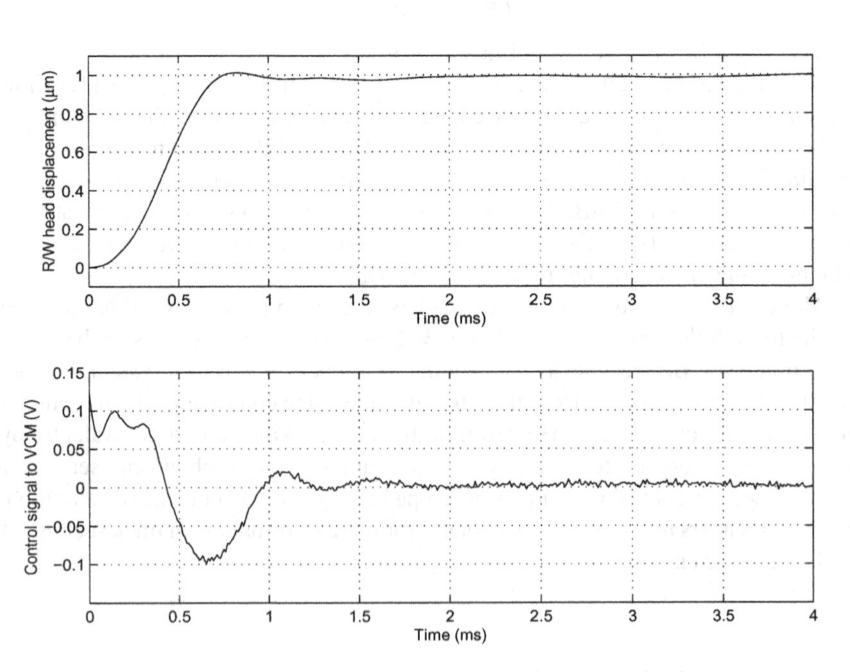

(b) h and u for the system with output disturbance and noise

Figure 11.2. Output responses and control signals of the CNF control system

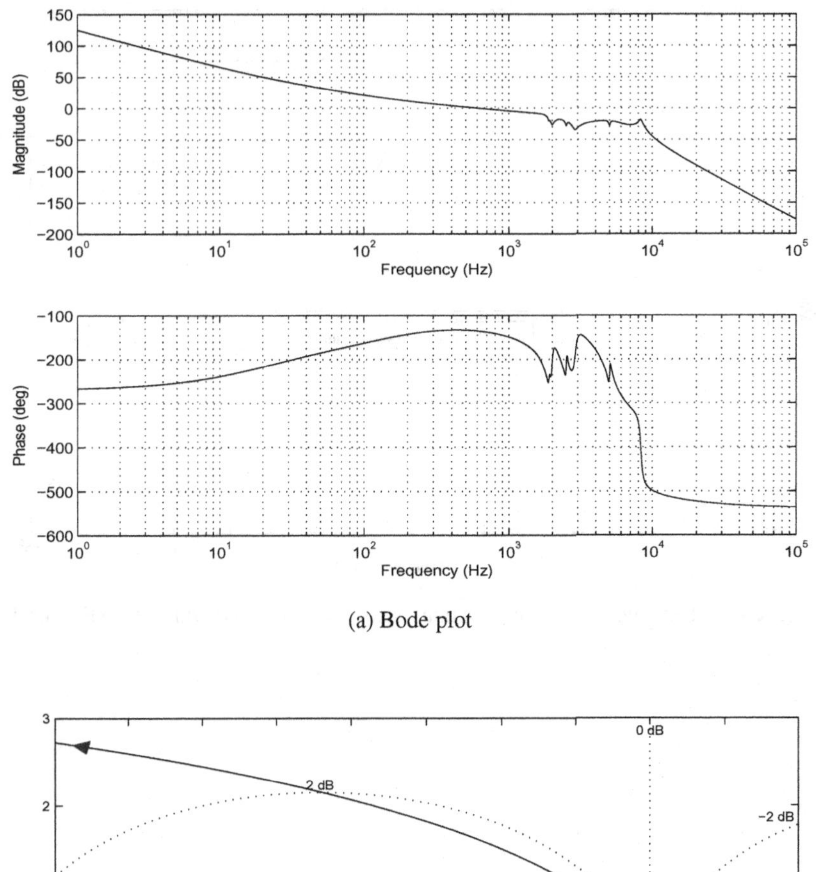

(a) Bode plot

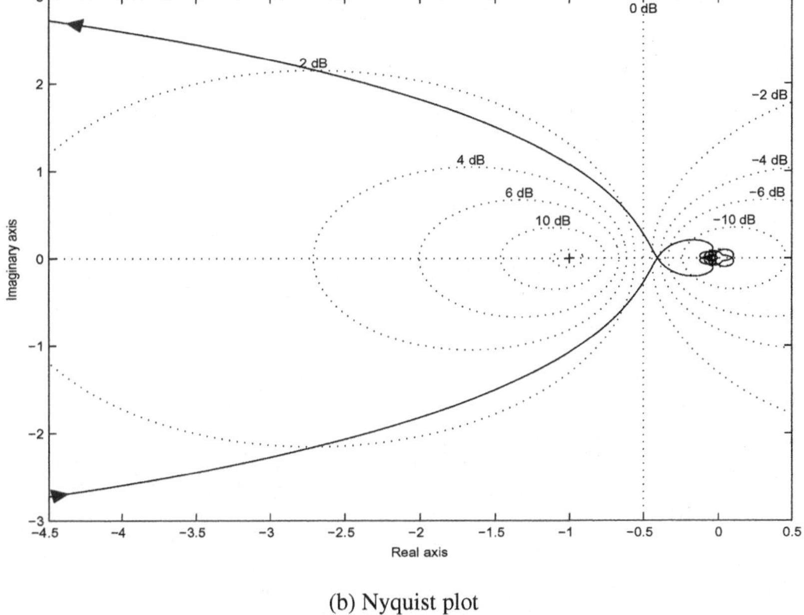

(b) Nyquist plot

Figure 11.3. Bode and Nyquist plots of the CNF control system

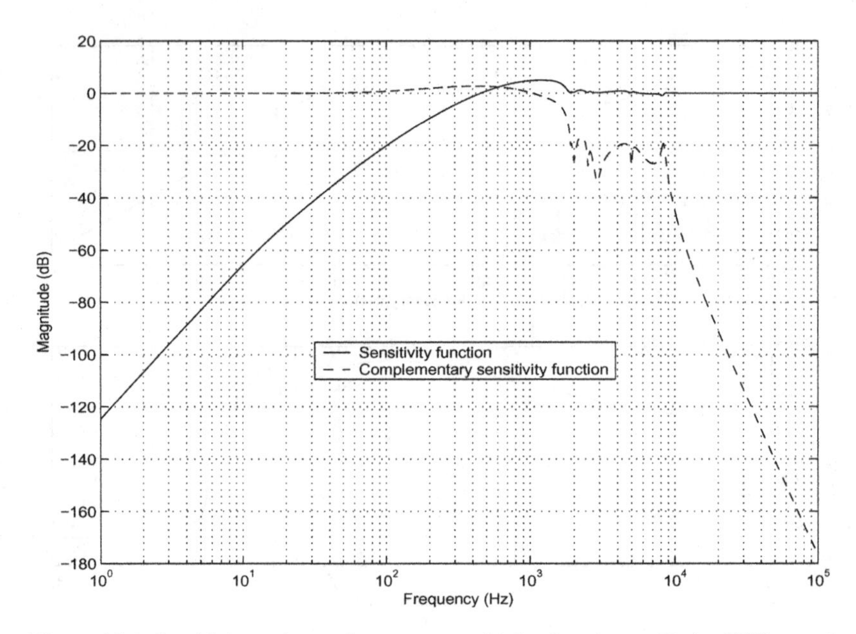

Figure 11.4. Sensitivity and complementary sensitivity functions with the CNF control

References

1. Franklin GF, Powell JD, Workman ML. Digital control of dynamic systems. 3rd edn Reading (MA): Addison-Wesley; 1998.
2. Fujimoto H, Hori Y, Yarnaguchi T, Nakagawa S. Proposal of seeking control of hard disk drives based on perfect tracking control using multirate feedforward control. Proc 6th Int Workshop Adv Motion Contr; Nagoya, Japan; 2000. p. 74–9.
3. Goh TB, Li Z, Chen BM, Lee TH, Huang T. Design and implementation of a hard disk drive servo system using robust and perfect tracking approach. IEEE Trans Contr Syst Technol 2001; 9:221-33.
4. Gu Y, Tomizuka M. Digital redesign and multi-rate control for motion control – a general approach and application to hard disk drive servo system. Proc 6th Int Workshop Adv Motion Contr; Nagoya, Japan; 2000. p. 246–51.
5. Hara S, Hara T, Yi L, Tomizuka M. Two degree-of-freedom controllers for hard disk drives with novel reference signal generation. Proc American Contr Conf; San Diego, CA; 1999. p. 4132–6.
6. Huang Y, Messner WC, Steele J. Feed-forward algorithms for time-optimal settling of hard disk drive servo systems. Proc 23rd Int Conf Ind Electron Contr Instrum; New Orleans, LA; 1997. p. 52–7.
7. Ho HT. Fast bang-bang servo control. IEEE Trans Magn 1997; 33:4522–7.
8. Ishikawa J, Tomizuka M. A novel add-on compensator for cancellation of pivot nonlinearities in hard disk drives. IEEE Trans on Magn 1998; 34:1895–7.
9. Iwashiro M, Yatsu M, Suzuki H. Time optimal track-to-track seek control by model following deadbeat control. IEEE Trans Magn 1999; 35:904–9.
10. Pao LY, Franklin GF. Proximate time-optimal control of third-order servomechanisms. IEEE Trans Automat Contr 1993; 38:560–80.
11. Patten WN, Wu HC, White L. A minimum time seek controller for a disk drive. IEEE Trans Magn 1995; 31:2380–7.
12. Takakura S. Design of a tracking system using n-delay two-degree-of-freedom control and its application to hard disk drives. Proc 1999 IEEE Int Conf Contr Appl; Kohala Coast, HI; 1999. p. 170–5.
13. Wang L, Yuan L, Chen BM, Lee TH. Modeling and control of a dual actuator servo system for hard disk drives. Proc 1998 Int Conf Mechatron Technol; Hsinchu, Taiwan; 1998. p. 533–8.
14. Weerasooriya S, Low TS, Mamun AA. Design of a time optimal variable structure controller for a disk drive actuator. Proc Int Conf Ind Electron Contr Instrum; Hawaii; 1993. p. 2161–5.

15. Yamaguchi T, Soyama Y, Hosokawa H, Tsuneta K, Hirai H. Improvement of settling response of disk drive head positioning servo using mode switching control with initial value compensation. IEEE Trans Magn 1996; 32:1767–72.
16. Zhang DQ, Guo GX. Discrete-time sliding mode proximate time optimal seek control of hard disk drives. IEE Proc Contr Theory Appl 2000; 147:440–6.
17. Chang JK, Ho HT. LQG/LTR frequency loop shaping to improve TMR budget. IEEE Trans Magn 1999; 35:2280–82.
18. Hanselmann H, Engelke A. LQG-control of a highly resonant disk drive head positioning actuator. IEEE Trans Ind Electron 1988; 35:100–4.
19. Weerasooriya S, Phan DT. Discrete-time LQG/LTR design and modeling of a disk drive actuator tracking servo system. IEEE Trans Ind Electron 1995; 42:240–7.
20. Chen BM, Lee TH, Hang CC, Guo Y, Weerasooriya S. An H_∞ almost disturbance decoupling robust controller design for a piezoelectric bimorph actuator with hysteresis. IEEE Trans Contr Syst Technol 1997; 7:160–74.
21. Hirata M, Atsumi T, Murase A, Nonami K. Following control of a hard disk drive by using sampled-data H_∞ control. Proc 1999 IEEE Int Conf Contr Appl; Kohala Coast, HI; 1999. p. 182-6.
22. Li Y, Tomizuka M. Two degree-of-freedom control with adaptive robust control for hard disk servo systems. IEEE/ASME Trans Mechatron 1999; 4:17–24.
23. Hirata M, Liu KZ, Mita T, Yamaguchi T. Head positioning control of a hard disk drive using H_∞ theory. Proc 31st IEEE Conf Dec Contr; Tucson, AZ; 1992. p. 2460–1.
24. Kim BK, Chung WK, Lee HS, Choi HT, Suh IH, Chang YH. Robust time optimal controller design for hard disk drives. IEEE Trans Magn 1999; 35:3598–607.
25. Teo YT, Tay TT. Application of the l_1 optimal regulation strategy to a hard disk servo system. IEEE Trans Contr Syst Technol 1996; 4:467–72.
26. Du CL, Xie LH, Teoh JN, Guo GX. An improved mixed H_2/H_∞ control design for hard disk drives. IEEE Trans Contr Syst Technol 2005; 13:832–9.
27. Chen R, Guo G, Huang T, Low TS. Adaptive multirate control for embedded HDD servo systems. Proc 24th Int Conf Ind Electron Contr Instrum; Aachen, Germany; 1998. p. 1716–20.
28. McCormick J, Horowitz R. A direct adaptive control scheme for disk file servos. Proc 1993 American Contr Conf; San Francisco, CA; 1993. p. 346–51.
29. Weerasooriya S, Low TS. Adaptive sliding mode control of a disk drive actuator. Proc Asia-Pacific Workshop Adv Motion Contr; Singapore, 1993. p. 177–82.
30. Workman ML. Adaptive proximate time optimal servomechanisms [PhD diss]. Stanford University; 1987.
31. Internet websites: www.storagereview.com; www.pcguide.com; www.storage.ibm.com; www.mkdata.dk; 2001.
32. Porter J. Disk drives' evolution. Presented at 100th Ann Conf Magn Rec Info Stor; Santa Clara, CA; 1998.
33. Mamun AA. Servo engineering. Lecture Notes; Dept of Electrical & Computer Engineering, National University of Singapore; 2000.
34. Zhang JL. Electronic computer magnetic storing devices. Beijing: Military Industry Publishing House; 1981 (in Chinese).
35. Li Q, Hong OE, Hui Z, Mannan MA, Weerasooriya S, Ann MY, Chen SX, Wood R. Analysis of the dynamics of 3.5″ hard disk drive actuators. Technical Report. Data Storage Institute (Singapore); 1997.
36. Chang JK, Weerasooriya S, Ho HT. Improved TMR through a frequency shaped servo design. Proc 23rd Int Conf Ind Electron Contr Instrum; New Orleans, LA; 1997. p. 47–51.

37. Li Z, Guo G, Chen BM, Lee TH. Optimal control design to achieve highest track-per-inch in hard disk drives. J Inform Stor Proc Syst 2001; 3:27–41.
38. Sacks AH, Bodson M, Messner W. Advanced methods for repeatable runout compensation. IEEE Trans Magn 1995; 31:1031–6.
39. Bodson M, Sacks A, Khosla P. Harmonic generation in adaptive feedforward cancellation schemes. IEEE Trans Automat Contr 1994; 39:1939–44.
40. Guo L, Tomizuka M. High speed and high precision motion control with an optimal hybrid feedforward controller. IEEE/ASME Trans Mechatron 1997; 2:110–22.
41. Kempf C, Messner W, Tomizuka M, Horowitz R. A comparison of four discrete-time repetitive control algorithms. IEEE Contr Syst Mag 1993; 13:48–54.
42. Guo L. A new disturbance rejection scheme for hard disk drive control. Proc American Contr Conf; Philadelphia, PA; 1998. p. 1553–7.
43. Weerasooriya S. Learning and compensation for repeatable runout of a disk drive servo using a recurrent neural network. Technical Report; Magnetics Technology Center, National University of Singapore; 1995.
44. Jamg G, Kim D, Oh JE. New frequency domain method of nonrepeatable runout measurement in a hard disk drive spindle motor. IEEE Trans Magn 1999; 35:833–8.
45. Ohmi T. Non-repeatable runout of ball-bearing spindle-motor for 2.5″ HDD. IEEE Trans Magn 1996; 32:1715–20.
46. Abramovitch D, Hurst T, Henze D. An overview of the PES Pareto method for decomposing baseline noise sources in hard disk position error signals. IEEE Trans Magn 1998; 34:17–23.
47. Zeng S, Lin RM, Xu LM. Novel method for minimizing track seeking residual vibrations of hard disk drives. IEEE Trans Magn 2001; 37:1146–56.
48. Mah YA, Lin H, Li QH. Design of a high bandwidth moving-coil actuator with force couple actuation. IEEE Trans Magn 1999; 35:874–8.
49. McAllister JS. The effect of disk platter resonances on track misregistration in 3.5″ disk drives. IEEE Trans Magn 1996; 32:1762–6.
50. Hanselmann H, Mortix W. High-bandwidth control of the head positioning mechanism in a winchester disk drive. IEEE Contr Syst Mag 1987; 7:15–9.
51. Guo G. Lecture notes in servo engineering. Dept of Electrical and Computer Engineering, National University of Singapore; 1998.
52. Weaver PA, Ehrlich RM. The use of multirate notch filters in embedded servo disk drives. Proc American Contr Conf; San Francisco, CA; 1993. p. 4156–60.
53. Lin Z, Chen BM, Liu X. Linear systems toolkit. Technical Report, Dept of Electrical and Computer Engineering, University of Virginia, Charlottesville (VA); 2004.
54. Cheng G, Chen BM, Peng K, Lee TH. A Matlab toolkit for composite nonlinear feedback control. Proc 8th Int Conf Contr Automat Robot Vision; Kunming, China; 2004. p. 878–83.
55. Cheng G, Chen BM, Peng K, Lee TH. A Matlab toolkit for composite nonlinear feedback control. Available online at http://hdd.ece.nus.edu.sg/~bmchen/.
56. Chen BM, Lee TH, Venkataramanan V. Hard disk drive servo systems. New York (NY): Springer; 2002.
57. Ljung L. System identification: theory for the user. Englewood Cliffs (NJ): Prentice Hall; 1987.
58. Sinha NK, Kuszta B. Modeling and identification of dynamic systems. New York (NY): Van Nostrand Reinhold Company; 1983.
59. Eykhoff P. System identification – parameter and state estimation. New York (NY): John Wiley; 1981.

60. Hsia TC. System identification. Lexington (MA): Lexington Books; 1977.
61. Sage AP, Melsa JL. System identification. New York (NY): Academic Press; 1971.
62. Dammers D, Binet P, Pelz G, Vobkarper LM. Motor modeling based on physical effect models. Proc 2001 IEEE Int Workshop on Behav Model Simulation; Santa Rosa, CA; 2001. p. 78–83.
63. Dubi A. Monte Carlo applications in systems engineering. New York (NY): John Wiley; 2000.
64. Evans M, Swartz T. Approximating integrals via Monte Carlo and deterministic methods. London: Oxford University Press; 2000.
65. Mikhailov GA. Parametric estimates by the Monte Carlo method. Utrecht, The Netherlands: V.S.P. International Science; 1999.
66. Rake H. Step response and frequency response methods. Automatica 1980; 16:519–26.
67. Eykhoff P. Trends and progress in system identification. New York (NY): Pergamon Press; 1981.
68. Chen J, Gu G. Control oriented system identification: an H_∞ approach. New York (NY): John Wiley & Sons; 2000.
69. Gong JQ, Guo L, Lee HS, Yao B. Modeling and cancellation of pivot nonlinearity in hard disk drives IEEE Trans Magn 2002; 38:3560–5.
70. Liu X, Chen BM, Lin Z. On the problem of general structural assignments of linear systems through sensor/actuator selection. Automatica 2003; 39:233–41.
71. Chen BM, Lin Z, Shamash Y. Linear systems theory: a structural decomposition approach. Boston: Birkhäser; 2004.
72. Sannuti P, Saberi A. A special coordinate basis of multivariable linear systems – Finite and infinite zero structure, squaring down and decoupling. Int J Contr 1987; 45:1655–704.
73. Saberi A, Sannuti P. Squaring down of non-strictly proper systems. Int J Contr 1990; 51:621–9.
74. Chen BM. Robust and H_∞ Control. London: Springer; 2000.
75. Rosenbrock HH. State-space and Multivariable Theory. New York (NY): John Wiley; 1970.
76. MacFarlane AGJ, Karcanias N. Poles and zeros of linear multivariable systems: a survey of the algebraic, geometric and complex variable theory. Int J Contr 1976; 24:33–74.
77. Commault C, Dion JM. Structure at infinity of linear multivariable systems: a geometric approach. IEEE Trans Automat Contr 1982; 27:693–6.
78. Pugh AC, Ratcliffe PA. On the zeros and poles of a rational matrix. Int J Contr 1979; 30:213–27.
79. Verghese G. Infinite frequency behavior in generalized dynamical systems [PhD diss]. Stanford University; 1978.
80. Owens DH. Invariant zeros of multivariable systems: a geometric analysis. Int J Contr 1978; 28:187–98.
81. Morse AS. Structural invariants of linear multivariable systems. SIAM J Contr 1973; 11:446–65.
82. Moylan P. Stable inversion of linear systems. IEEE Trans Automat Contr 1977; 22:74–8.
83. Scherer C. H_∞-optimization without assumptions on finite or infinite zeros. SIAM J Contr Optimiz 1992; 30:143–66.
84. Ziegler JG, Nichols NB. Optimum settings for automatic controllers. Trans ASME 1942; 64:759–68.
85. Ziegler JG, Nichols NB. Process lags in automatic control circuits. Trans ASME 1943; 65:433–44.

86. Franklin GF, Powell JD, Emami-Naeini A. Feedback control of dynamic systems. 3rd edn Reading (MA): Addison-Wesley; 1994.
87. Anderson BDO, Moore JB. Optimal control: linear quadratic methods. Englewood Cliffs (NJ): Prentice Hall; 1989.
88. Fleming WH, Rishel RW. Deterministic and stochastic optimal control. New York (NY): Springer-Verlag; 1975.
89. Kwakernaak H, Sivan R. Linear optimal control systems. New York (NY): John Wiley; 1972.
90. Saberi A, Sannuti P, Chen BM. H_2 optimal control. London: Prentice Hall; 1995.
91. Doyle J, Glover K, Khargonekar PP, Francis BA. State space solutions to standard H_2 and H_∞ control problems. IEEE Trans Automat Contr 1989; 34:831–47.
92. Chen BM, Saberi A, Sannuti P, Shamash Y. Construction and parameterization of all static and dynamic H_2-optimal state feedback solutions, optimal fixed modes and fixed decoupling zeros. IEEE Trans Automat Contr 1993; 38:248–61.
93. Zhou K, Khargonekar P. An algebraic Riccati equation approach to H_∞-optimization. Syst Contr Lett 1988; 11:85–91.
94. Chen BM, Saberi A, Bingulac S, Sannuti P. Loop transfer recovery for non-strictly proper plants. Contr Theory Adv Technol 1990; 6:573–94.
95. Zames G. Feedback and optimal sensitivity: Model reference transformations, multiplicative seminorms, and approximate inverses. IEEE Trans Automat Contr 1981; 26:301–20.
96. Limebeer DJN, Anderson BDO. An interpolation theory approach to H_∞ controller degree bounds. Linear Algebra Appl 1988; 98:347–86.
97. Doyle JC. Lecture notes in advances in multivariable control. ONR-Honeywell Workshop; 1984.
98. Francis BA. A course in H_∞ control theory. Berlin: Springer; 1987.
99. Glover K. All optimal Hankel-norm approximations of linear multivariable systems and their $\mathcal{L}_\infty$ error bounds. Int J Contr 1984; 39:1115–93.
100. Kwakernaak H. A polynomial approach to minimax frequency domain optimization of multivariable feedback systems. Int J Contr 1986; 41:117–56.
101. Kimura H. Chain-scattering approach to H_∞-control. Boston: Birkhäuser; 1997.
102. Zhou K, Doyle J, Glover K. Robust and optimal control. Englewood Cliffs (NJ): Prentice Hall; 1996.
103. Başar T, Bernhard P. H_∞ optimal control and related minimax design problems: a dynamic game approach. 2nd edn Boston: Birkhäuser; 1995.
104. Willems JC. Almost invariant subspaces: an approach to high gain feedback design – part I: almost controlled invariant subspaces. IEEE Trans Automat Contr 1981; 26:235–52.
105. Willems JC. Almost invariant subspaces: an approach to high gain feedback design – part II: almost conditionally invariant subspaces. IEEE Trans Automat Contr 1982; 27:1071–85.
106. Liu K, Chen BM, Lin Z. On the problem of robust and perfect tracking for linear systems with external disturbances. Int J Contr 2001; 74:158–74.
107. Chen BM, Lin Z, Liu K. Robust and perfect tracking of discrete-time systems. Automatica 2002; 38:293–9.
108. Lewis FL. Applied optimal control and estimation. Englewood Cliffs (NJ): Prentice Hall; 1992.
109. Athans M. A tutorial on LQG/LTR methods. Proc American Contr Conf; Seattle, WA; 1986. p. 1289–96.

110. Chen BM. Theory of loop transfer recovery for multivarible linear systems [PhD diss]. Pullman (WA): Washington State University; 1991.
111. Chen BM, Saberi A, Sannuti P. A new stable compensator design for exact and approximate loop transfer recovery. Automatica 1991; 27:257–80.
112. Doyle JC, Stein G. Multivariable feedback design: concepts for a classical/modern synthesis. IEEE Trans Automat Contr 1981; 26:4–16.
113. Goodman GC. The LQG/LTR method and discrete-time control systems. Technical report. MIT (MA): Report No.: LIDS-TH-1392; 1984.
114. Kwakernaak H. Optimal low sensitivity linear feedback systems. Automatica 1969; 5:279–85.
115. Matson CL, Maybeck PS. On an assumed convergence result in the LQG/LTR technique. Proc 26th IEEE Conf Dec Contr; Los Angeles, CA; 1987. p. 951–2.
116. Niemann HH, Sogaard-Andersen P, Stoustrup J. Loop Transfer Recovery: Analysis and Design for General Observer Architecture. Int J Contr 1991; 53:1177–203.
117. Saberi A, Chen BM, Sannuti P. Loop transfer recovery: analysis and design. London: Springer; 1993.
118. Stein G, Athans M. The LQG/LTR procedure for multivariable feedback control design. IEEE Tran Automat Contr 1987; 32:105–14.
119. Zhang Z, Freudenberg JS. Loop transfer recovery for nonminimum phase plants. IEEE Trans Automat Contr 1990; 35:547–53.
120. Chen BM, Saberi A, Ly U. Closed loop transfer recovery with observer based controllers: analysis. in Contr Dynam Syst (ed. Leondes CT). San Diego (CA): Academic Press; 1992; 51:247–93.
121. Chen BM, Saberi A, Ly U. Closed loop transfer recovery with observer based controllers: design. in Contr Dynam Syst (ed. Leondes CT). San Diego (CA): Academic Press; 1992; 56:295–348.
122. Chen BM, Saberi A, Berg MC, Ly U. Closed loop transfer recovery for discrete time systems. in Contr Dynam Syst (ed. Leondes CT). San Diego (CA): Academic Press; 1993; 56:443–81.
123. Hu T, Lin Z. Control systems with actuator saturation: analysis and design. Boston: Birkhäuser; 2001.
124. Kirk DE. Optimal control theory. Englewood Cliffs (NJ): Prentice Hall; 1970.
125. Venkataramanan V, Chen BM, Lee TH, Guo G. A new approach to the design of mode switching control in hard disk drive servo systems. Contr Eng Prac 2002; 10:925–39.
126. Itkis U. Control systems of variable structure. New York (NY): Wiley; 1976.
127. Yamaguchi T, Numasato H, Hirai H. A mode-switching control for motion control and its application to disk drives: Design of optimal mode-switching conditions. IEEE/ASME Trans Mechatron 1998; 3:202–9.
128. Salle JL, Lefschetz S. Stability by Liapunov's direct method. New York (NY): Academic Press; 1961.
129. LaSalle J. Stability by Liapunov's direct method with applications. New York (NY): Academic Press; 1961.
130. Lin Z, Pachter M, Banda S. Toward improvement of tracking performance – nonlinear feedback for linear systems. Int J Contr 1998; 70:1–11.
131. Turner MC, Postlethwaite I, Walker DJ. Nonlinear tracking control for multivariable constrained input linear systems. Int J Contr 2000; 73:1160–72.
132. Chen BM, Lee TH, Peng K, Venkataramanan V. Composite nonlinear feedback control: theory and an application. IEEE Trans Automat Contr 2003; 48:427–39.

133. Venkataramanan V, Peng K, Chen BM, Lee TH. Discrete-time composite nonlinear feedback control with an application in design of a hard disk drive servo system. IEEE Trans Contr Syst Technol 2003; 11:16–23.
134. He Y, Chen BM, Wu C. Composite nonlinear control with state and measurement feedback for general multivariable systems with input saturation. Syst Contr Lett 2005; 54:455–69.
135. He Y, Chen BM, Wu C. Composite nonlinear feedback control for general discrete-time multivariable systems with actuator nonlinearities. Proc 5th Asian Contr Conf; Melbourne, Australia; 2004. p. 539–44.
136. Lan W, Chen BM, He Y. Improving transient performance in tracking control for a class of nonlinear systems with input saturation. Syst Contr Lett 2006; 55:132–8.
137. He Y, Chen BM, Lan W. Improving transient performance in tracking control for a class of nonlinear discrete-time systems with input saturation. Proc 44th IEEE Conf Dec Contr; Seville, Spain; 2005. p. 8094–9.
138. Peng K, Chen BM, Cheng G, Lee TH. Modeling and compensation of nonlinearities and friction in a micro hard disk drive servo system with nonlinear feedback control. IEEE Trans Contr Syst Technol 2005; 13:708–21.
139. Chen BM, Zheng D. Simultaneous finite and infinite zero assignments of linear systems. Automatica 1995; 31:643–8.
140. Cheng G, Peng K, Chen BM, Lee TH. A microdrive track following controller design using robust and perfect tracking control with nonlinear compensation. Mechatron 2005; 15:933–48.
141. Iannou PA, Kosmatopoulos EB, Despain AM. Position error signal estimation at high sampling rates using data and servo sector measurements. IEEE Trans Contr Syst Technol 2003; 11:325–34.
142. Weerasooriya S, Low TS, Huang YH. Adaptive time optimal control of a disk drive actuator. IEEE Trans Magn 1994; 30:4224–6.
143. Xiong Y, Weerasooriya S, Low TS. Improved discrete proximate time optimal controller of a disk drive actuator. IEEE Trans Magn 1996; 32:4010–2.
144. Mizoshita Y, Hasegawa S, Takaishi K. Vibration minimized access control for disk drives. IEEE Trans Magn 1996; 32:1793–8.
145. Yamaguchi T, Nakagawa S. Recent control technologies for fast and precise servo system of hard disk drives. Proc 6th Int Workshop Adv Motion Contr; Nagoya, Japan; 2000. p. 69–73.
146. Tsuchiura KM, Tsukuba HH, Toride HO, Takahashi T. Disk system with sub-actuators for fine head displacement. US Patent No: 5189578; 1993.
147. Miu DK, Tai YC. Silicon micromachined SCALED technology. IEEE Trans Ind Electron 1995; 42:234–9.
148. Fan LS, Ottesen HH, Reiley TC, Wood RW. Magnetic recording head positioning at very high track densities using a microactuator based, two stage servo system. IEEE Trans Ind Electron 1995; 42:222–33.
149. Aggarwal SK, Horsley DA, Horowitz R, Pisano AP. Microactuators for high density disk drives. Proc American Contr Conf; Albuquerque, NM; 1997. p. 3979–84.
150. Ding, J, Tomizuka M, Numasato H. Design and robustness analysis of dual stage servo system. Proc American Contr Conf; Chicago, IL; 2000. p. 2605–09.
151. Evans RB, Griesbach JS, Messner WC. Piezoelectric microactuator for dual stage control. IEEE Trans Magn 1999; 35:977–82.
152. Fan LS, Hirano T, Hong J, Webb PR, Juan WH, Lee WY, *et al.* Electrostatic microactuator and design considerations for HDD application. IEEE Trans Magn 1999; 35:1000–5.

153. Guo L, Chang JK, Hu X. Track-following and seek/settle control schemes for high density disk drives with dual-stage actuators. Proc 2001 IEEE/ASME Int Conf Adv Intell Mechatron; Como, Italy; 2001. p. 1136–41.
154. Guo L, Martin D, Brunnett D. Dual-stage actuator servo control for high density disk drives. Proc 1999 IEEE/ASME Int Conf Adv Intell Mechatron; Atlanta, GA; 1999. p. 132–7.
155. Guo W, Weerasooriya S, Goh TB, Li QH, Bi C, Chang KT, *et al.* Dual stage actuators for high density rotating memory devices. IEEE Trans Magn 1998; 34:450–5.
156. Guo W, Yuan L, Wang L, Guo G, Huang T, Chen BM, *et al.* Linear quadratic optimal dual-stage servo control systems for hard disk drives. Proc 24th IEEE Ind Electron Soc Ann Conf; Aachen, Germany; 1998. p. 1405–10.
157. Hernandez D, Park SS, Horowitz R, Packard A. Dual-stage track-following design for hard disk drives. Proc American Contr Conf; San Diego, CA; 1999. p. 4116–21.
158. Horsley DA, Hernandez D, Horowitz R, Packard AK, Pisano AP. Closed-loop control of a microfabricated actuator for dual-stage hard disk drive servo systems. Proc American Contr Conf; Philadelphia, PA; 1998. p. 3028–32.
159. Hu X, Guo W, Huang T, Chen BM, Discrete time LQG/LTR dual-stage controller design and implementation for high track density HDDs. Proc American Contr Conf; San Diego, CA; 1999. p. 4111–5.
160. Kobayashi M, Horowitz R. Track seek control for hard disk dual-stage servo systems. IEEE Trans Magn 2001; 37:949–54.
161. Li Y, Horowitz R. Track-following controller design of MEMS based dual-stage servos in magnetic hard disk drives. Proc 2000 IEEE Int Conf Robot Automat; San Francisco, CA; 2000. p. 953–8.
162. Mori K, Munemoto T, Otsuki H, Yamaguchi Y, Akagi K. A dual-stage magnetic disk drive actuator using a piezoelectric device for a high track density. IEEE Trans Magn 1991; 27:5298–300.
163. Schroeck SJ, Messner WC. On controller design for linear time-invariant dual-input single-output systems. Proc American Contr Conf; San Diego, CA; 1999. p. 4122–6.
164. Semba T, Hirano T, Hong J, Fan LS. Dual-stage servo controller for HDD using MEMS microactuator. IEEE Trans Magn 1999; 35:2271–3.
165. Suthasun T, Mareels I, Mamun AA. System identification and control design for dual actuated disk drive. Contr Eng Prac 2002; 12:665–76.
166. Takaishi K, Imamura T, Mizasgita Y, Hasegawa S, Ueno T, Yamada T. Microactuator control for disk drive. IEEE Trans Magn 1996; 32:1863–6.
167. Du CL, Guo GX. Lowering the hump of sensitivity functions for discrete-time dual-stage systems. IEEE Trans Contr Syst Technol 2005; 13:791–7.
168. Canudas de Wit C, Lischinsky P. Adaptive friction compensation with partially known dynamic friction model. Int J Adapt Contr Signal Proc 1997; 11:65–80.
169. Canudas de Wit C, Olsson H, Aström KJ, Lischinsky P. A new model for control of systems with friction. IEEE Trans Automat Contr 1995; 40:419–25.
170. Canudas de Wit C, Olsson H, Aström KJ, Lischinsky P. Dynamic friction models and control design. Proc American Contr Conf; San Francisco, CA; 1993. p. 1920–6.
171. Olsson H, Aström KJ. Observer-based friction compensation. Proc 35th IEEE Conf Dec Contr; Kobe, Japan; 1996. p. 4345–50.
172. Dahl PR. Solid friction damping of mechanical vibrations. AIAA J 1976; 14:1675–82.
173. Ge SS, Lee TH, Ren SX. Adaptive friction compensation of servomechanisms. Int J Sys Sci 2001; 32:523–32.
174. Maria HA, Abrahams ID. Active control of friction-driven oscillations. J Sound Vibra 1996; 193:417–26.

175. Abramovitch D, Wang F, Franklin G. Disk drive pivot nonlinearity modeling – Part I: frequency domain. Proc American Contr Conf; Baltimore, MD; 2004. p. 2600–3.
176. Ishikawa J, Tomizuka M. Pivot friction compensation using an accelerometer and a disturbance observer for hard disk drives. IEEE/ASME Trans Mechatron 1998; 3:194–201.
177. Wang F, Abramovitch D, Franklin G. A method for verifying measurements and models of linear and nonlinear systems. Proc American Contr Conf; San Francisco, CA; 1993. p. 93–7.
178. Wang F, Hurst T, Abramovitch D, Franklin G. Disk drive pivot nonlinearity modeling – Part II: time domain. Proc American Contr Conf; Baltimore, MD; 1994. p. 2604–7.
179. Chang HS, Baek SE, Park JH, Byun YK. Modeling of pivot friction using relay function and estimation of its functional parameters. Proc American Contr Conf; San Francisco, CA; 1999. p. 3784–9.
180. Liu X, Liu JC. Analysis and measurement of torque hysteresis of pivot bearing in hard disk drive applications. Tribology Int 1999; 32:125–30.
181. Low TS, Guo W. Modeling of a three-layer piezoelectric bimorph beam with hysteresis. J Microelectromech Syst 1995; 4:230–7.
182. Chang TP. Seismic response analysis of nonlinear structures using the stochastic equivalent linearization technique [PhD diss]. New York (NY): Columbia University; 1985.
183. Peng K, Venkataramanan V, Chen BM, Lee TH. Design and implementation of a dual-stage actuator HDD servo system via composite nonlinear feedback approach. Mechatron 2004; 14:965–88.
184. Hwang CL, Lin CH. A discrete-time multivariable neuro-adaptive control for nonlinear unknown dynamic systems. IEEE Trans Syst Man Cyb B 2000; 30:865–77.
185. Adriaens HJMTA, de Koning WL, Banning R. Modeling piezoelectric actuators. IEEE/ASME Trans Mechatron 2000; 5:331–41.
186. Cruz-Hernandez JM, Hayward V. Phase control approach to hysteresis reduction. IEEE Trans Contr Syst Technol 2001; 9:17–26.
187. Cheng HM, Ewe MTS, Chiu GTC, Bashir R. Modeling and control of piezoelectric cantilever beam micro-mirror and micro-laser arrays to reduce image banding in electrophotographic processes. J Micromech Microeng 2001; 11:487–98.
188. Guo G, Chen R, Low TS, Wang Y. Optimal control design for hard disk drive servosystems. IEE Proc–Contr Theor Appl 2002; 149:237–42.
189. Ewe MTS, Grice JM, Chiu GTC, Allebach JP, Chan CS, Foote W. Banding reduction in electrophotographic processes using a piezoelectric actuated laser beam deflection device. J Imaging Sci Technol 2002; 46:433–42.
190. Lin CL, Jan HY, Shieh NC. GA-based multiobjective PID control for a linear brushless DC motor. IEEE/ASME Trans Mechatron 2003; 8:56–65.
191. Hwang CL, Jan C. Optimal and reinforced robustness designs of fuzzy variable structure tracking control for a piezoelectric actuator system. IEEE Trans Fuzzy Syst 2003; 11:507–17.
192. Jan C, Hwang CL. A nonlinear observer-based sliding-mode control for piezoelectric actuator systems: Theory and experiments. J Chinese Inst Engr 2004; 27:9–22.
193. Huang YC, Lin DY. Ultra-fine tracking control on piezoelectric actuated motion stage using piezoelectric hysteretic model. Asian J Contr 2004; 6:208–16.
194. Hwang CL, Jan C. Nano trajectory control of multilayer low-voltage PZT render actuator systems. Asian J Contr 2004; 6:187–98.
195. Hwang CL, Chen YM. Discrete sliding-mode tracking control of high-displacement piezoelectric actuator systems. J Dynam Syst–Trans ASME 2004; 126:721–31.

196. Hwang CL, Chen YM, Jan C. Trajectory tracking of large-displacement piezoelectric actuators using a nonlinear observer-based variable structure control. IEEE Trans Contr Syst Technol 2005; 13:56–66.
197. Ikhouane F, Manosa V, Rodellar J. Adaptive control of a hysteretic structural system. Automatica 2005; 41:225–31.
198. Cruz-Hernandez JM, Hayward V. Position stability for phase control of the Preisach hysteresis model. Trans Canadian Soc Mech Eng 2005; 29:129–42.
199. Hwang CL, Jan C. State-estimator-based feedback control for a class of piezoelectric systems with hysteretic nonlinearity. IEEE Trans Syst Man Cyb A 2005; 35:654–64.
200. Ikhouane F, Rodellar J. On the hysteretic Bouc–Wen model. Nonlinear Dynam 2005; 421:63–78.
201. Moheimani SOR, Vautier BJG. Resonant control of structural vibration using charge-driven piezoelectric actuators. IEEE Trans Contr Syst Technol 2005; 13:1021–35.
202. Caughey TK. Derivation and application of the Fokker–Planck equation to discrete nonlinear dynamic systems subjected to white random excitation. J Acoust Soc Am 1963; 35:1683–92.
203. Crandall ST. Perturbation techniques for random vibration of nonlinear systems. J Acoust Soc Am 1963; 35:1700–05.
204. Lyon RH. Response of a nonlinear string to random excitation. J Acoust Soc Am 1960; 32:953–60.
205. Booton, Jr., RC. Nonlinear control systems with random inputs. IRE Trans Circuit Theory 1954; CT–1:9–18.

Index